Materials Characterization for Systems Performance and Reliability

SAGAMORE ARMY MATERIALS
RESEARCH CONFERENCE PROCEEDINGS

Recent volumes in the series:

Materials Characterization for Systems Performance and Reliability

Edited by
James W. McCauley
Army Materials and Mechanics Research Center
Watertown, Massachusetts
and
Volker Weiss
Syracuse University
Syracuse, New York

PLENUM PRESS • NEW YORK AND LONDON

Library of Congress Cataloging in Publication Data

Sagamore Army Materials Research Conference (31st: 1984: Lake Luzerne, N.Y.)
  Materials characterization for systems performance and reliability.

  (Sagamore Army Materials Research Conference proceedings; 31st)
  Bibliography: p.
  Includes index.
  1. Materials — Congresses. 2. Reliability (Engineering) — Congresses. I. McCauley, James
W. II. Weiss, Volker, 1930–   . III. Title. IV. Series: Sagamore Army Materials Research
Conference. Sagamore Army Materials Research Conference proccedings; 31st.
TA401.3.S23   1984   620.1′1   85-19118
ISBN 0-306-42095-3

Proceedings of the 31st Sagamore Conference, entitled Materials Characterization for Systems
Performance and Reliability, held August 13–17, 1984, at Lake Luzerne, New York

© 1986 Plenum Press, New York
A Division of Plenum Publishing Corporation
233 Spring Street, New York, N.Y. 10013

Printed in the United States of America

31st SAGAMORE CONFERENCE COMMITTEE

Chairman

JAMES W. McCAULEY

Army Materials and Mechanics Research Center

Vice-Chairman

WENZEL E. DAVIDSOHN

Army Materials and Mechanics Research Center

Program Director

VOLKER WEISS

Syracuse University

Conference Administrator

WILLIAM K. WILSON

Syracuse University

Conference Coordinator

KAREN A. KALOOSTIAN

Army Materials and Mechanics Research Center

Program Coordinator

MARY ANN HOLMQUIST

Syracuse University

Poster Coordinator

CATHERINE A. BYRNE

Army Materials and Mechanics Research Center

Steering Committee

ALFRED L. BROZ

Army Materials and Mechanics Research Center

GORDON A. BRUGGEMAN

Army Materials and Mechanics Research Center

DAVID R. CLARKE

IBM Corporation

DONALD G. GROVES

National Research Council

LUCY HAGAN

U.S. Army Material Command

GARY L. HAGNAUER

Army Materials and Mechanics Research Center

PHILLIP A. PARRISH

U.S. Army Research Office

RUSTUM ROY

The Pennsylvania State University

PREFACE

The Sagamore Army Materials Research Conferences have been held in the beautiful Adirondack Mountains of New York State since 1954. Organized and conducted by the Army Materials and Mechanics Research Center (Watertown, Massachusetts) in cooperation with Syracuse University, the Conferences have focused on key issues in Materials Science and Engineering that impact directly on current or future Army problem areas. A select group of speakers and attendees are assembled from academia, industry, and other parts of the Department of Defense and Government to provide an optimum forum for a full dialogue on the selected topic. This book is a collection of the full manuscripts of the formal presentations given at the Conference.

The emergence and use of nontraditional materials and the excessive failures and reject rates of high technology, materials intensive engineering systems necessitates a new approach to quality control. Thus, the theme of this year's Thirty-First Conference, "Materials Characterization for Systems Performance and Reliability," was selected to focus on the need and mechanisms to transition from defect interrogation of materials after production to utilization of materials characterization during manufacturing.

The guidance and help of the steering committee and the dedicated and conscientious efforts of Ms. Karen Kaloostian, Conference Coordinator, and Mr. William K. Wilson, and Ms. Mary Ann Holmquist are gratefully acknowledged. The continued active interest and support of Dr. Edward S. Wright, Director, AMMRC; Dr. Robert W. Lewis, Associate Director, AMMRC; and COL L. C. Ross, Commander/ Deputy Director, AMMRC; are greatly appreciated.

James W. McCauley
Conference Chairman

CONTENTS

SECTION III

MATERIALS CHARACTERIZATION IN MANUFACTURING I: STARTING MATERIALS

Chairman: Phillip A. Parrish

SECTION IV

MATERIALS CHARACTERIZATION IN MANUFACTURING II: IN-PROCESS

Chairman: Alfred L. Broz

SECTION V

MATERIALS CHARACTERIZATION IN MANUFACTURING III: FINAL PRODUCT

Chairman: Ram Kossowsky

SECTION VI

IN-SERVICE AND PERFORMANCE MATERIALS CHARACTERIZATION ISSUES

Chairman: Gordon A. Bruggeman

SECTION VII

MANAGEMENT CONSIDERATIONS

Chairman: Donald G. Groves

SECTION VIII

SPECIFIC EXAMPLES

Chairman: James W. McCauley

PANEL MEMBERS

Charles Craig, U.S. Department of Energy
Robert Green, The John Hopkins University
Rustum Roy, The Pennsylvania State University
John Wachtman, Jr., Rutgers University
Ram Kossowsky, Wetinghouse R&D Center
Volker Weiss, Syracuse University
Harry Light, U.S. Army Material Development
        and Readiness Command

MATERIALS CHARACTERIZATION: DEFINITION, PHILOSOPHY AND OVERVIEW OF CONFERENCE

James W. McCauley

United States Army Laboratory Command
*Army Materials Technology Laboratory
Watertown, MA   02172

INTRODUCTION

In the 21st century, the Army will require materials with maximum achievable performance and high reliability at the lowest possible cost. However, current problems associated with excessive reject rates and failures of prototype and fielded high technology systems suggest that dramatic changes must take place given current trends and future requirements. These problems combined with the phase-out of manufacturing arsenals and associated procurement laws necessitate the use of unambiguous specifications which include the full projected life of the system. Moreover, the emergence and actual use of nontraditional, advanced materials like composites, polymers and high performance ceramics and the need for man-rating merely exacerbates the problem.

Further impacting the situation is the much increased competition in the international high technology market place. Requirements for long term reliability seem to be moving up in consummer's priorities. Increased usage of automated manufacturing (CAD/CAM) techniques is only part of the solution. It, however, could compound the quality problem if proper procedures are not utilized. First, our thinking must change from failure analysis or screening quality (accept/reject criteria) to failure prevention. This will demand management techniques similar to those of Deming[1] and Crosby[2]. Second, a technology base must be generated to unambiguously identify (interrogate) advanced materials, in all stages of production and usage in a quantifiable way. This is a prerequisite for the use of artificial intelligence in the manufacturing process.

*Formerly the Army Materials and Mechanics Research Center

MATERIALS CHARACTERIZATION

Any engineering system requires certain material properties
for performance. In an ideal world, any extrinsic property of a
material is a function of its intrinsic characteristics. Therefore,
by identifying the unique characteristics of the material the
properties are also defined in an unambiguous way. The converse is
not true. Identification of a property does not uniquely define a
material. This is the fundamental ambiguity of advanced materials
and is beautifully illustrated by a series of polyurethanes[3]
(Table 1). It can be seen that it is much more advantageous to
identify the polyurethane by a microstructural feature like wt%
hard segment than by using any of the mechanical properties. In
the real world the unique characteristics of any material not only
consist of chemistry and physical structure of its constitutive
components, but also include defects.

Materials characterization has been very succinctly defined by
a National Academy of Sciences Materials Advisory Board Report[4].
"Characterization describes those features of the composition and
structure (including defects) of a material that are significant
for a particular preparation, study of properties, or use, and
suffice for reproduction of the material." The unique signature of
any material therefore is[5]:

$$P = f(c,M,PD) \tag{1}$$

c = chemistry, phases
M = Microstructure (physical structure)
PD = Processing Defects

Therefore, any property of a material can be unambiguously defined
by its unique signature. Using this simple concept, a material's
full life cycle can be schematically illustrated as shown in
Figure 1. A variety of concepts are illustrated in this figure.
In order to insure certain performance requirements, performance
specifications should be replaced with material specifications
and design allowables obtained via a materials science and engineering approach: materials science being the relation between material
characteristics and properties, while materials engineering is the
relation of properties to performance criteria. Figures of merit
have typically been used to relate a required performance to a set
of properties. In theory, therefore, materials can be tailored to
meet specific requirements. The right side of this chart defines
the technological pushers, while the left side defines the pullers.

2

Materials characterization is critical in defining the unique signature of the processed material (c,M,PD) and also for assessing the service effects. Chemistry, microstructure, and processing defects may change as a result of the service environment, besides the formation of new defects (SD). The service related signature

Table I. Properties of a Series of Polyurethanes

Sample#	Wt % Hard Segment	Initial Modulus (MPa)	Ultimate Elongation (%)	Fracture Stress (MPa)	Toughness (Jm-3)
I	10	0.7	700	1.7	8.3×10^7
II	21	5.5	800	7.8	3.3×10^8
III	32	3.4	1400	22	2.1×10^9
IV	43	130	1200	33	2.6×10^9
V	55	260	700	31	1.7×10^9
VI	66	520	400	33	1.2×10^9
VII	77	1400	40	38	1.6×10^8

may, therefore, be defined as follows:

$$P' = f(\Delta c, \Delta M, \Delta PD, SD) \tag{2}$$

Δc = changed composition
ΔM = changed microstructure
ΔPD = changed processing defects
SD = new service-related defects

ADVANCED MATERIALS

It is being demonstrated everyday that many advanced materials are viable candidates for present engineering systems which require enhanced performance. However, the properties of these new materials, in many cases, depend not only on defects but on subtle, often minute, differences in their chemistry and microstructure.

For example, the creep rates of emerging high performance ceramics like various types of sialons often vary by orders of magnitude[6] depending on the characteristics of a grain boundary phase that is 10-1000Å in thickness[7]. Recent work on zeolites[8] demonstrate that the Si/Al ratio in a single crystal about 1.0μm in diameter can vary from 20:1 to 40:1 in a distance of 0.5μm. Apparent subtle variations in certain structural fibers have also been observed. Mechanical testing[9] of Kevlar 49 epoxy composites exhibit large variability in creep rupture experiments. Failure times at identical stress levels vary by up to over three orders of magnitude. Besides the variation in intrinsic fiber characteristics the interfacial region between fiber and matrix in composites also has a profound effect on properties. With the advent of new materials, joining techniques are relying more and more on advanced concepts over traditional welding and mechanical fastening. Optimum bonding in these cases depends critically on the nature (chemistry and structure) of the joining faces.

All of the preceding factors necessitate new directions in materials R&D. The first aspect involves the incorporation of quality criteria into the full manufacturing cycle, similar to what many Japanese manufacturers and a smaller number of American manufacturers have done (and Deming suggested many years ago). The second aspect involves derivation of quantitative relationships between properties and unique signature data for generations of usable design allowables and material specifications. The third phase involves a merging of traditional NDT and materials characterization into a unified field (nondestructive characterization) for final product quality control and field evaluation of environmental degradation and contamination.

4

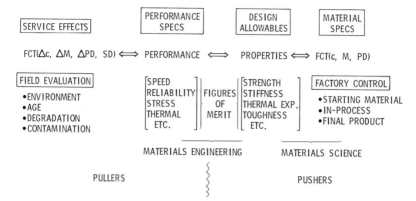

Figure 1. Full Life Cycle Materials Characterization

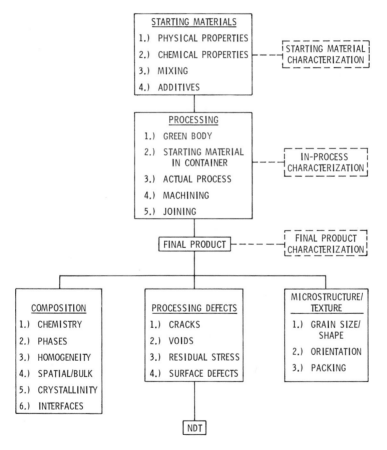

Figure 2. Characterization in Manufacturing

MATERIALS CHARACTERIZATION IN MANUFACTURING

The most important aspect of future R&D should be in establishing materials characterization as an integrated part of manufacturing material-intensive systems. Figure 2 illustrates this. The
key areas of focus involve the evaluation of techniques to fully
characterize starting materials and material during processing.
The techniques should be amendable to automation and computerization
for feedback control. Powder characterization and green body
characterization will be critical. For in-process characterization,
the relation between measurable characteristics or properties and
desired properties via contact or noncontact sensors is the key
aspect for this part of manufacturing. Sensor technology must be
advanced as well as quantitative relationships developed between
measurable characteristics and desired properties.

Obviously, the determination of the full unique signature
of a material may be cost or technically prohibitive. However, the
concept provides a sound quantitative framework for the derivation
of data bases suitable for computer interrogation in an artificial
intelligence mode if desired. Two routes are possible: either
direct measurements of unique signature or measurement of other
characteristics (like density, sonic velocity, etc.) which are a
direct function of the unique signature and can be carried out
nondestructively.

Full unique signature determination may not be as intimidating
as it appears. Of course, if complete signature determination is
carried out, the costs and time required may be prohibitive. However, only certain applications would require this--the cost and
performance trade-offs would not allow for full component characterization in most cases. Utilization of easier-to-measure characteristics like density, sonic velocities, X-ray fluorescent chemistry,
X-ray radiographic, NMR or ultrasonic images, etc., may be all that
is required. However, these measurable characteristics must be
related to the unique signature characteristics: composition, grain
size, grain boundary characteristics, defects, etc.

In order to accomplish a major part of these goals, traditional
NDT must work hand-in-hand with materials characterization. By
doing this it will be possible to go beyond defect screening and
use the emerging NDT techniques for unique signature determination.
With future miniaturization and computerization, laboratory-based

characterization procedures should evolve to be field- and factory-adaptable. In the 21st Century even portable suction-cup type vacuum systems will be available for various particle and photon-beam surface and bulk analysis.

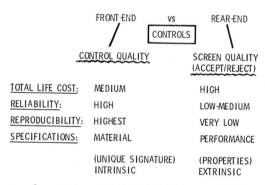

Figure 3.   Material Reliability and Reproducibility via Full-Cycle Materials Characterization

RELIABILITY

A material must be uniform within a single piece or item and also be reproducible from batch to batch. As previously defined[10] uniformity is the control of variations in the character (unique signature) within a piece, while reproducibility is the control of variations from piece to piece in a repetitive process. Reliability, therefore, is the uniformity and reproducibility of a constant (desired) property value continuously. Quality, on the other hand, is the reliable conformance to requirements.

The use of unambiguous material specifications consisting of unique chemistry, physical structure, and defect signature is a failsafe solution to enhanced reliability. Moreover, these same concepts applied in the field to assess service-related changes in the unique signature allows for reliable life predictions. Traditional nondestructive testing must merge with laboratory-based materials characterization concepts into a unified field of nondestructive characterization. Whereas, with traditional materials (mostly metals), the detection of flaws or processing defects by NDT sufficed for high reliability, in the 21st Century, materials properties will be controlled primarily by their unique chemical and physical structure. There are at least five options available to assure the reliability of materials:

    a. Measure Final Product Signature
      • Actual measurement of property on manufactured component.
      • Measurement of property on identically manufactured component.
      • Estimation of property from a data base of identical materials.
      • Estimation of property from non- or semidestructive property evaluation via unique signature.

    b. Control Final Product Signature
      • Complete manufacturing process control.

In the short term, focus should be on measuring the final product signature, and in the middle-to-far term on controlling the final product signature via starting material and process control. By changing from accept/reject criteria based on defects of manufactured material to full-cycle quality control in starting materials, during processing, and on the final product, total life-

8

cycle costs will be dramatically reduced and reliability will be substantially improved as illustrated in Figure 3. For high performance ceramic materials Bowen[11] has determined that 50 + 25% of their total manufacturing costs are due to excessively high reject rates.

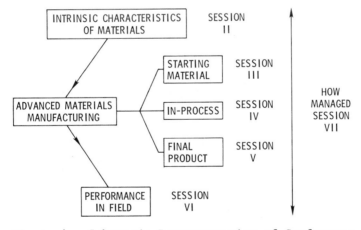

Figure 4.   Schematic Representation of Conference

9

For all practical purposes, destructive property measurement on a manufactured component is unrealistic. Measurement on an identically manufactured component is possible, but not unambiguous. For design purposes data bases are critical, but without unique signature data there would be no absolute way of material verification. Some major efforts have been performed to develop mechanical property data bases for potential use as design allowables[12]. These data, although extremely useful, are not yet sufficient for design use. Further, since there is virtually no correlation to any unique signature data (since this was not required) the data cannot be used for generating material specifications or process refinement. The conclusions are simple. Material reliability (and also manufacturing costs) will be assured by quantitative process control using full-cycle characterization in manufacturing and user verification of unique signatures.

SUMMARY

There are both technical and managerial implications of an increased awareness and useage of materials characterization in the emergence of advance materials:

    a. Technical:
        • Enhanced focus in materials science and engineering
        • Importance of unique signatures
        • Design data bases which include signature data sensor development
        • Field and factory adaptable equipment
        • Evolution of characterization standards and standardized procedures.

    b. Managerial:
        • Extension from processing defect characterization (NDT) to chemical and structural interrogation of advanced materials.
        • Extreme cost of prototype equipment requires new/innovative methods of implementation.
        • Expand traditional support role to include focussed and separate, but integrated programs.
        • Set of mechanisms for the development of characterization standards and calibration of test methods.

CONFERENCE FORMAT

The conference has been designed to address these concepts in a systematic way illustrating the criticality and usefulness of characterization in materials science and engineering and advanced materials manufacturing. First, overviews will be presented on advanced materials processing and the emerging concepts of full-cycle materials characterization and nondestructive characterization.

Examples from all advanced materials will be presented in all of the sessions. In the first regular session papers will be presented illustrating the subtle effects of chemistry and micro-structure on materials. The next four sessions form the core of the conference dealing with all of the aspects of materials charac-terization in manufacturing and in-service. Finally, and perhaps most importantly, the critical issue of management implementation is addressed in the last formal session. Figure 4 is a schematic representation of the entire conference.

REFERENCES

1. W. E. Deming, "Quality, Productivity, and Competitive Position," MIT C&R for Adv. Eng. Study, Cambridge, MA (1982).
2. P. B. Crosby, Quality is Free, New American Library, NY (1979).
3. A. L. Chang, R. M. Briber, E. L. Thomas, R. J. Zdrahala and F. E. Critchfield, "Morphological Study of the Structure Developed During the Polymerization of a Series of Segmented Polyurethanes," Polymer, Vol. 23, pp. 1060-1068 (1982).
4. Characterization of Materials, NAS Materials Advisory Board, MAB-229-M, March, (1967).
5. J. W. McCauley, "Structural and Chemical Characterization of Processed Crystalline Ceramic Materials," in Characterization of Materials in Research, Ceramics and Polymers; Edited by J. J. Burke and V. Weiss, Syracuse University Press, Syracuse, NY, pp. 175-209 (1975).
6. M. H. Lewis, G. R. Heath, S. M. Winder and R. J. Lumby, "High Temperature Creep and Fracture of $\beta'-Si_3N_4$ Ceramic Alloys."
7. D. R. Clarke, "The Microstructure of Nitrogen Ceramics," in Progress in Nitrogen Ceramics, ed. by F. L. Riley, Martinus Nijhoff Publishers, Boston, pp. 341-358 (1983).
8. C. E. Lyman, Analytical Electron Microscopy of Zeolites in Proceedings of the 41st Annual Meeting of the Electron Micro-scopy Society of America, pp. 344-345 (1983).
9. F. P. Gerstle, Jr. and S. C. Kunz, "Prediction of Long-Term Failure in Kevlar 49 Composites," in Long-Term Behavior of Composites, ASTM STP 813, T. K. O'Brien, ed., Philadelphia, PA pp. 263-292 (1983).
10. Ceramic Processing, NAS Materials Advisory Board, Pub. 1576 (1968).
11. H. K. Bowen, personal communication.
12. D. C. Larsen, J. W. Adams and S. A. Bortz, Survey of Potential Data for Design Allowable MIL-Handbook Utilization for Structural Silicon-Base Ceramics, Final Report, Contract DAAG46-79-C-0078, AMMRC, Watertown, MA December (1981).

MATERIALS PROCESSING SYSTEMS CONTROL AND COMPETITION

George B. Kenney

Materials Processing Center
Massachusetts Institute of Technology
Cambridge, Massachusetts 02139

INTRODUCTION

Technological and, subsequently, economic leadership in communication, transportation, information processing, and other high tech systems is directly dependent upon advances in materials processing and manufacturing technologies. It is for this reason that the firms, industries, academic institutions, and governments of various nations are competing to develop and promote advanced materials technologies. At the heart of their efforts are systems performance and reliability which can be either constrained or enhanced by the quality of available materials. For example, the quality of the silica in fiber optics, gallium arsenide in electronic circuitry, or advanced ceramics in adiabatic engines determine how well these high tech systems will function. In this context, it is important to recognize that control of the microstructure of a material is critical in determining its engineering properties and subsequent performance in use. Ultimately, technical and economic competitive advantage is determined by the efficiency with which scientific information based on materials characterization and analysis is generated and applied to process control and commercialization of technology.

MATERIALS, MATERIALS PROCESSING, AND INTERNATIONAL COMPETITION

Materials processing is the engineering field that seeks to control the structure, properties, shape, and performance of materials in an economic and socially cost-effective manner. It lies at the heart of the broader field of materials science and engineering, linking basic science to societal need and experience, Figure 1.

13

For thousands of years, long before the field of materials science was dreamt of, materials processing was an essential part of society, practiced at its best by skilled artisans who worked with materials to advance their utility and aesthetic appeal. For illustrations, we can turn to such well-known examples as the pottery, textiles, and cast arts of Asia Minor of 5000 years ago, the beautiful and functional Japanese samurai swords, medieval iron-making, and the ubiquitous American blacksmith.

The knowledge of these artisans was empirical, and their processing can best be described as "materials carftsmanship". They achieved properties and performance through processing, but lacked the basic science and understanding available to modern processers, Figure 1, and without the essential modern concept that properties and performance are controlled through control of structure. Today, the crucial role of structure control at all levels (submicroscopic, microscopic, and macroscopic) and structure-property relationships is understood, as are the essential linkages of materials processing to the scientific base and to societal needs. Largely as a result of this understanding, materials processing has become a field critical to current national needs. Materials processing is central, for example, to the development and economic production of sophisticated new materials for the energy, transportation, electronic, and aerospace technologies. It is within this context that the importance of materials characterization, in production process design and control for subsequent systems performance and reliability, cannot be over-emphasized.

The direct dependence of high technology systems performance and reliability on advanced materials has been well documented. One of the best current examples is certainly that of the space shuttle, protected by ceramic tiles from the tremendous temperatures experienced upon reentry into the earth's atmosphere. Similarly, current electronic and computer industries are based upon computer chips made from silicon single crystals. Another technology, optical telecommunications, depends on an advanced material just making its debut, glass fibers. These and other established advanced materials and their production processes have been carefully studied, designed, and characterized.

The linkage between the development and application of materials characterization and testing procedures for product quality control and improved reliability is well defined and discussed extensively in following sessions of this conference. What is not well defined or fully appreciated is the relationship between advances in materials processing and competition, both national and international, in advanced technology systems. It is precisely this

Table 1
High Technology Ceramics Markets (1980 – $M)*

	Japan[+]	United States
Electronic Applications		
Ceramic Powders	130	120
Electronic IC Packages/Substrates	540	340
Capacitors	325	425
Piezoelectrics	145	50
Thermistor/Varistors	125	75
Ferrites	380	100
Gas/Humidity Sensors	5	40
Translucent Ceramics	20	25
Structural Ceramics		
Cutting Tools	125	750
Structural Ceramics	120	130
TOTALS	$1,915	$2,055

+   Reference 1.
*   Excludes fibers, nuclear fuels, spark plugs.

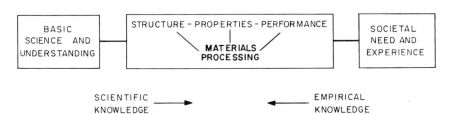

Figure 1.   Materials processing as part of materials science
and engineering

relationship upon which the Japanese government has founded its much publicized, 10-year, $450 million, national initiative.

In 1981, Japan's Ministry of Industry and Trade, through its Agency of Industrial Science and Technology, founded the Industrial Base Technology Development Project, Figure 2. The specific goal of this project is the "development of base technologies for the next generation of industries." Clearly, Japan's current half-billion-dollar national initiative, which is to run through the decade of the 1980's, is directed towards the industrial position and competitiveness of Japanese industries of the 1990's, as opposed to current industrial sectors. This truly represents long-range industrial planning at the national level. In this context, Japan has identified and selected three areas -- new materials, biotechnology, and computers -- as being central to their long-range industrial and economic strategy.

New materials selected for specific study are fine ceramics, special function polymers, special metals, and composites. Advanced ceramics are fundamental to the expanding electronics industry and to future adiabatic power plant systems. Special polymers include the emerging electrically conductive polymers. Special metallic materials include metastable, rapidly solidified alloy systems. Composites include polymer as well as ceramic and metal matrix systems.

Successful development for these materials systems will require the application of existing and development of new characterization techniques to define and quantify appropriate processing technologies consistent with performance and systems reliability requirements. Technological leadership in any of these materials systems has profound international competitive and economic implications.

ADVANCED CERAMICS AS A CASE STUDY

Advanced ceramics are an excellent example of an emerging materials technology with profound implications for dependent high technology systems and subsequent international trade and economic competition. It has been suggested that advanced ceramics will be to the late 1980's and 1990's what plastics were to the 1960's. The current status of the advanced ceramic markets is presented in Table 1.

As of 1980, the world market for advanced ceramics was $4.12 billion. Of this, $1.9 billion was provided by Japanese suppliers while $2.1 billion was provided by U.S. suppliers. European sources account for roughly an additional $150 million, primarily as cutting

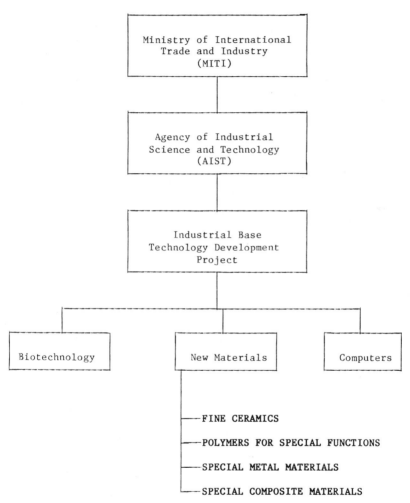

Figure 2.   Development of base technologies for the next generation of industries.

tools. Table 1 does not include figures for large vertically integrated captive producers in both the U.S. and Japan.

Electronics applications account for approximately 88 percent
and 70 percent of the total market value of Japanese and U.S. production of fine ceramics, respectively. While this electronic
market offers significant future growth potential, it is dwarfed by
the potential size and impact of fine ceramics in heat and wear
resistant applications, which include engine components. At
present, there exists no established technology for the mass production, testing, and application of fine ceramics in high-temperature engines. However, as part of the concerted Japanese research
and development effort, several Japanese companies are independently
testing critical engine components. Several advanced ceramic parts
such as swirl chambers and glow plugs for diesel engines are
currently in volume usage, and even ceramic turbocharger rotors were
introduced in selected 1984 Japanese vehicles.

The Japanese government and industry are intent upon developing
the full spectrum of technology necessary to extend electronics
applications and to commercialize mechanical applications of fine
ceramics. If successful, this initiative would further enhance the
competitive position of Japan's electronics industry and also secure
technological leadership in the next generation of energy efficient,
oil- and gas-fueled power plants in automotive, industrial, and
utility applications. Leadership in efficient, energy conversion
systems based upon fine ceramic technology would provide Japanese
industry with a subsequent production cost advantage, minimizing
Japanese dependence on foreign imports of petroleum products and
strategic materials such as cobalt and chromium. Furthermore,
extensive foreign markets could be secured for Japanese products
such as automobiles, internal combustion engines, and turbines.
Clearly, fine ceramics offer Japan an opportunity to minimize its
critical dependence on foreign materials and petroleum while
expanding its economic growth and export market share, upon which
the health of the Japanese economy depends.[1]

Consider the following situation. Advanced ceramics have been
demonstrated to be technically viable in adiabatic engine applications. For example, a turbocharged ceramic diesel, road tested by
the U.S. Army, demonstrated a 33 percent increase in fuel economy
over the equivalent all-metallic engine.[2] In cutting tool applications, advanced ceramics have also increased cutting tool life at
higher cutting speeds and demonstrated 220 percent overall improvements in productivity.[3] Advanced ceramics are also ultimately
viable in turbine blade applications. Consider, if you will,
ceramic-based power turbines operating at better than 50 percent
thermal efficiency to drive advanced ceramic cutting tools to

18

Figure 3. Typical greenware structure of an advanced ceramic component (1-5 μm particles).

produce finished metallic parts for automobiles powered by turbo-compounded ceramic engines. These engines can deliver up to 45 percent better fuel economy than the best currently available engines. Clearly, dominance in advanced ceramics technology has profound national and international competitive implications. Advanced ceramics also represent for the Japanese a rare opportunity to utilize domestically available raw materials such as silica, readily available in common sand. In this light, it is easy to appreciate the motivation behind Japan's national initiative with respect to advanced ceramics.

In 1981, Japan established its Fine Ceramics Technology Research Association, with the primary goals of promoting joint technological development and effective and early entry into the marketplace. The specific research and development objectives are to:

- improve the reliability and reproducibility of ceramic components,
- establish a high temperature ceramic engine system, and
- promote basic research of fine ceramic materials.

The first of these objectives, reliability and reproducibility, addresses the critical shortcoming of advanced ceramics. Improvements in this area will of necessity require enhanced characterization of this material's microstructure and subsequent quantification of the structure-processing relationship.

Japan's collaborative effort is well underway as witnessed by the rapid development of the Fine Ceramic Office. Established early in 1981, it had a membership exceeding 170 companies by October of 1982. This office considers ceramics to be "a new generation of materials to replace metals."

CURRENT STATE OF THE ART IN CERAMICS

A ceramic component is typically manufactured in four discrete steps. First, a dry powder is produced from raw materials. This powder is then slip cast, injection molded, tape cast, or pressed into a desired "greenware" shape. This formed piece is next fired or sintered to consolidate the powder into a monolithic piece.

During the sintering operation this part typically shrinks in size by roughly 14 percent. The last step is the machine finishing of the sintered piece to the final, desired shape.

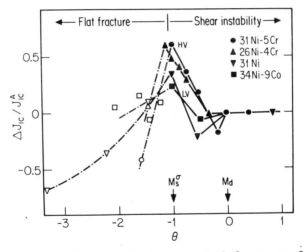

Figure 4. The effects of structural defects on finished ceramic components.

Table 2
Manufacturing Costs of High Value
Added Materials

	Structural Ceramics	Advanced Metals and Polymers
Raw Materials	5-10%	30-60%
Mixing, Shaping	5-10%	10-20%
Firing	10-15%	10-20%
Finishing	30-50%	5%
Rejection[+]	40-60%	5-10%

[+] Cost of rejection varies, 25-75%, depending on process and component

Machine finishing the sintered ceramic piece is necessary because the greenware piece distorts during firing due to its non-uniform packing density. This machine finishing introduces micro-cracks into the surface. These cracks act as stress concentrators and contribute to failures in service. The root of the problem is the irregular packing of the powder particles in the greenware piece which introduces voids and other defects at the front end of the production process which are carried through to the final finished part. These voids, defects, surface cracks, and other defects represent the collection of stress concentration factors which generate the reliability and reproducibility problem exhibited by current advanced ceramic components.

Figure 3 illustrates the microstructure of a high quality greenware piece produced by current commercial advanced ceramic standards. The irregularly sized powder particles have agglomerated and leave a rather nonuniform distribution of voids throughout the piece. During firing, the agglomerated, high density regions sinter or densify rapidly, causing the voids to grow in some cases. These voids are not completely eliminated during the sintering process and represent stress concentration centers in the final piece. In addition, the nonuniform densification of the piece during sintering, which causes the piece to distort, also creates internal stresses which further enhances the stress concentration factors.

The typical state-of-the-art advanced ceramic piece contains sintered defects greater than 50 μm in size. It is these internal defects, surface cracks, and other stress concentration factors which contribute to the high Wyball distribution of the mechanical properties of advanced ceramics. What this means in practical terms is that individual ceramic components exhibit a tremendous distribution of performance. Unfortunately, the engineer must carefully consider this Wyball distribution and design the component based on a guaranteed performance well below the tested range of performance.

Figure 4 illustrates the problems associated with current advanced ceramic processing technologies. It is these defects that also contribute to the high cost of advanced ceramic components, Table 2. With reject rates as high as 75 percent, it is obvious why reproducibility and subsequent reliability are major factors in ongoing ceramics research and development programs.

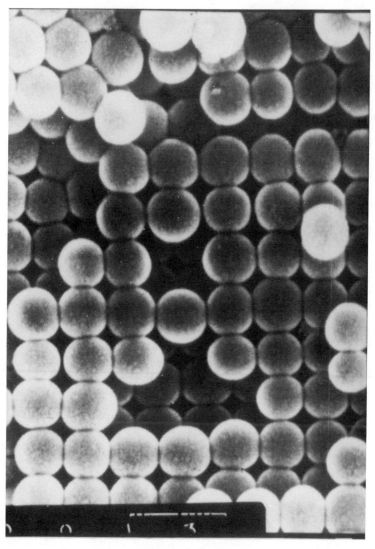

Figure 5. Uniformly packed, submicron (0.3 μm), monosized powders.

One such program is that of the Ceramics Processing Research Laboratory at Massachusetts Institute of Technology (MIT) which is directed by Professor H. Kent Bowen. In the late 1970's, Professor Bowen concluded that the solution to the reproducibility problem in ceramics was the control of the entire ceramics processing system, from powder preparation through sintering, to minimize defect size and population. The proposition, simply put, was to:

a) produce an ideal powder,
b) pack this powder uniformly, and
c) sinter the greenware part into a net final shape requiring no finish machining.

The ideal ceramic powder would be cubic and pack together neatly to nearly full density in the green state. Since it is not possible to produce cubic particles, monosized, unagglomerated, spherical powders were selected as the nearest substitute. After several years of effort, this MIT research team had developed the technology necessary to produce such oxide powders by precipitation from alkoxide solution.[4-6] This same research team went on to develop the expertise necessary to pack these monosized powders into well-ordered greenware structures, Figure 5. Current research efforts are directed towards sintering these unique greenware structures to full density. The early results of these efforts are very encouraging. For example, extremely good densification is possible at much lower sintering temperatures and at much shorter sintering times. Quantification of electrical, magnetic, mechanical, and other properties is under way with very promising early results.

The tremendous progress made to date by MIT's Ceramics Processing Research Laboratory team is directly attributable to improvements in process control. From the very beginning, each step in this process development effort has required meticulous care with respect to testing procedures and characterization. In many instances, several complementary testing and characterization procedures are required to quantify single parameters due to the lack of existing standardized characterization procedures for ceramics. In essence, the advances made to date can be measured by the progress made in identifying, developing, and applying the appropriate testing and characterization procedures necessary to quantify the pertinent process control variables.

As stated earlier, the improved materials required to meet future societal needs depend on innovative processing technology. These can only be developed based on new scientific knowledge and data quantified by appropriate testing and characterization procedures.

25

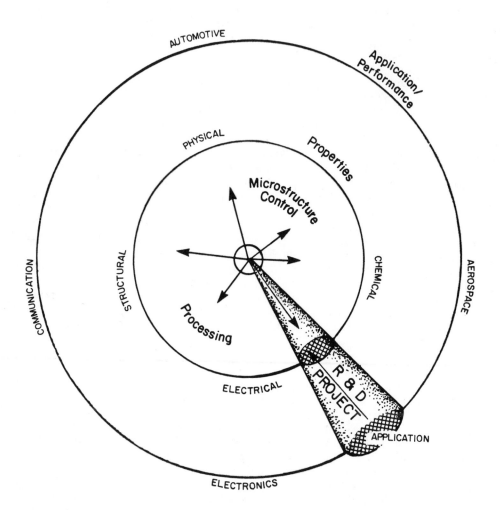

Figure 6. Technology leverage points

TECHNOLOGY LEVERAGE POINTS AND INTERNATIONAL COMPETITION

The fact that the structure of a material determines its properties, in turn determining its performance in applications, is well known to materials engineers and scientists. It is also well known that the structure of every material is sensitive and specific to the process by which it is produced. What is not fully appreciated is the full potential impact and importance of materials process technology and control on structure and performance in a generic materials sense. In fact, process control of microstructure can represent a very potent technology leverage point in any materials system.

As illustrated in Figure 6, processing determines the microstructure of engineering materials, which in turn determines its physical, chemical, electrical, structural, and all other properties. These properties thereby establish the performance of the material in various uses, which might include automotive, aerospace, electronic, communication, or any other applications.

Typically, the materials engineer is asked to improve the performance of a particular material in a specific application. The materials engineer or scientist then defines a research and development project which is designed to enhance a partial property of the material in question, Figure 6. This generally involves modifying the chemistry/alloying of the system or somehow modifying the process itself by which the material is made. By either design or happenstance, the microstructure of the material is altered and, hopefully, its properties and performance are improved. Underlying all of this is a tremendous opportunity in materials development.

It is clear from Figure 6 that generic control of the microstructure of a material provides direct control of all its properties. Microstructural control provides the opportunity to tailor the properties of a material to any particular application. Hence, the materials technology leverage point is defined. By focusing upon microstructural control of a material, or the core of Figure 6, one can take advantage of a multiplicity of application opportunities. The option is a collection of seemingly unrelated research and development projects with the same material geared towards the same series of applications.

The market implications of advanced ceramics as a technology leverage point are illustrated in Figure 7. Technological leadership in the processing of advanced ceramics to provide enhanced properties leverages this generic class of materials into a multitude of seemingly unrelated markets.

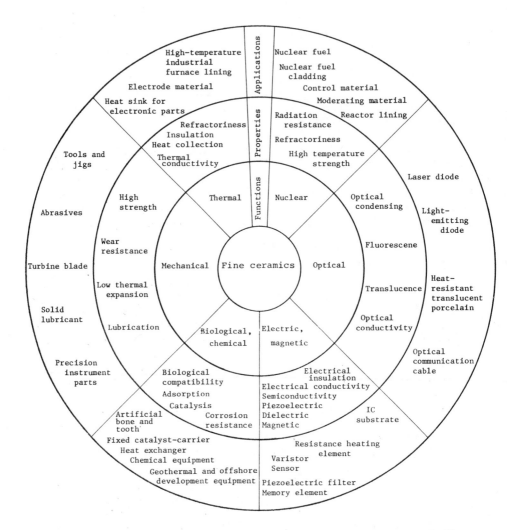

Figure 7. Functions, properties, and applications of fine ceramics.[7]

The potential domestic and international economic impact of these materials based technology leverage points is tremendous. For example, the net impact of U.S. versus foreign dominance in advanced structural ceramics for automotive applications has been estimated to be $280 billion in gross national product and 250,000 domestic jobs between 1985 and 2005. If a foreign country was to dominate advanced ceramic technology in this application alone, the domestic loss would exceed $100 billion in projected gross national product and over 100,000 existing jobs.[8]

Advanced ceramics is but one example of where advances in materials process technology and control represents a potential technology leverage point. However, in all cases, the dependence of advances in materials process and control technology upon the development and application of appropriate materials and process characterization and testing procedures is self-evident.

Given the economics and international competitive issues at stake, this Sagamore Conference on materials characterization and testing is most appropriate and should serve as the foundation for ongoing efforts in this area. The opportunities are clear. Our continued leadership in high technology systems depends upon advances in materials science, engineering, and processing. Future advances are in turn based on materials characterization and testing technology. Time alone will be the judge of whether or not we seize the leadership opportunity in this important, fundamental technology.

REFERENCES

1. G.B. Kenney and H.K. Bowen, "High Tech Ceramics in Japan: Current and Future Markets," Ceramic Bulletin of the American Ceramic Society, Vol. 62, No. 5, (1983).
2. S. Robb, "Cummins Successfully Tests Adiabatic Engine," Bulletin of the American Ceramic Society, Vol. 62, No. 2, (1983).
3. S. Robb, "Ford, GTE Announce Si_3N_4 Cutting-Tool Developments," Bulletin of the American Ceramic Society, Vol. 62, No. 2, (1983).
4. H.K. Bowen, "Ceramics as Engineering Materials: Structure-Property-Processing," Materials Research Society Symposium Proceedings, Vol. 24, Elsevier Publishing Co., (1984).
5. E.A. Barriner, R. Brook, and H.K. Bowen, "The Sintering of Monodisperse TiO_2," in: Sintering and Heterogeneous Catalysis, eds. G.C. Kuczynski, A.E. Miller, and G.A. Sargent, Plenum Publishing Corp., (1984).

6. E.A. Barringer and H.K. Bowen, "Formation, Packing, and Sintering of Monodisperse TiO_2 Powder," Communication of the American Ceramic Society, December, (1982).
7. The Fine Ceramics Office - Its Background and Activities, Ministry of International Trade and Industry, Fine Ceramics Office, November, (1983).
8. L.R. Johnson, A.P.S. Teotia, and L.G. Hill, "A Structural Ceramics Research Program: A Preliminary Economic Analysis," Center for Transportation Research, ANL/CNSV-38 Argonne National Laboratory, March, (1983).

NONDESTRUCTIVE MATERIALS CHARACTERIZATION

Robert E. Green, Jr.

Center for Nondestructive Evaluation
The Johns Hopkins University
Baltimore, MD 21218

ABSTRACT

In materials production industries it has become increasingly evident that nondestructive techniques are required in order to monitor and control as many stages in the manufacturing process as possible. Ideally it would be desirable to have nondestructive measurement techniques which are capable of monitoring the production process; the material stability during transport, storage, and fabrication; and the rate of degradation of materials during their in-service life. Although, historically, nondestructive techniques have been used to detect flaws in a qualitative manner, in recent years increased attention has been given to the development of more sophisticated techniques for the quantitative nondestructive characterization of materials in order to assure that the devices and structures fabricated from these materials function properly and safely throughout their intended service life. Several such innovative nondestructive materials characterization techniques are described in the present paper. Finally, some comments will be made with regard to the role to be played by nondestructive materials characterization in the manufacture of advanced materials in the 21st century.

INTRODUCTION

As clearly pointed out by McCauley[1], in order to have advanced materials possessing optimum properties and high in-service reliability for ever increasing high-technology applications, particularly to meet the requirements of the 21st century, "Traditional nondestructive testing must merge with laboratory-based materials characterization concepts into a unified field of

nondestructive or semidestructive property (unique signature)
evaluation." Materials characterization has been defined by the
National Academy of Sciences Materials Advisory Board[2] as
follows: "Characterization describes those features of the
composition and structure (including defects) of a material that
are significant for a particular preparation, study of properties,
or use, and suffice for reproduction of the material."

Traditionally the vast majority of materials characterization
techniques have been destructive, e.g. chemical compositional
analysis, metallographic determination of microstructure, tensile
test measurement of mechanical properties, etc. Also, traditionally, nondestructive testing techniques have been primarily
limited to the detection of macroscopic defects, particularly
cracks, in structures and devices which have already been constructed and most often have already been in service for a
relatively extended period of time. Following these conventional
nondestructive tests, it has been standard practice to use somewhat arbitrary accept-reject criteria as to whether or not the
structure or device should be removed from service. The present
unfavorable status of a large segment of American industry, e.g.
metals production, coupled with the premature failure of
structures and devices, e.g. bridges and automobiles, dramatically
show that our traditional approaches must be drastically modified
if we are to be able to meet future industrial and military needs.

The development of high quality low porosity ceramic materials
capable of performing reliably at high temperatures, the optimization of metal-matrix and polymeric composites, and the improvement
of metallic superalloys can only result from proper application of
nondestructive materials characterization techniques to monitor
and control as many stages in the production process as possible.
A number of innovative nondestructive materials characterization
techniques are described in the present paper. Some of these have
already been developed to a high degree of sophistication and
successfully applied to specific materials characterization
problems, while others are still in the developmental stage and
the range of their applicability has not been fully defined.
Suggestions will also be given as to sources of new innovative
nondestructive materials characterization techniques which will
play an ever increasing role in the manufacture of high quality
advanced materials in the 21st century.

RAPID X-RAY DIFFRACTION IMAGING

Despite the wide use of electro-optical systems for remote
intensified viewing of x-ray radiologic and radiographic images,
the use of such systems for viewing x-ray diffraction images has
been much more limited. However, rapid viewing and recording of
x-ray diffraction images affords a valuable tool for nondestructive

characterization of materials. In order to construct a system for such applications consideration must be given to optimizing both x-ray generation and detection in the wavelength regime suitable for diffraction.

In 1971, the present author published a review paper[3] which described the design and performance characteristics of electro-optical systems for direct display of x-ray diffraction images. Subsequently, Green[4] in 1977, and Winter and Green[5] in 1985, published other papers related to this subject. All of the systems reported may be grouped into two generic categories depending on the techniques used to permit rapid viewing and recording of the x-ray diffraction images, namely the direct method and the indirect method. In the direct method, an x-ray sensitive vidicon television camera directly converts the x-ray image into an electronic charge pattern on a photoconductive target, which is read out by a scanning electron beam and displayed as a visible image on a television monitor. In the indirect method, the x-ray diffraction image is converted into a visible light image by a fluorescent screen. This visible light image is then optically coupled either by a lens or a fiber-optic faceplate to the input photocathode of a low-light-level electro-optical device. Depending on the type of electro-optical device used, the output image may either be viewed directly or displayed on a television monitor. There are two categories of high gain electro-optical detectors used with the indirect x-ray diffraction imaging method. In the first category fall those systems which use a low-light-level television tube as a detector, display the diffraction image on a television monitor, and then use a video-tape or video-cassette recorder to make a permanent record of the image. In the second category are those systems which use a low-light-level image intensifier tube as a detector and either use a camera to photograph the image or an auxiliary television camera to display the image on a monitor and to record it on a video recorder.

Most of the research of the present author has been conducted using image intensifier systems since the modular approach is the most practical and it has been shown that a multiple-stage image intensifier system coupled to an external fluorescent screen is the most sensitive and only truly instantaneous system.

Figure 1 shows schematic drawings of one type of commonly used first generation portable x-ray image intensifier and a more recently developed second generation miniature x-ray image intensifier. With the first generation device an x-ray image incident on the fluorescent screen scintillator is converted to a visible light image. This visible light image travels through the fiber optic input faceplate and is converted into an electron image at the photocathode. The electrons constituting this image are accelerated by an electrostatic field and focused by a conical electron lens onto the output phosphor screen where they are converted into a stronger visible image. This intensified visible image is transmitted through the output fiber-optic faceplate of the first stage and through the imput fiber-potic faceplate of the second stage where it is amplified a second time and similarly for the third stage. Three stages of amplification are such that individual x-ray photons can be detected, since the luminous flux gain of top quality three stage image intensifier tubes is more than one million.

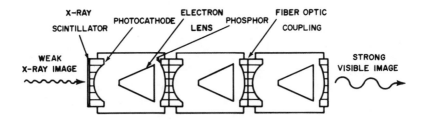

PORTABLE IMAGE X-RAY INTENSIFIER

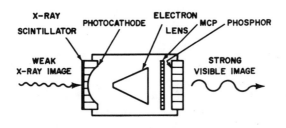

MINIATURE IMAGE X-RAY (MINIX) INTENSIFIER

Figure 1. Schematic drawings of first generation (upper) and second generation (lower) x-ray image intensifier tubes for rapid display of x-ray diffraction images.

The newer second generation miniature x-ray image intensifier operates in essentially the same fashion as the first generation device except for the extremely important addition of a microchannel plate. The electron image from the photocathode is focused by the electron lens upon the input face of the microchannel plate. In the microchannel network the electrons are multiplied many times and upon emerging strike the output phosphor where they are converted into a strong visible image. Although the gain of this type intensifier tube is only several hundred thousand, one can afford to sacrifice some gain when using high intensity x-ray generators. Moreover, the small size of this type tube makes it ideal for attachment to goniometers.

These electro-optical systems have been used in the author's laboratory to orient single crystals, to study crystal lattice rotation accompanying plastic deformation, to measure the rate of grain boundary migration during recrystallization annealing of cold-worked metals, to determine the physical state of exploding metals, to monitor the amorphous to crystalline phase transformation of rapidly solidified metals, to rapidly measure residual stress, to study the dynamics of structural phase transitions in ferroelectric crystals, and to record topographic images of lattice substructure and defects.

Figure 2 shows a schematic illustration of the application of a dynamic x-ray diffraction system for viewing and recording real time Debye-Scherrer x-ray diffraction patterns during the transformation of rapidly solidified amorphous metals to the more stable crystalline state[6]. As shown in the figure, a stream of molten metal is ejected from an orifice in a non-reactive container onto a rotating drum chill block. When the proper experimental conditions are present the molten metal solidifies sufficiently rapidly so that the distribution of atoms prevalent in the liquid state are retained in the solid state. However, since the normal equilibrium state of such rapidly quenched materials is crystalline, many materials will transform to this equilibrium state rapidly. By using the dynamic x-ray diffraction system shown and by changing the distance between the spinning drum chill block and the collimated x-ray beam, all intermediate transformation states can be readily monitored and the associated transformation times determined. With the large amount of research devoted to development of amorphous metallic alloys, which possess chemical and physical properties superior to their crystalline counterparts, dynamic x-ray diffraction systems could play a major role in assessing the stability of these materials and provide feedback to optimize and control their manufacture. In fact, such systems can be used in a similar manner to control the processing of more conventional materials.

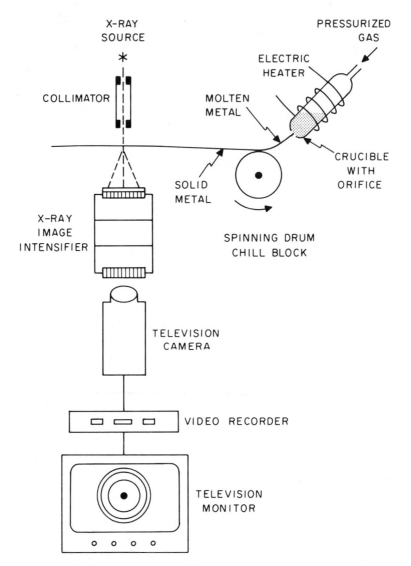

Figure 2. Application of a dynamic x-ray diffraction system for viewing and recording real-time Debye–Scherrer x-ray diffraction patterns during the transformation of rapidly solidified amorphous metals to the more stable crystalline state[6].

Figure 3 serves to show how a similar x-ray imaging electro-optical system can be combined with an asymmetric crystal topographic system[7] to display substructure in a copper single crystal. These photographs, which show surface reflection topographs taken at slightly different glancing angles yield quantitative information about the degree of lattice strain and microorientation in the crystal. Figure 3(c) indicates the best glancing angle to use for a topograph is coverage of a large area of the copper crystal is desired. On the other hand, Figures (a), (b), or (d) are to be preferred for obtaining most information about one particular sub-grain. Observation of the topographic image on the output of the image intensifier tube while rocking the crystal yields a series of topographs (topographical analog to a rocking curve) each with good sensitivity to strain. This yields instantaneous information about the distribution of strain over the entire area of the crystal.

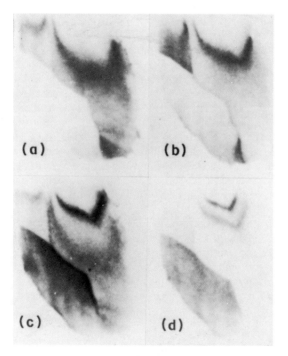

Figure 3.   Surface reflection x-ray topographs taken at slightly different glancing angles from a copper single crystal containing sub-grains. The images reproduced were instantaneously visible on the x-ray image intensifier output[7]. (a) initial setting, (b) 84 sec rotation, (c) 120 sec additional rotation, (d) 234 sec additional rotation.

A review paper was published in 1973 which traced the development of generation and detection systems for flash x-ray diffraction and summarized the state-of-the-art of such systems[8]. A comparative evaluation was presented of flash x-ray diffraction systems using generators which rely on increased electron beam current and those which rely on higher voltage. A detailed description of the most rapid flash x-ray diffraction developed to date was given. This system used a Field Emission Fexitron single channel 300kV pulsed x-ray generator incorporating an x-ray tube with a beryllium output window. A first generation three-stage image intensifier tube, as described above, was used as a detector. Using this system Laue transmission x-ray diffraction patterns of single crystals and Debye-Scherrer diffraction patterns of polycrystalline aggregates were obtained with exposure times of 30 nanoseconds. A second review was presented in 1977, which was directly concerned with the application of flash x-ray diffraction systems to materials testing[9]. The x-ray diffraction patterns obtained using such systems have yielded detailed information about microstructural alterations caused by explosions, heat pulses, and shock waves.

Recently a flash x-ray diffraction system based on a commercially available pulse generator incorporating a specially designed demountable field emission tube was applied to the investigation of structural phase transformations in ferroelectric crystals caused by polarity switching and electrically initiated temperature jumps[10]. A photograph showing the main components of this system is reproduced in Figure 4. An illustration of the use of this system is a study of the ferroelectric domain switching in gadolinium molybdate (GMD). A strong electric field was applied to a thin plate shaped GMD crystal. By reversing the polarity of the field, an interchange of the intensity of the (560) and (650) reflections was observed. Figure 5 shows typical x-ray diffraction patterns obtained from a GMD crystal in the two oppositely polarized states. The position of reflection (650) was first recorded at the location indicated by the black arrow in Figure 5a. After the field reversal this reflection was recorded at the location indicated by the white arrow in Figure 5b, where it replaced the (560) reflection which was too weak to be recorded.

Continuing improvements in electro-optical technology suggest that such systems will find ever increasing application in monitoring a variety of materials processing applications. Included in these improvements are third generation image intensifier tubes incorporating gallium arsenide -- aluminum gallium arsenide photocathodes, solid state charge coupled and charge injection television cameras with digital signal output, and perhaps most promising of all computer controlled digital signal processing systems.

Figure 4. Main components of flash x-ray diffraction system[10].
(1) Cable connection to x-ray pulse generator,
(2) Remote tubehead, (3) Tubehead flange, (4) Vacuum
discharge tube endplate, (5) Vacuum pump, (6) Tube-
vacuum pump connection pipe, (7) Specimen mount
goniometer, (8) X-ray image intensifier, (9) Intensifier power supply, (10) Polaroid camera.

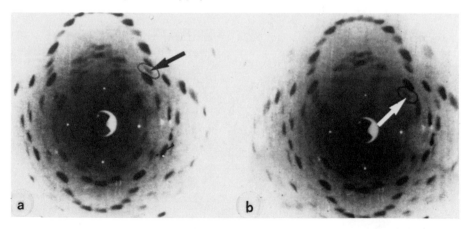

Figure 5. X-ray diffraction patterns from GMD crystal[10].
(a) Before domain switching, (b) After domain
switching. Arrows indicate positions of (650)
reflection.

ULTRASONIC VELOCITY AND ATTENUATION MEASUREMENTS

Ultrasonic techniques offer very useful and versatile non-destructive methods for investigating the microstructure and associated mechanical properties of materials. Although historically ultrasonic techniques have been used to make such evaluations in a qualitative manner, current efforts are directed at development of more sophisticated ultrasonic techniques capable of quantitative sizing of flaws permitting determination of material response using fracture mechanics analysis.

The use of ultrasonic waves as nondestructive probes has as a prerequisite the careful documentation of the propagational characteristics of the ultrasonic waves themselves[11]. Since in nondestructive evaluation applications it is not desirable for ultrasonic waves to alter the material through which they pass, it is necessary to work with very low amplitude waves, which normally are regarded to obey linear elasticity theory. In general, three different linear elastic waves may propagate along any given direction in an anisotropic material. These three waves are usually not pure modes since each wave generally has particle displacement components both parallel and perpendicular to the wave normal. However, one of these components is usually much larger than the other; the wave with a large parallel component is called quasi-longitudinal, which the waves with a large perpendicular component are called quasi-transverse. Also of great importance to elastic wave propagation in anisotropic materials, the direction of the flow of energy per unit time per unit area, the energy-flux vector does not, in general, coincide with the wave normal as it does in the isotropic case, i.e. the ultrasonic beam exhibits refraction even for normal incidence.

Here is a good place to call attention to the fact that, although many investigators draw a close analogy between electromagnetic (light) wave propagation and elastic (sound) wave propagation in solid materials, great caution should be exercised in doing so. The behavior of anisotropic materials with respect to propagation of elastic waves is much more complicated than the propagation of electromagnetic waves, since the material constitutive equations required to properly describe elastic waves are of higher order tensor quantities than those required to describe electromagnetic waves[12].

Although geometrical effects can cause energy to be lost from the ultrasonic beam, such losses are not indicative of intrinsic loss mechanisms associated with the microstructure. Once proper precautions are taken to either eliminate or control these geometrical effects, ultrasonic attenuation measurements serve as a very sensitive indicator of internal loss mechanisms caused by microstructures and microstructural alterations in metallic

materials[1 3]. This sensitivity derives from the ability of
ultrasonic waves to interact with volume defects such as cracks,
microcracks, included particles, and precipitates; surface
defects such as grain boundaries, interphase boundaries, and
magnetic domain walls; and dislocation line defects. Moreoever,
ultrasonic waves are sensitive to the interaction of point
defects such as impurity atoms and vacancies with dislocations.

Klinman and Stephenson[1 4] conducted an experimental study
of longitudinal ultrasonic wave attenuation versus grain size
measurements for both laboratory and commercially produced plates
of plain carbon-manganese ferrite-pearlite steel. Figures 6 and
7 show their experimental results. The curves drawn through the
data points are least squares fits to the Rayleigh scattering
relation. In order to determine the effect of frequency
measurements were made at both 5 and 10MHz on the laboratory
produced steel plate. As the results show in Figure 6, it is
evident that for this material the higher frequency established
a definite relationship between ultrasonic attenuation and
metallographically measured grain size, while the lower
frequency did not. However, comparison between the 10MHz data
in Figure 6 with the data in Figure 7, also measured at 10MHz, shows
the much larger scatter in data obtained with the factory
produced steel plate.

Figure 8 demonstrates the ability of ultrasonic attenuation
measurements to detect extremely small microstructural alterations
during fatigue testing and hence give early warning of fatigue
induced failure. This figure was taken from the work of Joshi and
Green[1 5] who used ultrasonic attenuation measurements as a
continuous monitor of fatigue damage in polycrystalline steel
specimens which were cycled in reverse bending as cantilever beams
to fracture at 30Hz. Note that the ultrasonic attenuation change
indicated microstructural alterations, probably microcrack
formation, prior to detection of an additional ultrasonic pulse
on the A-scan display as is used in conventional nondestructive
ultrasonic testing to detect crack formation.

Gefen and Rosen[1 6] performed innovative studies of the
relationship between the attenuation of ultrasonic waves and
changes in the precipitate structures occurring during the aging
process in age hardening alloys. Figure 9 from their work shows
the variation in ultrasonic attenuation of 10MHz longitudinal
waves in Duraluminum 2024 specimens during isothermal aging at the
temperatures indicated. Gefen and Rosen attributed the observed
ultrasonic attenuation maxima to relaxation processes caused by
the growth of precipitates and the accompanying structural and
coherency changes. They pointed out that Duraluminum 2024 should
exhibit two maxima in the ultrasonic attenuation due to the
formation of GPB and GP zones, respectively. The maxima in

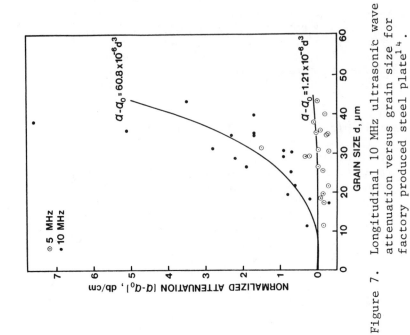

Figure 7. Longitudinal 10 MHz ultrasonic wave attenuation versus grain size for factory produced steel plate[14].

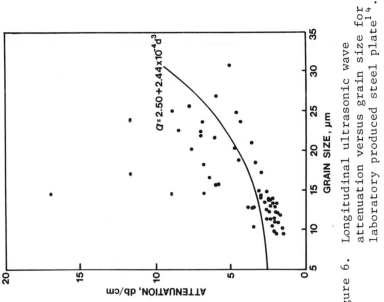

Figure 6. Longitudinal ultrasonic wave attenuation versus grain size for laboratory produced steel plate[14].

42

attenuation occurs at aging times when the interaction of the
ultrasonic waves with the strain field around the precipitates is
also a maximum. This was confirmed by additional experiments
which showed that at higher ultrasonic frequencies the maximum
interaction occurred for smaller precipitates at shorter aging
times.

In addition to the examples cited here a large amount of
work has been conducted on the scattering of ultrasonic waves by
grains in polycrystalline aggregates in the Rayleigh, stochastic,
and diffusive frequency regimes. Moreover, in recent years
considerable attention, both experimental and theoretical, has
been given to the use of ultrasonic wave scattering from both
simulated and real defects in an effort to determine their size,
shape, and orientation. Although this effort has only met with
partial success, work in this area is continuing.

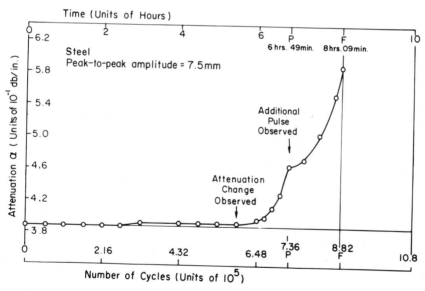

Figure 8. Longitudinal ultrasonic wave attenuation versus number
of fatigue cycles for polycrystalline steel specimen[15].

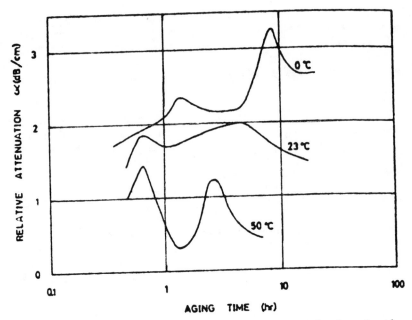

Figure 9. Variation of ultrasonic attenuation during isothermal
aging of Duraluminum 2024 at different temperatures[16].

Nonlinear effects associated with ultrasonic wave propagation
in materials may also be used to advantage for nondestructive
materials characterization[17]. Although x-ray diffraction
techniques have historically been the nondestructive method most
often used in actual practice to measure residual stress (strain),
they are not optimally suited for field applications partially
because the necessary equipment is heavy and bulky and, perhaps
more importantly, they suffer from the fact that they only serve
to determine the state of stress (strain) in a surface layer of
a material, while in many practical cases a knowledge of the bulk
stresses (strains) is desired. Ultrasonic residual stress
(strain) measurements have been made far less frequently,
probably because of lack of familiarity with nonlinear elasticity
theory and the experimental techniques necessary to obtain the
required ultrasonic velocity measurement accuracy. Most ultrasonic measurements of residual stress (strain) have been based on
stress induced acoustical birefringence of a homogeneous isotropic

44

solid. When shear (transverse) waves are propagated through a homogeneous isotropic solid at right angles to the direction of applied or residual stress, the shear wave with particle displacement parallel to the direction of the stress propagates with a faster velocity than the shear wave with particle displacement perpendicular to the direction of the stress. Other ultrasonic techniques can also be used to make such measurements. For example, five of the equations derived originally by Hughes and Kelly[18] relate an elastic wave velocity to a stress component in the material. These equations can be used either singly or in appropriate combination to determine stress components from experimental measurements of various ultrasonic wave velocities. Moreover, additional features of nonlinear elastic wave propagation such as harmonic generation, temperature dependence of ultrasonic velocity, change in velocity as a function of applied magnetic field, or ultrasonic wave dispersion have been proposed for residual stress determinations. Details of most of these measurement techniques as well as a survey of theoretical considerations may be found in a recent review article[19].

The major difficulty associated with reliable ultrasonic residual stress (strain) measurement is that the change in elastic wave velocities in solid materials due to residual stress is small and other factors which can cause greater velocity changes may mask residual stress effects. The prime factor in this regard is the fact that all real structural materials are initially anisotropic and not isotropic. The solidification and/or forming and heating processes associated with metal preparation does not permit a random distribution crystallographic orientations among the grains of the polycrystalline aggregate and often does not even permit a uniformity of grain size. Therefore, real polycrystalline metals possess a "texture" and this texture strongly influences the mechanical properties of the metal including ultrasonic wave propagation. Thus, the ultrasonic residual stress measurement problem becomes one of measuring changes in anisotropy before and after stressing rather than the commonly treated theoretical problem of stress-induced anisotropy of an originally isotropic solid.

Additional problems arise with ultrasonic wave propagation in inhomogeneous materials. The presence of a single bounding surface complicates the propagational characteristics of ultrasonic waves in solid materials and can lead to erroneous interpretation of velocity and attenuation measurements. The presence of many bounding surfaces, such as may occur in composites, complicates the propagational characteristics even more and, except in a few special cases, the problems have not been solved analytically. However, solution of these problems will permit proper ultrasonic measurements to be an invaluable tool in characterizing composite materials.

OTHER INNOVATIVE TECHNIQUES

A number of other techniques have been developed or greatly improved over the past 10 years which also have a high probability of playing an important role in nondestructive characterization of materials during the various stages of their manufacturing process in the 21st century. Unfortunately space limitations do not permit more than mention of a number of these techniques and a brief description of a few of them.

X-ray

High resolution real-time radiographic imaging includes the use of microfocus x-ray generators, new high resolution fluorescent screens, slit scanners utilizing fluorescent coatings coupled to fiber optic bundles, and fluorescent fiber optics coupled to image intensifiers. New developments in this area can permit high sensitivity detection without the normally associated loss in resolution. With higher resolution imaging devices, signal digitizing, and computer processing and storage, film can be eliminated, removing a major inspection cost and storage problem.

Computerized axial tomography is a method of image reconstruction utilizing multiple images. These images are then processed to form a digital view of a slice of the object. While utilized primarily in the medical field this technique has the potential to supplant film radiography in industrial applications. With the continued improvement in small dedicated computers practical industrial tomography becomes feasible. For mass produced items the time for reconstruction becomes a problem. Research is proceeding to reduce this reconstruction time.

Rapid residual stress determination systems have been undergoing development for some time. Properly applied x-ray diffraction techniques can accurately determine near surface stress (strain) characteristics of crystalline materials. The use of image intensifier and position sensitive proportional counter detectors coupled with computer processing of the experimental data has permitted such measurements to be made very rapidly. Work has been proposed to directly measure internal stress distributions at greater depths below the surface by using high energy x-ray sources.

Acoustic

Laser beam ultrasound generation and detection affords the opportunity to make truly non-contact ultrasonic measurements. Incorporation of scanning techniques would greatly increase the capability of testing large structures without the present necessity of either immersing the test object in a water tank or using water squirter coupling. In addition measurements can be made on hot bodies and in hostile corrosive environments. Laser beam interferometric detectors can also be used for acoustic emission monitoring in similar environments and on moving machinery such as turbines. More work needs to be conducted on both the generation and detection techniques as well as scanning methods for optimization as far as size, cost, and ruggedness for field operation is concerned. Figure 10 taken from the work of Rosen[19] shows a schematic of such a non-contact system which used a Nd:YAG pulsed laser to generate Rayleigh waves at the surface of a variety of metal specimens. The surfaces tested had been microstructurally modified by either electron beam or high-power laser irradiation to either form an amorphous layer on the crystalline bulk, to produce different layers of transformed phases, or to induce case hardening of steel components. The Rayleigh waves were detected in a non-contact manner using a dual laser beam interferometer.

Although it has been the general practice to use large solid-state pulsed laser to generate acoustic waves, recently Dewhurst[20] constructed a hand-held laser generator of ultrasonics based on some components from a laser rangefinding device. A photograph of this device is reproduced in Figure 11. Operating at a wavelength of 1.06 microns, the hand-held laser has provided optical pulses on a single shot basis with energies of 3-10 milli-joules in 10 nanoseconds. This optical output was sufficient to generate both surface and bulk ultrasonic waves with an aluminum specimen.

Optical

High resolution optical holographic interferometry may be used to measure the surface topography of an object with great precision. Current optical holographic interferometric techniques routinely measure surface displacements of test objects only to one-half the wavelength of the laser light used (approximately 3000 Angstroms for helium-neon). On the other hand, optical laser beam interferometric point probes routinely measure surface displacements to better than a fraction of an Angstrom. One method of getting around the limited dynamic range problem associated with contour holographic interferometry of complex surfaces is to use heterodyne techniques to interpolate between fringes. Figure 12 shows a top view of the tibial component of an artificial knee prosthesis using a double exposure holographic contouring technique. This hologram was obtained by using a different reference beam to record each image, and only when both reference beams illuminate the hologram will an interferogram be observed. If one of the two reference beams is shifted to a frequency slightly different from the other beam, then the previously observed

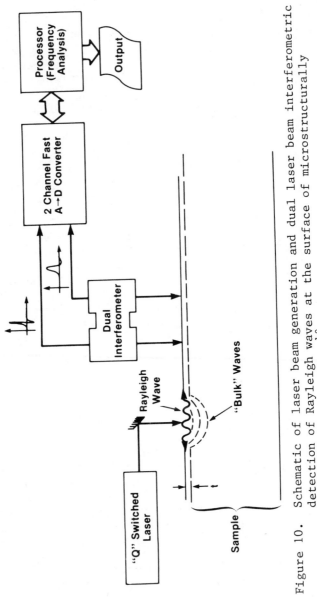

Figure 10. Schematic of laser beam generation and dual laser beam interferometric detection of Rayleigh waves at the surface of microstructurally modified metal surfaces[19].

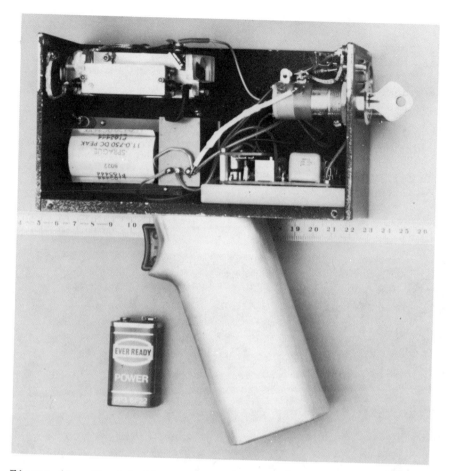

Figure 11. Hand-held laser generator of ultrasonic pulses[20].

stationary interference pattern appears to precess over the reconstructed image of the object. At shift frequencies on the order
of 100 kHz, the fringes move at such a rate as to be invisible to
the human eye. If, however, a lens is used to project the interferogram image onto a photodetector, then the effect of the
passing fringes is to generate a sinusoidal output current from
the photodetector which varies at the 100 kHz difference frequency.

If a second detector is placed next to the first in the image plane, it too will produce a 100 kHz sinusoidal output but perhaps at some phase difference relative to the first signal. In fact, if the distance between the two photodetectors is exactly one half of the fringe spacing observed in the stationary fringe pattern, then the phase difference between the two signals will be 180 degrees. Therefore, by using an electronic phasemeter to measure the phase difference between the two photodetector signals, one may determine the number of fringes between any two points on the image to some fraction of a fringe. Under the proper conditions, phase determinations of better than 0.4 degree may be made. This permits interpolation between fringes to 1/1000 of a fringe.

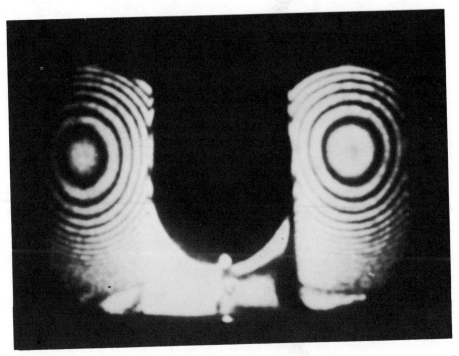

Figure 12. Holographic interferogram of the top view of a tibial component of an artificial knee prosthesis[21].

Figure 13 shows the results of using this type of heterodyne analysis on one of the surfaces of the knee prosthesis. In this case, one detector remained fixed in the image plane, while the other detector was scanned over the imaged surface. The phase

difference between these two signals was then plotted as a
function of the scanned detectors position. Note the gouge which
is readily discernable in Figure 13 but not at all apparent from the
static contour hologram shown in Figure 12.

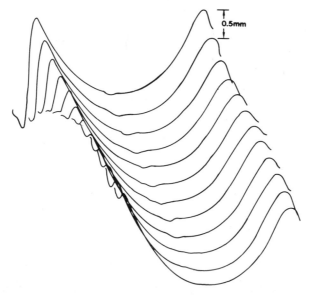

Figure 13.   Heterodyne analysis of one surface of the
knee prosthesis depicted in Figure 12
showing gouge[21].

Fiber optic sensor technology has markedly advanced in recent
years due primarily to developments in communication technology.
In addition to this application fiber optic sensors have been used
to detect pressure and temperature changes and acoustic waves.   A
promising area of application is for permanent assessment of the
structural integrity of relatively large composite structures such
as airframes by embedding optical fibers in the initial composite
lay-up and after assembly either continuously or periodically
monitor the transmission of light through the fiber.   Not only
would broken or highly stressed fibers indicate major damage
sites, but some evidence has accumulated to show that interfacial
problems between fiber and matrix may also be assessed.

Thermal

Vibrothermography is a technique which uses a real-time heat imaging device such as a pyroelectric vidicon or a video-thermographic camera to observe the heat patterns generated at defects by mechanical vibration of the material under test. This technique has been successfully applied to locate fatigue cracks and microstructural inhomogeneities in metals and a variety of structural defects and damage sites in complex composite laminates and structures. Figure 14 shows a vibrothermograph of an E-glass epoxy gate rotor seal from a hydraulic pump. Hot spots (white areas) are visible at the base of the teeth indicating regions where damage had developed locally during use of the pump. In order to make the vibrothermograph the rotor was mounted in a mechanical shaker and vibrated at approximately 13 kHz. The hot spots are 3-3.5°C above ambient temperature and are due to local matrix crazing and small regions of delamination.

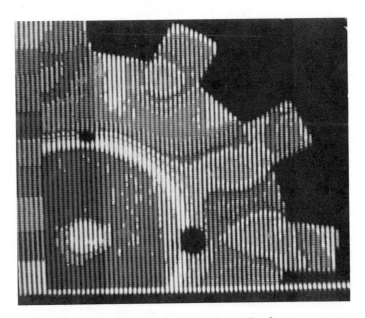

Figure 14. Vibrothermograph of an E-glass epoxy
gate rotor seal from a hydraulic
pump[22].

Photothermal imaging or photoacoustic microscopy, as
initially developed, is a technique where laser beam scanning of
a test object placed in a closed gas-filled container causes
changes in the gas pressure in direct proportion to thermal
property changes in the surface layers of the test object.
Recording of the pressure changes as a function of position of
the laser pump beam permits imaging of surface and sub-surface
microstructural features and defects in the test object. More
recent developments have permitted elimination of the gas-filled
container by use of a second probe laser beam which either
detects surface displacements of the test object due to localized
thermal expansion or changes in the refractive index of the air
just above the sample surface. Figure 15 illustrates the way
this "mirage effect" caused by the pump laser beam interacting
with the test object causes deflections of the probe laser beam
transversing the index of refraction altered region. Another
modification of this technique uses a chopped electron beam in a
scanning electron microscope to excite elastic and thermal waves
at the top surface of a test specimen. The propagation of these
waves through the test specimen is modified by the mechanical and
thermal characteristics of the material inhomogeneities through
which they pass. These modified waves are detected by a
piezoelectric crystal coupled to the bottom side of the test
specimen. By displaying and recording the output of the piezoelectric crystal as a function of position of the scanning electron
electron beam an "electron-acoustic" image of the test specimen
can be obtained. Alteration of the energy of the electron beam
and chopping frequency permits different layers in the test
specimen to be imaged as desired. Figure 16 compares a standard
scanning electron image of a sheet of polycrystalline aluminum
with an electron-acoustic image of the same identical region of
the test specimen. The electron-acoustic image reveals the grain
structure not detected by the secondary electron image. Figure 17
shows electron-acoustic magnitude and phase images of a COS/MOS
integrated circuit chip. Note that the contrast mechanisms
involved in these two images are different revealing different
sub-surface structures. There is no question that these emerging
thermal wave imaging techniques will make a marked impact on
nondestructive materials characterization.

Others

In addition to the techniques already mentioned there is no
doubt that improved eddy current and other electrical and magnetic
techniques will play a major role in nondestructive materials
characterization in the future. A number of electronic and
nuclear techniques, such as nuclear magnetic resonance and electron
spin resonance, which have hitherto been regarded as pure laboratory research tools, will find ever increasing practical application. There will also be continued improvements in computer

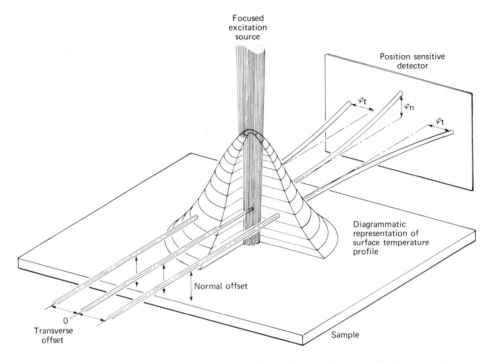

Figure 15. Schematic of photothermal imaging technique using the "mirage effect"[23]

controlled data acquisition and processing systems. Finally, it is suggested that since coupled physical phenomena, such as piezoelectricity, photoelasticity, and thermoelectricity, have proven to be very useful for making nondestructive measurements in the past, it is highly probable that in the 21st century other coupled phenomena will be the basis for new nondestructive materials characterization techniques.

 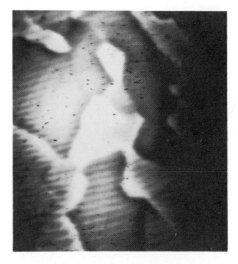

SECONDARY ELECTRON IMAGE   ACOUSTIC MAGNITUDE IMAGE

Figure 16. Comparison of conventional scanning electron microscope image of high purity polycrystalline aluminum sheet with electron-acoustic image[24].

 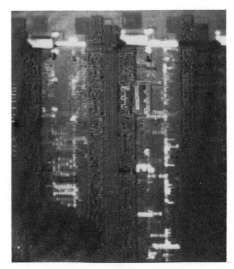

18 kV, 42X, 209.03 kHz, MAGNITUDE   18 kV, 42X, 209.03 kHz, PHASE

Figure 17. Electron-acoustic magnitude and phase images of a COS/MOS integrated circuit chip[24].

SUMMARY

Consideration has been given to several innovative nondestructive materials characterization techniques including rapid x-ray diffraction imaging, ultrasonic velocity and attenuation measurements, high resolution real-time radiographic imaging, computerized axial tomography, rapid residual stress determination, laser beam ultrasound generation and detection, high resolution optical holographic interferometry, fiber optic sensors, vibrothermography and photothermal imaging. Suggestions and examples have been given as to the role to be played by these techniques in the manufacture of advanced materials in the 21st century.

ACKNOWLEDGEMENTS

Thanks are due Professor Robert B. Pond, Sr. (The Johns Hopkins University), Dr. Masao Kuriyama (National Bureau of Standards), Professor Dov Rabinovich (Weizmann Institute, Israel), Drs. Ron Klinman and Ed Stephenson (Bethlehem Steel), Professor Moshe Rosen (The Johns Hopkins University), Dr. Richard J. Dewhurst (University of Hull, England), Professor James W. Wagner (The Johns Hopkins University), Professor Edmund G. Henneke, II (Virginia Polytechnic Institute and State University), and Dr. John C. Murphy (The Johns Hopkins University Applied Physics Laboratory) for supplying figures and technical information. Special thanks are due Mrs. Harriett Horton for her skill in typing the manuscript and to Mrs. Sydney Green for mounting the figures.

REFERENCES

1. J. W. McCauley, "The Role of Characterization in Emerging High Performance Ceramic Materials," Ceramic Bulletin, Vol. 63, No. 2, pp. 263-265 (1984).
2. "Characterization of Materials," NAS Materials Advisory Board, MAB-229-M, March (1967).
3. R. E. Green Jr., "Electro-Optical Systems for Dynamic Recording of X-ray Diffraction Images," Advances in X-ray Analysis, Vol. 14, pp. 311-337, Plenum Press, NY (1971).
4. R. E. Green Jr., "Direct Display of X-ray Topographic Images," Advances in X-ray Analysis, Vol. 20, Plenum Press, NY pp. 221-235 (1977).
5. J. M. Winter and R. E. Green Jr., "Rapid Imaging of X-ray Topographs," to be published in Applications of X-ray Topographic Methods to Materials Science, Plenum Press, NY (1985).
6. C. Beczak, R. B. Pond Sr., and R. E. Green Jr., The Johns Hopkins University (unpublished work).
7. W. J. Boettinger, H. E. Burdette, M. Kuriyama and R. E. Green Jr., "Asymmetric Crystal Topographic Camera," Rev. Sci. Instrum. 47, pp. 906-911 (1976).

8. J. A. Dantzig and R. E. Green Jr., "Flash X-ray Diffraction System," Advances in X-ray Analysis, Vol. 16, pp. 229- 241 (1973).

9. R. E. Green Jr., "Applications of Flash X-ray Diffraction Systems to Materials Testing," Proceedings of Flash Radiography Symposium, The American Society for Nondestructive Testing, Columbus, Ohio, pp. 151-164 (1977).

10. R. E. Green Jr. and D. Rabinovich, "Exploration of Structural Phase Transformations by Flash X-ray Diffraction," to be published in Proceedings of Flash X-ray Symposium, The American Society for Nondestructive Testing, Columbus, Ohio (1985).

11. R. E. Green Jr., Ultrasonic Investigation of Mechanical Properties, Vol. III Treatise on Materials Science and Technology, Academic Press, NY (1973).

12. E. G. Henneke II., and R. E. Green Jr., "Light Wave – Elastic Wave Analogies in Crystals," J. Acoust. Soc. Amer. 45, pp. 1367-1373 (1969).

13. R. E. Green Jr., "Effect of Metallic Microstructure on Ultrasonic Attenuation," Nondestructive Evaluation: Microstructural Characterization and Reliability Strategies, Metallurgical Society of AIME, Warrendale, PA pp. 115-132 (1981).

14. R. Klinman and E. Stephenson, "Relation Between Mechanical Properties, Grain Size, and Ultrasonic Attenuation in Plain Carbon Steel," Research Department, Bethlehem Steel Corporation, Bethlehem, PA (1980).

15. N. R. Joshi and R. E. Green Jr., "Ultrasonic Detection of Fatigue Damage," Engineering Fracture Mechanisms 4, pp. 577-583 (1972).

16. Y. Gefen and M. Rosen, "Behavior of Ultrasonic Attenuation During Aging of Duraluminum 2024," Materials Science Engineering 8, pp. 246-247 (1971).

17. D. S. Hughes and J. L. Kelly, "Second-Order Elastic Deformation of Solids," Phys. Rev. 92, pp. 1145-1149 (1953).

18. Y. -H. Pao, W. Sachse and H. Fukuoka, "Acoustoelasticity and Ultrasonic Measurements of Residual Stress," Physical Acoustics, Vol. XVII, pp. 61-143 (1984).

19. M. Rosen, Center for Nondestructive Evaluation, The Johns Hopkins University, Baltimore, MD (Private Communication).

20. R. J. Dewhurst, "A Hand-Held Laser-Generator of Ultrasonic Pulses," Nondestructive Testing Communications 1, pp. 93-103 (1983).

21. J. W. Wagner, "Heterodyne Holographic Contouring for Wear Measurements of Orthopedic Implants," Proceedings of the Second International Congress on Applications of Lasers and Electro-Optics, Laser Institute of America, Toledo, OH (1983).

22. E. G. Henneke II., Engineering Science and Mechanics Department, Virginia Polytechnic Institute and State University, Blacksburg, VA (Private Communication).

23. L. C. Aamodt and J. C. Murphy, "Thermal Effects in Photothermal Spectroscopy and Photothermal Imaging," J. Appl. Phys. 54, pp. 581-591 (1983).

24. J. W. Maclachlan, J. C. Murphy, R. B. Givens and L. C. Aamodt, "Microstructure Characterization by Thermal Wave Imaging," to be published in Proceedings of In-Process Nondestructive Characterization and Process Control Symposium, ASM Metals Congress, Detroit, Michigan (1984), American Society for Metals, Metals Park, OH (1985).

COMPOSITIONAL CHARACTERIZATION OF DIELECTRIC OXIDES

D. M. Smyth

Lehigh University
Materials Research Center, #32
Bethlehem, PA 18015

ABSTRACT

Many oxides containing transition metal elements are used in electronic and optical applications that require that they be insulating or transparent. The chemical criteria for such behavior are reviewed, and the strong effect of even very small changes in composition are emphasized. In some cases compositional changes in the ppm range can cause changes in electrical conductivity by many orders of magnitude. It is also indicated that in addition to oxides that are insulating after processing in air, there is another diverse family of oxides that are insulating only after chemical reduction. Very little is known about the properties of these materials but there is no reason to exclude them from possible applications.

INTRODUCTION

There are increasing applications for metallic oxides in electronic and electrooptic devices. A basic aspect of the characterization of candidates for such applications is whether they are insulating and transparent, or conducting and opaque. Among the oxides of and transition metals, this distinction can depend upon very subtle compositional differences. The same metallic oxide can be insulating or semiconducting, depending on impurity content or on differences in the metal/oxygen ratio in the range of one part in 10^5. This poses a major challenge in compositional control and materials characterization. It also suggests that an unexplored world of properties remains to be investigated. Our perceptions are biased by the normal ambient of our world, a fiercely oxidizing atmosphere by any chemical standard. Some compounds are

59

insulators and others are conductors when equilibrated with air. Under more reducing conditions, however, the behavior may be reversed, and there has been very little attention devoted to the properties of oxides in the highly reduced state. This article will review the basic compositional distinctions between insulating and conducting oxides, and will describe how these properties can be controlled by very minor compositional adjustments.

THE ELECTRICAL CONDUCTIVITY OF METALLIC OXIDES

Metallic oxides can be divided into two major classes. The main group metals generally form oxides corresponding to a single oxidation state, e.g. MgO, SrO, Al_2O_3, etc. In such compounds, excitation of an electron from the valence band, derived from oxygen 2p states, to the conduction band, derived from cation s states, corresponds to a reversal of the original transfer of electrons from metal atoms to nonmetal atoms to form the ionic constituents of the crystalline structure. Such a reverse transfer is very unfavorable. The consequence of this in electronic terms is that the band gaps of such compounds are wide, 6-12 eV, and electronic defects are unfavorable. As a result, impurity ions whose charge differs from that of the ion they replace (aliovalent impurities) are compensated by ionic defects rather than by electrons or holes, e.g. $Ca_{Na}^{\cdot}$ in NaCl produces equal numbers of V_{Na}'. (In this Kröger-Vink notation[1], the subscript indicates the location of the species in the structure, and the superscript gives its effective charge relative to the ideal lattice. Thus $Ca_{Na}^{\cdot}$ is a Ca^{++} ion substituted for a Na^+ ion, giving an extra positive charge ($\cdot$) to the site, while V_{Na}' is a vacant Na^+ site that is thus missing the usual positive charge.) Because the constituents of such compounds are limited to a single oxidation state, deviations from the ideal stoichiometric compositions are extremely limited. Changes in the metal/nonmetal ratio would correspond to partial oxidation or reduction toward another oxidation state, and also involve a form of electronic disorder.

In transition metal oxides, other adjacent oxidation states exist for the cation. These correspond to different occupancies of the d orbitals, and the availability of electronic states in the solid crystals that are derived from the atomic d orbitals results in smaller band gaps, typically 3-5 eV. Such crystals represent a much more hospitable environment for electronic defects, but they are not usually equally hospitable to electrons and holes.

If the cation is in its highest stable oxidation state, and lower oxidation states exist, e.g. TiO_2 can be reduced to Ti_2O_3, and Nb_2O_5 to Nb_2O_4, the compound can be partially reduced toward the lower oxidation states and conduction electrons are tolerated by the system; it becomes a semiconductor. The loss of oxygen

60

can be symbolized by the equilibrium reaction

$$O_o \rightleftharpoons 1/2O_2 + V_o^{\cdot\cdot} + 2e' \tag{1}$$

The departure of oxygen atoms leaves behind empty oxygen sites in the lattice, as well as the electrons that had been combined with oxygen atoms in the form of oxide ions. The introduction of more highly charged impurities (donor impurities) has a similar effect in that oxygen is lost from the solid solution and conduction electrons are left behind. As an example, solution of Nb_2O_5 up to ~ 0.5 atom % in TiO_2 results in a loss of oxygen

$$Nb_2O_5 \rightarrow 2Nb_{Ti}^{\cdot} + 4O_o + 1/2O_2 + 2e' \tag{2}$$

Nb_2O_5 has too many oxygens per cation to substitute directly for TiO_2, so molecular oxygen is lost and electrons are left behind. This corresponds to chemical reduction in that $Ti^{+4} + e^-$ is equivalent to Ti^{+3}. Thus it is well-known that TiO_2 can be reduced or donor-doped to a semiconducting state. In its oxidized state, however, and in the absence of donor impurities, TiO_2 is an excellent insulator. This class of oxides is referred to as reduction-type semiconductors.

The oxides of transition metal elements in their lowest oxidation state correspond to the other side of the compositional

$$1/2O_2 \rightleftharpoons O_o + V_{Ni}'' + 2h^{\cdot} \tag{3}$$

where $h^{\cdot}$ is a hole (unoccupied electron state) in the valence band. Since $Ni^{+2} + h^{\cdot}$ is equivalent to Ni^{+3}, an achievable oxidation state, holes are easily tolerated in the electronic structure, and NiO equilibrated in air has an excess of oxygen and is a p-type semiconductor. This can be accentuated by the addition of an acceptor dopant, such as Li_2O, that gives room in the lattice for the addition of more oxygen

$$Li_2O + 1/2O_2 \rightarrow 2Li_{Ni}' + 2O_o + 2h^{\cdot} \tag{4}$$

NiO that has been highly reduced or donor-doped, on the other hand, is an insulator. This class of oxides is referred to as oxidation-type semiconductors.

The oxidation-type and reduction-type semiconductors thus vary markedly in their conductivity-composition relationships. The former are conductors when oxidized and insulators when reduced, while the reverse is true for the latter. We tend to categorize

the properties according to the behavior that results from processing in air, and to forget that these properties can be
drastically changed by equilibration with a reducing environment.

THE SIGNIFICANCE OF NATURALLY-OCCURRING IMPURITIES

$BaTiO_3$ is a typical reduction type semiconductor and has been
extensively studied in our laboratory[2,3,4]. The relationship
between nonstoichiometry and conductivity is best demonstrated
by the equilibrium electrical conductivity, as shown in Figure 1.
Under these high temperature conditions, the oxide maintains
thermodynamic equilibrium with the surrounding gas phase. Oxygen
enters or leaves the oxide as the oxygen partial pressure, PO_2, is
varied, and the electron and hole concentrations vary accordingly.
This results in the characteristic changes in the conductivity
as a function of PO_2 seen in Figure 1. In the region of low PO_2,
oxygen is leaving the crystal with decreasing PO_2 according to
Equation (1), leaving behind oxygen vacancies and electrons. The
conductivity thus increases with decreasing PO_2. In the region of
high PO_2, the oxide is taking up a stoichiometric excess of oxygen
with increasing PO_2. The entering oxygen atoms become oxide ions
at the expense of the electron content of the valence band, and
the hole concentration and p-type conductivity increase accordingly with increasing PO_2. Since there are no oxidizable species
in $BaTiO_3$, and the structure has no room for excess oxygen, an
oxygen excess region would not be expected if we were dealing
with an ideal $BaTiO_3$ crystal. In real oxides, however, the
impurity content is usually in the 10-100 ppm range (atomic substitution). The impurities in $BaTiO_3$ are predominantly of lower
charge than the ions they replace in the crystal, because of
natural elemental abundances. Thus such acceptor impurities as
Na'_{Ba}, Al'_{Ti}, and Fe'_{Ti} are much more likely than donor impurities
such as $Nb^{\cdot}_{Ti}$ or $La^{\cdot}_{Ba}$. In fact the total of all possible acceptor
impurities, plus those impurities having the same charge as the
host ions, e.g. Sr_{Ba} or Zr_{Ti}, plus oxygen and silicon, accounts
for over 99.7% by weight of the earth's crust. The bias toward
a naturally-occurring excess of acceptor impurities is thus very
strong. Acceptor impurity oxides have less oxygen per cation
than the host oxide they replace, e.g. Na_2O for BaO or Al_2O_3 for
TiO_2, and result in a corresponding number of oxygen vacancies.
A real crystal of $BaTiO_3$ with its acceptor impurity content,
therefore, does not have a completely filled oxygen sublattice,
and the partial filling of the impurity-related vacancies by
oxygen from the gas phase results in a slight stoichiometric
excess of oxygen at high PO_2. The oxidation reaction is then

$$V^{\cdot\cdot}_o + 1/2O_2 \rightleftharpoons O_o + 2h^{\cdot} \qquad (5)$$

and the p-type conductivity increases with increasing PO_2. A
comparison of the temperature dependences of the conductivities

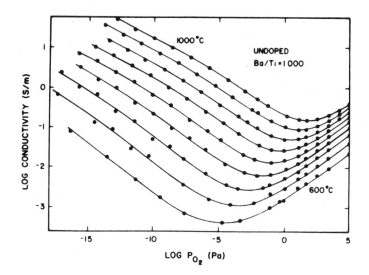

Figure 1. Equilibrium electrical conductivity of polycrystalline BaTiO$_3$ from 600-1000°C at 50°C intervals. 10^5 Pa $\approx$ 1 atm. Reproduced from[4] with permission from the authors and the American Ceramic Society.

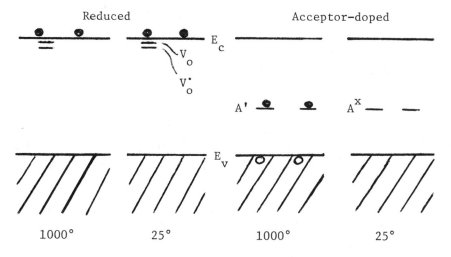

Figure 2. Schematic band diagrams for BaTiO$_3$. V$_o$ and V$_o^{\cdot}$ represent oxygen vacancy levels; A' and A^x are occupied and empty acceptor levels. $\bullet$ = electron in the conduction band, o = hole in the valence band.

63

in the high and low P_{O_2} regions indicates that it is energetically much easier to add a stoichiometric excess of oxygen than to create a stoichiometric deficiency. This is consistent with the ease of filling existing vacancies that result from the impurity content. Under normal conditions, only a small fraction of these extrinsic vacancies are ever filled by oxidation. The conductivity behavior shown in Figure 1 is quantitatively accounted for by Equations (1) and (3) treated according to the principles of dilute solution thermodynamics[2,3].

Over the temperature range covered in Figure 1, the electrons and holes created by reduction or oxidation, respectively, are all free to contribute to the electrical conduction. The n-type and p-type conductivities are thus similar in magnitude. The situation for samples cooled to room tememprature is quite different, however. Because of the available lower oxidation states of Ti, electrons are easily tolerated in the structure; $Ti^{+4} + e^-$ is equivalent to Ti^{+3}. Therefore, when a reduced sample is cooled from the equilibration temperature, the electrons remain free to conduct even far below room temperature. Such material is black and semiconducting, and is not particularly useful. This is true for both reduced and donor-doped $BaTiO_3$, since the latter is effectively reduced by oxygen loss according to Equation (2), for donor concentrations up to about 0.5%.

When an oxidized p-type sample is cooled down, however, the behavior is quite different. Since a hole combined with any of the normal constituents of $BaTiO_3$ does not correspond to any achievable oxidation state, the electronic structure of $BaTiO_3$ does not tolerate free holes at lower temperatures. During cooling, the holes begin to be increasingly trapped by the acceptor centers somewhere between 600°C and room temperature

$$A' + h^{\cdot} \rightleftharpoons A^x \qquad\qquad (6)$$

where A' is a generalized acceptor impurity center, and A^x is the neutral combination of an acceptor and a trapped hole. At room temperature the hole concentration is reduced to miniscule levels and the conductivity has dropped by a factor of about 10^{10}. Such material is a light-colored insulator and is the basic ingredient in ceramic capacitors for which current world production exceeds 10^{10} units per year.

In semiconductor terms, this striking asymmetry is shown in the form of a simplified band diagram in Figure 2. $V_o^{\cdot\cdot}$ represents two donor levels located very close to the conduction band, while A' is an acceptor level about 1 eV above the valence band. At room temperature, thermal energy is sufficient to supply the energy necessary to keep the electrons out of the shallow donor levels, but is insufficient to keep the electrons in the acceptor levels,

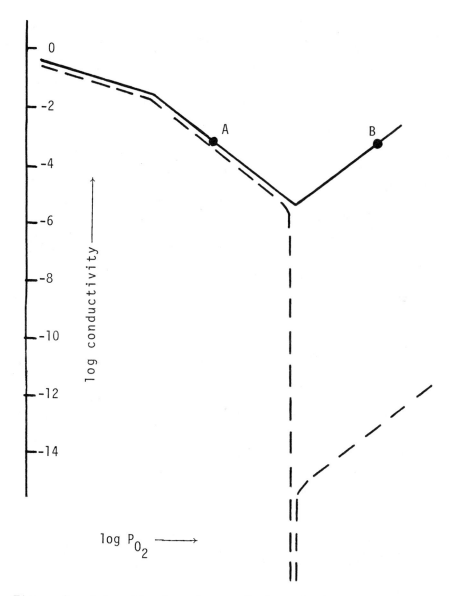

Figure 3. Schematic dependence of electrical conductivity of BaTiO$_3$ on P$_{O_2}$. Solid line: measured at 1000°C; dashed line: quenched and measured at 25°C.

and they fall back into the valence band, annihilating holes in the process.

From another point of view, the conductivities of $BaTiO_3$ equilibrated at various values of P_{O_2} are compared schematically in Figure 3 for samples maintained at 1000°C and for those subsequently cooled to room temperature with no compositional change. The asymmetry can be dramatized by considering two identical samples, one equilibrated at point A, and the other at point B. Their conductivities are identical in magnitude at 1000°C, but one is n-type due to oxygen deficiency (A), while the other is p-type due to an oxygen excess (B). Their compositions differ by only about 5 ppm in oxygen content. After cooling to room temperature, the conductivity of the n-type material remains about the same, but that of the p-type sample has decreased by a factor of 10^{10} as the holes have been trapped by the acceptor impurity centers. This represents an enormous compositional leverage on an easily measurable property.

In order for $BaTiO_3$ to be useful as a cermaic dielectric, _all_ of the following conditions must be fulfilled. It must have a significant excess of impurities of the acceptor type. The resulting oxygen vacancies must be partially filled by equilibration in air to give an oxygen excess, p-type material, and the holes must be immobilized to give an insulating material at normal ambient temperatures. Any deviation from these requirements would give a semiconducting material. Fortunately, all of these conditions are fulfilled naturally, and we consider $BaTiO_3$ to be an insulator. This effect of the naturally occurring acceptor impurity content can be generalized for reduction-type semiconductors as a class, including oxides containing Ti, Nb, Ta, W, etc.[5]. As a result, there are numerous applications for such compounds that depend on their insulating character or transparency.

THE OXIDATION TYPE SEMICONDUCTORS

As described earlier, there is another class of oxides, the oxidation-type semiconductors, that are generally ignored for applications that require them to be insulators or transparent. This class includes oxides containing transition metal ions in their lowest available oxidation states that can be oxidized toward a higher achievable oxidation state. These materials are generally black semiconductors when oxygen-excess or acceptor-doped and p-type, but are light-colored insulators when reduced or donor-doped, the reverse of the situation for reduction-type semiconductors. Because they are easily oxidized, such compounds as NiO, CoO, MnO, etc., are invariably oxygen-excess and p-type after equilibration in air, and are thus semiconductors of limited applicability. There is no restriction against equilibrating them under reducing conditions, however, to give oxygen-deficient, n-type materials that will generally be insulators. This will open up a wide range of materials for possible applications that have previously been required because of the tendency to attribute to materials the properties that result from processing in our normal, oxidizing atmosphere. Property changes can also be achieved by proper doping. Thus, while NiO can be made highly conducting by doping with an acceptor such as Li_2O, as shown in Equation (4); with a donor dopant such as Cr_2O_3, the situation is quite different.

$$Cr_2O_3 \longrightarrow 2Cr_{Ni}^{\bullet} + V_{Ni}'' + 3O_o \tag{7}$$

Under reducing conditions the extra oxygen may be lost

$$Cr_2O_3 \longrightarrow 2Cr_{Ni}^{\bullet} + 2O_o + 1/2O_2 + 2e' \tag{8}$$

At room temperature the electrons will be trapped by the donor centers

$$Cr_{Ni}^{\bullet} + e' \rightleftharpoons Cr_{Ni}^{x} \tag{9}$$

giving an insulating material.

Some years ago, a so-called "nickel capacitor" appeared on the market. This consisted of a piece of nickel foil that had been thermally oxidized to give a coating of NiO that would serve as the dielectric layer. A metallic electrode was applied to the NiO surface, leads were attached, and the device packaged. This gave a simple, inexpensive capacitor having the structure Ni-NiO-metal. In order to function as a useful dielectric, the NiO must be uniformly insulating; however, the thermally formed oxide is quite nonuniform. The side of the layer in contact with metallic nickel is, by definition, in its maximum state of reduction, but the outer layer that was in contact with air during the thermal oxidation is oxidized and semiconducting. There is a steep gradient of conductivity across the film and this leads to terrible properties as a capacitor. It is necessary to heat the oxidized foil in a reducing atmosphere to reduce the outer surface of the NiO to an insulating state. This is easily done, and the resulting electroded capacitor has quite good properties. It is inefficient, however, because of the low dielectric constant of NiO, and it remained on the market only briefly. In spite of its lack of commercial success, this serves as a specific example of the reduction of an oxidation-type semiconductor to an insulating state for a specific application. It seems likely that comparable opportunities could be exploited if a broader search for appropriate materials were made.

CONCLUSION

Transition metal oxides are usually highly asymmetric in their electrical conductivity depending on whether they have an excess or a deficiency of oxygen relative to the ideally stoichiometric composition. Near the compositional transition, differences of a few ppm in oxygen content or impurity concentration can result in changes in conductivity by several orders of magnitude. The reduction-type semiconductors are typically in an oxygen-excess, insulating state after equilibration with air, and we have a wide knowledge of their properties in terms of ferroelectricity, piezoelectricity, electrooptic activity, etc. We know much less about the properties of the oxidation-type semiconductors, which must be reduced to be insulating or transparent. It would be surprising if there were not a number of useful materials in this latter class. We cannot afford to restrict our outlook only to those properties that result from equilibration with our own specific atmosphere.

ACKNOWLEDGEMENT

The support of the Ceramics Program of the Division of Materials Research, National Science Foundation, is gratefully acknowledged.

REFERENCES

1. F. A. Kröger and H. J. Vink. In "Solid State Physics" Vol. 3, F. Seitz and D. Turnbull, eds., p. 307, New York, Academic (1956).

2. N.-H. Chan, R. K. Sharma, and D. M. Smyth, J. Am. Ceram. Soc. 64, 556 (1981).

3. N.-H. Chan, R. K. Sharma, and D. M. Smyth, J. Am. Ceram. Soc. 65, 167 (1982).

4. N.-H. Chan and D. M. Smyth, J. Am. Ceram. Soc. 67, 285 (1984).

5. D. M. Smyth, Prog. in Solid State Chem. 15, 145 (1984).

HIGH MODULUS POLYMERS FROM FLEXIBLE CHAIN THERMOPLASTICS

Roger S. Porter

Department of Polymer Science and Engineering
Materials Research Laboratory
University of Massachusetts Amherst, MA 01003

ABSTRACT

A review is provided for the field of high modulus fibers. Emphasis is placed on research for the development of high tensile moduli from flexible chain polymers, with polyethylene used as an example. Advances in drawing techniques and in the characterization of draw are also reviewed. These developments with flexible chain polymers are compared to the recent U.S. advances in the achievement of high modulus fibers from the rigid rod polymers.

- Recent Developments in High Modulus Polymers
- Preparation and Characterization of Drawn Thermoplastics
- Deformation of Low Density Polyethylenes
- Deformation of High Density Polyethylene
- Gas Permeation with CO_2 as a Probe for Amorphous Orientation

RECENT DEVELOPMENTS IN HIGH MODULUS POLYMERS

Perspective

The world position of nations is currently influenced by technology. In this contemporary competition, military material has become the science and engineering of materials.

With the competition over steel an established example, we may consider the prospects for the two other major classes of materials, ceramics and polymers. The latter class of materials, polymers, is also being challenged in Japan. The following stark example is an excerpt from a translation of "Nikkei Sangyo Shimbum, 7/27/82."

The Ministry of International Trade and Industry (MITI), the synthetic fiber industry, and academic institutions are to engage in the joint development and practical application of "the third generation fiber" which will have more than twice the high tenacity, high modulus, low elongation and other properties of the present fibers. MITI has designated as a government subsidized project, effective fiscal year 1983, this next generation research/development program. The industrial infrastructure plans to allocate 3 billion - 5 billion yen in funds, with practical application targeted at five years in the future. The new fiber is expected to replace nylon and carbon fiber and expand the area of fiber demand. MITI and the industry, therefore, look forward to the contribution of this development project as a conclusive factor for the revitalization of the fiber industry now suffering from the structural recession, and for increasing the product value-addition.

At present, the closest fiber to this third generation fiber is DuPont's Aramid (also known as "Kevlar®49"), with a high tenacity of ~28 grams/denier ~3.6 GPa, the world's most tenacious (commercial), but with a maximum modulus inferior to carbon fiber and polyethylene.

This is the world's first development project in the area of new materials, which is of equal importance with electronics and biotechnology. It is particularly uniquely Japanese, because the research project is to be conducted jointly by government, industry and academic circles under the government subsidy program.

Flexible Polymers, the Polyethylene Example

Polymer researchers have approached the problem of making the strongest possible polymers in two diverse ways: (1) by chemically constructing polymers with rigid and linear backbone chains and (2) by processing existing conventional polymers in ways that a permanent transformation of the internal structure and properties occurs. Chemical construction of rigid macromolecules has been approached by syntheses leading to para-substituted aromatic rings in the polymer backbone. In general, these polymers cannot be processed by means of the common polymer-processing techniques; however, some industrial examples, viz, Kevlar and X-500, have been processed into fibers (of very high strength).

In the second category, polymers are converted into a highly oriented chain conformation with substantially increased modulus by drawing from a dilute flowing solution, gel state, or by extruding a supercooled melt, or by solid-state extruding or drawing below the melting point under controlled conditions.

New and successful drawing techniques have been recently developed by workers in the U.S. (at UMass) and Japan. It has been found possible to ultradraw single crystal mats of ultrahigh molecular weight polyethylene (UHMWPE). By the principal deformation technique of solid state coextrusion, draw has been achieved even at room temperature, and at up to 130°C, i.e. to below the melting point. Moreover, a stable extrudate results which exhibits a high crystal orientation. Multiple drawing by repeated coextrusion at 110°C produced an extrudate of UHMWPE with a draw ratio (DR) of 110 and a tensile modulus of 100 GPa. An even higher DR was achieved by a combination of solid-state coextrusion followed by tensile drawing at controlled rate and temperature. The maximum achieved for the present by this drawing combination was a DR of 250. This superdrawn sample gave a tensile modulus of 222 GPa. This is about twice the highest previously reported room temperature experimental value (110 GPa) for polyethylene. Figure 1 summarizes these new results and provide a comparison with previous achievements by solid state drawing of high density polyethylenes.

Rigid-Rod Polymers

Various molecular architectures and processing procedures have been employed to achieve high modulus and high strength in polymeric fibers. Linear, flexible chain polymers, semi-rigid chain polymers and extended or rigid chain polymers all have been manufactured into fibers with high tensile mechanical properties. Additionally, carbon and graphitic fibers produced from polymeric precursors exhibit some of the highest performance characteristics of commercially available materials to date.

Carbon and graphitic fibers have been extensively investigated over the last two decades owing to their high temperature stability and exceptional mechanical properties. Commercially available fibers possess tensile moduli of up to 690 GPa along with tensile strengths of 2.2 GPa[7]. While such fibers have amply demonstrated high tensile modulus and strength, they are quite brittle which may limit their use in various applications. Also, in order to produce carbon and graphitic fibers extreme processing conditions are required, resulting in high production costs and hence final material cost. Additionally, the high electrical conductivity of these fibers is not always desired in high performance applications. While carbon fibers have demonstrated very desirable high performance characteristics there still exists a need for other high performance materials. Work on new materials is continuing in many areas and a sizeable amount of this activity is concerned with extended chain or rigid rod polymers possessing high performance characteristics.

Fibers produced from lyotropic liquid crystalline solutions of extended chain polymers have not only achieved desirable high performance characteristics but have become commercially successful engineering materials through the development of conventional wet spinning techniques for their manufacture. Both Monsanto[8] and

DuPont[9] have had success in developing high modulus/high strength
fibers based on wholly aromatic polymers which possess a rod-like
character deriving from steric effects; however, only DuPont has
pursued commercial development (Kevlar®). This success in producing
high performance fibers from extended chain macromolecules has
encouraged further investigation of rigid rod polymers. A sizeable
research effort sponsored by the U.S. Air Force Wright-Patterson
Materials Laboratory and the U.S. Air Force Office of Scientific
Research (Ordered Polymers Research Program)[10] is currently
investigating the high performance nature of novel rigid macromolecules.

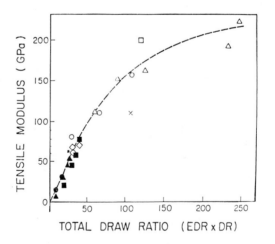

Figure 1. Solid State Extrusion Draw of Single Crystal Mats of
UHMWPE to an Extrusion Draw Ratio (EDR) Followed by
Conventional Draw (DR) at 110°C.

PREPARATION AND CHARACTERIZATION OF DRAWN THERMOPLASTICS

Perspective

Recently there has been a considerable interest in the development of thermoplastics with high tensile modulus. This has been
pursued by introducing a high degree of chain orientation and
extension by the techniques of solid state drawing[1] and extrusion[2].

In the original studies of the crystallization of an oriented
high density polyethylene (HDPE) melt using an Instron capillary
rheometer[3], it was found that polyethylene extrudates with a
remarkably high modulus (up to 70 GPa) could be obtained[4]. The
process was modified later to a solid state extrusion by Capiati,
et al.[5], in that melt crystallized morphologies of HDPE were
extruded below their melting point through conical dies.

The effect of the processing conditions on the properties of
HDPE was studied by various workers and for more details the
reader is referred to the original papers[6-8] and an extensive
review by Zachariades, et al.[9].

In this review we shall discuss some of the developments that
have extended the range of the solid state extrusion technique for
HDPE and also allowed the extrusion of materials that were previously found difficult to process by conventional solid state
extrusion.

In early extrusion experiments melt crystallized morphologies
of high density polyethylene (Alathon 7050 M_W = 59,000) were
prepared by melting at 160^0C and subsequent pressure. These
morphologies were extruded through a conical brass die, with an
included entrance angle of 20^0, in a 3/8" bore Instron capillary
rheometer under constant extrusion pressure.

Figure 1 shows that as the draw ratio increases the apparent
viscosity increases and hence the extrusion rate drops rapidly.
The rapid increase in viscosity is thought to be related to the
strain hardening observed in a conventional tensile test at high
elongation[10]. The extrusion rate is increased with high extrusion pressure; however, there is an upper limit to the pressure.
Over this limit stick-slip occurs and a spiral fractured extrudate
is produced that is similar to spiral products found in shear
fracture of polymeric melts during extrusion[11]. The deep shear
parabolic profile in the extrudates suggests that stick-slip may
arise from shear fracture of the extrudate[10,12,13].

The range of moduli and extrusion draw ratios that can be
achieved by this process is shown in Figure 2. The moduli are
measured at 0.1% strain and a strain rate of 2.0 x 10^{-5} sec^{-1} and

the extrusion draw ratio (EDR) are determined by the ratio of the cross-sectional areas of the extrusion dies at the exit and the entrance. At lower extrusion temperatures, higher moduli were obtained, presumably because of the greater efficiency of deformation. Efficiency of deformation is the ratio of the bulk deformation to the deformation of the individual polymer chains. However, the maximum available draw ratio is also dependent on temperature such that the maximum modulus is observed for samples extruded at 132^0C and a draw ratio of 40. The physical properties of HDPE rod extrudates are listed and compared with the properties of its single crystal in Table 1.

Various workers have studied the properties of highly oriented HDPE[14-16] and proposed morphological models to account for the high modulus of these morphologies[17-19]. The PE unit cell has a modulus of > 240 GPa[20,21]. In the oriented morphologies obtained by extrusion the PE chain crystallizes in the trans-trans conformation so that only bending and extension of the carbon-carbon bonds may occur. Thus the crystallized chain has a very high modulus in the chain direction (Table 1):

Recently, the range of the conventional extrusion process has been extended to the processing of new morphologies under a wider range of processing conditions by the split billet technique[22]. According to this technique melt crystallized morphologies of HDPE with longitudinal free surfaces (Figure 3) were extruded through a conical brass die. This split billet technique – with longitudinal free surface extrusion – proceeds at a higher rate and more importantly to a higher maximum draw ratio without fracture. This was thought to be due to a stress relief which may result from the free surface in the center of the billet[22]. The increase in extrusion draw ratio using the split billet technique is shown in the processing map illustrated in Figure 4. The lines in Figure 4 join points of maximum extrusion draw ratio that can be continuously extruded. Since above 138^0C the extrudate melts at the die exit, a region of extrudability can be drawn. The dashed line shows the extrudability limits of the split billet technique for processing HDPE (M_w = 59,000).

Push-Pull Extrusion

A further enhancement of the extrusion rate was achieved by applying a pull load on the emerging extrudate from the die[23]. The effect of a pull load on the extrusion rate of HDPE (M_w = 59,000) split billet extrudates obtained under constant extrusion pressure at different extrusion temperatures is shown in Figure 5. The effects of push (extrusion) force and pull force on the extrusion rate are superimposable. The push and pull forces are related by

$$\text{Push force} = \lambda \text{x (Pull force)}$$

where λ is a proportionality factor arising from the hydrostatic transmission of forces through the polymer on the extrusion die. From this relationship the pressure at the base of the conical die has been evaluated and estimated to be similar to the tensile strength of the extrudate at the extrusion temperature. This correspondence between the pressure at fracture and the tensile strength may be explained by the fact that the material is being elongated in the die as the extrusion proceeds.

Also, in conventional extrusion there is a maximum pressure that can be applied without stick-slip. This imposes a restriction on the maximum extrusion rate and draw ratio. In push-pull extrusion forces up to the tensile strength of the material can be applied and the true maximum extrusion draw ratio can be obtained at a given temperature. These data added on the processing map (Figure 4) indicate clearly that the push-pull technique extends considerably the range of the extrusion process. The mechanical properties of the extrudates prepared by the push-pull extrusion technique are independent of the nature of the applied pressure.

Coextrusion

During the development of the split billet technique, it was realized that by placing a polymer film of HDPE between the billet halves it could be extruded with the outside halves to a draw ratio that was not obtainable using a slit die[22]. The film in the center of the billet is forced to follow the two outside billet halves because of the compression in the conical die and friction between the components. Two important features of the split billet coextrusion are: a) The film and the substrates (billet halves) may undergo the same strain, if the total stress in the film and substrates are similar. The total stress in the film may be controlled by its cross-sectional area and b) if the film in the center of the billet has the same size as the flat surface of the split billet it deforms with a deep shear profile and undergoes severe shear stress whereas if it occupies only the center of the billet it undergoes extensional deformation and consequently may be drawn to a much higher draw ratio.

The versatility of the split billet coextrusion technique was demonstrated with the preparation of highly oriented thin films of the semicrystalline poly(vinylidene fluoride)[24,25] and polyamides[26] as well as the amorphous polystyrene and polyethylene tere-phthalate[27,28] by coextruding within HDPE split billet.

The poly(vinylidene fluoride) thin films were obtained at $120\,^{0}C$ and were of a maximum extrusion draw ratio 7. The amorphous poly(ethylene terephthalate) thin films were coextruded at $100\,^{0}C$

to an EDR of 12. The atactic polystyrene films were coextruded at
126°C to EDR > 20 and had a microfibril morphology[29].

Nylon thin films could also be processed by the split billet
technique under reasonably modest conditions although the existence
of hydrogen bonding makes their processing feasible under vigorous
conditions only. The nylon films were treated prior to extrusion
with anhydrous ammonia which is absorbed by the sample quantitatively and acts as a volatile plasticizer in the material[30]. The
ammonia was retained within the sample during extrusion by inserting
the treated nylon films into an HDPE split billet and pressurizing
the assembly in the extruder at some pressure (>~ 5.5 MPa) above
the vapor pressure of the ammonia at the extrusion temperature
(<~ 95°C). For nylon 6, coextrusions were performed through conical
dies of EDR=12 in the temperature range of 65-95°C, i.e. at least
125°C below its melting point. On exiting the extrusion die, the

Table 1:   Comparison of Physical Properties of Drawn HDPE Rods
            with PE Single Crystals

Physical Property*	Polyethylene Single Crystal	Ultraoriented PE Rod
Tensile modulus, GPa	320, 285, 240	70
Strain to fracture, %	6	3
Tensile strength, GPa	13	0.4
Linear Expansion Co-efficient, 10^{-5} °C^{-1}	(a) axis + 22 (b) axis + 3.8 (c) axis - 1.2	-0.9 $\pm$ 0.1 Fiber axis
Melting point, °C	145.5	139 ($10°C$ min^{-1} heating rate)
Crystallinities		85%
Crystal orientation function	1.0	0.996 $\pm$ 0.002
Birefringence	0.059[0]	0.0632

*Mechanicals are for crystal (C) axis direction.

[0]Calculated for crystal (C) axis direction[32].

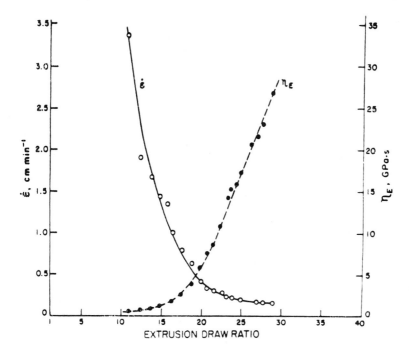

Figure 1. The Relation Between Viscosity, Extrusion Rate and Draw Ratio for HDPE (M_W = 59,000).

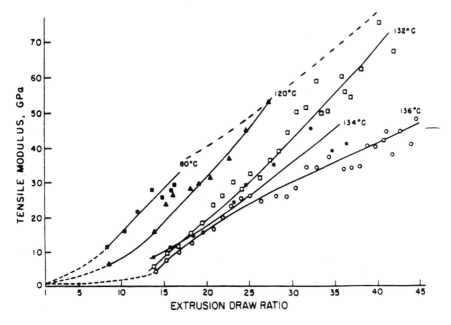

Figure 2. Variation of Modulus with Extrusion Draw Ratio at a Series of Different Extrusion Temperatures and Extrusion Press = 0.23 GPa. The Maximum Draw Ratio is shown by the dashed line (Ref. 14).

77

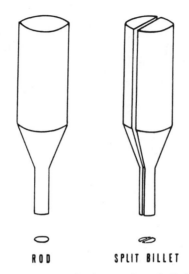

ROD   SPLIT BILLET

Figure 3.   A schematic of the Rod and Split Billet.

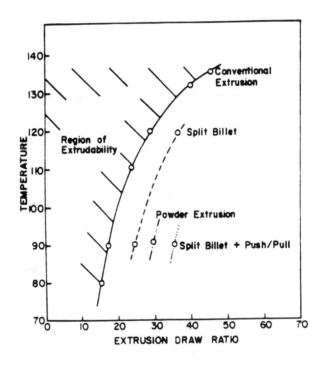

Figure 4.   Processing Map for Alathon 7050, Under Extrusion Pressure
of 0.23 GPa.   Lines connect points of maximum obtainable
Extrusion draw ratio at given temperature.

ammonia volatilized quantitatively from the extruded films. The tensile modulus of these films was ~ 13 GPa and is significantly higher than the modulus of untreated higher nylon analogues deformed in the solid state[30,31].

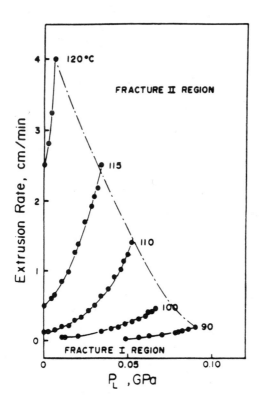

Figure 5. The effect of the Addition of Pull Load on Extrusion Rate under Extrusion Pressure of 0.23 GPa at a Series of Different Extrusion Temperatures. The dashed line connects the points of fracture at the given temperature.

DEFORMATION OF LOW DENSITY POLYETHYLENES

Perspective

The crystalline (solid) state extrusion of several thermo-plastics has been studied extensively over this past decade for the purpose of producing anisotropic morphologies. Despite the fact that low density polyethylene (LDPE) is the largest volume thermo-plastic in the world[1], it has been utilized in only a few such studies: note the solid state and hydrostatic extrusions by Buckley and Long[2] and by Alexander and Wormell[3] respectively. The advent of linear LDPE resins has opened new opportunities for studying the effect of short as well as long branches in solid state extrusion. Thus, the purpose of this study is not only to evaluate the property changes achieved through uniaxial extrusion draw of three polyethylenes, but also to compare the linear and long branched LDPE polymers and also LDPE with prior evaluations of HDPE[4-15]. Consideration will also be given to the differences and similarities between solid-state extrusion and cold drawing in inducing high uniaxial orientation. Four principal methods of characterization were used: thermal analysis, density, tensile and birefringence measurements[16-31]. The variables of draw were extrusion temperature and draw ratio. Additional properties such as die swell and transparency are considered. The LDPE samples studied are described in Table 1.

Our goal is to produce flawless extrudates at the highest possible draw ratios (DR) for the extrusion temperatures studied. At high extrusion ratio, the back pressure becomes so large that instabilities take place, giving rise to irregularities in the extrudates. The ratio between the entrance and the exit cross-sectional areas of the die expresses the extent of deformation and is referred to as the draw ratio (DR). The final product, re the extrudate, is characterized by three parameters: DR, the temperature of extrusion, T_{ex}, and the plunger velocity.

Results

Figures 1 and 2 intercompare the melting point as a function of extrusion draw ratio for the three polymers extruded at 80^0C. The melting point for samples A and C drops at EDR > 2.0 and then remains constant. For B, MP decreases regularly with extrusion draw. There is, though, little temperature dependence of MP.

The melting of low density polyethylenes as a function of extrusion draw differs markedly from that of HDPE for which the melting point is reported to increase modestly with draw[9-11, 32-39]. Buckley and Long[2] also observed no appreciable change in the melting points for LDPE.

The tensile modulus, E, varies similarly with EDR for all three low density polyethylenes. After going through a minimum at near EDR = 2, the modulus increases markedly with extrusion draw. The overall increase is up to 4.5, 2.5 and 4.0 times that of the original isotropic polymers for A, B, and C respectively. There is also a minor but clear dependence of modulus on extrusion temperature. At higher extrusion draw ratios and for the lower crystallinity polymers, A and B, E decreases with increasing extrusion temperature. Polymer A (Figure 3) shows the most rapid increase in modulus with draw. Although undrawn A has a lower modulus, it reaches a higher value than its linear counterpart B. That C reaches higher values than the former stems more from its higher ductility at these conditions than from its higher crystallinity.

The effect of extrusion draw on tensile modulus for LDPE differs from that for HDPE[11]. In the latter case, the modulus increases slowly at draw ratios less than 10-15 whereas at higher draw ratios, the modulus increases rapidly and linearly with extrusion draw. The minimum in modulus at low EDR was not observed, in accord with Buckley and Long[2], who also found only a slight increase in tensile modulus on solid-state extrusion of LDPE. Our results also compare well with those for cold-drawn LDPE[40-42] including the anomalous pattern of the minumum modulus which seems to be unique to low density polyethylene[43]. The highest moduli attained are shown in Table 2 which also includes the highest moduli reported in literature[42-48].

An explanation for the minumum has been given by Frank, et al.[45] on the basis of two mechanisms: c-shear process and twin boundary migration. The c-axis shear mechanism is related to the mobility of the structure arising from an appreciable branch content which also gives rise at room temperature to a low shear modulus on planes along and perpendicular to the draw direction[46]. Ward[48] showed that the mechanical anisotropy of LDPE is well predicted by the aggregate model.

Birefringence has been chosen to assess the extent of orientation during solid state extrusion of low density polyethylenes because it may be directly related to the permanent strain[42]. Birefringence is the difference between refractive indices along and perpendicular to the draw direction. As the chain becomes more oriented, birefringence Δn increases as defined by the following equation[42]:

$$\Delta n = \Delta n_{max}(1 - \frac{3}{2} \overline{\sin^2\theta})$$

where Δn_{max} is the maximum birefringence of full orientation and θ the angle between the chain axis and the draw direction. According to the above equation, Δn initially rises sharply with increasing draw ratio and then turns plateaus at high draw[48]. This is indeed what we observe for our low density polyethylenes, see Figure 4.

From Figure 4 we can see that within precision, A and B are indistinguishable, whereas C, of higher crystallinity, reaches higher values of birefringence: 0.068 ± 0.009. This value may be higher than any other previously reported for PE but the large uncertainty limits the significance of this result. In any case, this value is comparable with those obtained for ultradrawn HDPE fibers[11]: 0.062 ± 0.002. The highest value in birefringence for polymer A is 0.046 ± 0.004 at EDR = 4.9 and extrusion temperature $T_{ex} = 22^0C$ which is comparable to that of cold-drawn LDPE[47,49,50]. Also, both extruded and cold drawn LDPE show the same pattern in birefringence change with draw. Therefore, it seems that birefringence is not influenced by long branching at least not within the precision of our results. There is a small but clear extrusion temperature dependence of birefringence in the case of polymer A but not in that of B or C perhaps because of the large uncertainty.

Table 1: Characteristics of Polyethylene Samples Studied

Sample	Type	Grade	Manufacturer	ρ, g/cm^3.	MI*	$\bar{M}_n$ **	Crystallinity (%) ***
A	LDPE	Alathon 20	DuPont	0.920	1.9	32,000	49
B	LLDPE	FW 1290	CdF Chimie	0.920	0.8	36,000	49
C	LLDPE	FW 1180	CdF Chimie	0.935	1.2	34,000	59

*ASTM D 1238, Melt Index.

**Calculated from Equation 1.

***Calculated from density.

82

The higher birefringence at lower draw temperature may be explained in terms of higher draw efficiency. The lower the temperature of draw, the higher the fraction of energy input that is stored elastically[23].

Table 2: Presently Achievable Tensile Moduli of Polyethylene[+]

Sample	Density (g/cm^3)	EDR$_{max}$	E(GPa)	Reference
A	0.920	4.9	0.73	--
B	0.920	5.5	0.46	--
C	0.935	6.9	1.5	--
LDPE*	---	6.0	0.83	Hadley et al.[44]
LDPE*	0.915	6.0	<0.75	DeCandia et al.[41]
LLDPE*	0.914	8.0	1.1	DeCandia et al.[41]
HDPE**	---	40	70	Zachariades et al.[11]

*Cold drawn.

**Solid state extruded.

[+]Highly drawn by a very specialized technique, UHMWPE, with a special initial morphology, was found to exhibit ultra high tensile modulus, i.e. 222 GPa[a].

[a]Reference: T. Kanamoto, A. Tsuruta, K. Tanaka, M. Takeda and R. S. Porter, Polym. J., 15, 327 (1983).

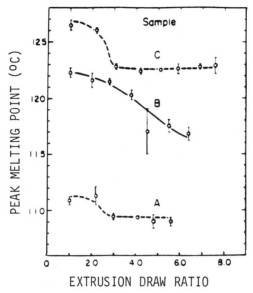

Figure 1.   Peak Melting Versus EDR for Samples A, B and C Extruded at 80°C.

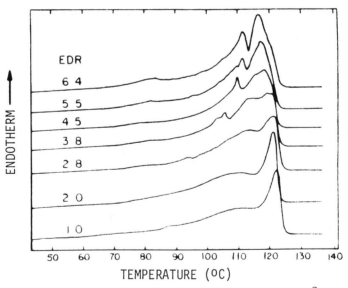

Figure 2.   DSC Endotherms for Sample B Extruded at 80°C at Indicated EDR.

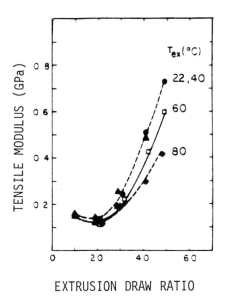

EXTRUSION DRAW RATIO

Figure 3.   Tensile Modulus Versus EDR for Sample A. T_{ex} = 22 (▲);
40 (●); 60 (□); 80°C (♦).

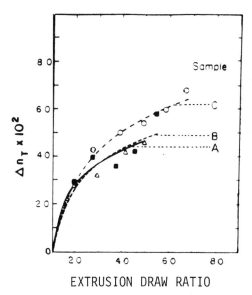

EXTRUSION DRAW RATIO

Figure 4 .   Birefringence Versus EDR for Samples A (△), B (■) and
C (○) Extruded at Romm Temperature.

DEFORMATION OF HIGH DENSITY POLYETHYLENE

Perspective

The plastic deformation of semicrystalline polymers has been extensively studied[1-3]. Many structural models[4-7] have been proposed to explain the remarkable enhancement of the mechanical properties of semicrystalline polymers on drawing. These models assume either crystal continuity (continuous crystals) or taut tie molecules, connecting longitudinally crystallites as a structural element, providing an efficient axial force transmission. The microstructure of crystalline components in highly-drawn, ultrahigh modulus polyethylene has been studied by many methods including x-ray diffraction[5,8], electron microscopy[9], nitric acid etching plus gel permeation chromatography[10,11], and by Raman spectroscopy[12,13]. Each technique has given important information about the existence of chain-extended crystalline components in highly-drawn high density polyethylene (HDPE). The important role of this component on the thermal and morphological properties has been discussed in a previous paper[14].

Recently, Ward, et al.[5] estimated the crystal length distribution of the chain-extended crystalline components in highly-drawn HDPE based on x-ray and modulus data. More recently, Thomas, et al.[9] directly determined the crystal length distribution in HDPE extrudates using the dark field electron microscopy. These studies show that the chain-extended crystalline component increases systematically with increasing draw ratio (DR). They ascribe the remarkable enhancement of the mechanical properties to the growth of this component which improves crystal continuity. Indeed, we have shown that a room temperature tensile moduli of up to 222 GPa can be achieved by super-drawing (DR ~ 250) of single crystal mats of ultrahigh molecular weight polyethylene[15]. The films of highest tensile modulus are highly-crystalline, $\geq$ 90%, the result of a significant growth in crystalline tie molecules. Nevertheless, the growth of this component during deformation has not been well understood.

Results

In previous papers[8,14,16,17], we have reported on the changes in the morphology and properties of HDPE as a function of EDR. The structural parameters of the extrudates prepared at 90°C are shown in Table 1. The features of change with EDR may be summarized:

1. Tensile modulus increases steadily with EDR.
2. The density (viz crystallinity) increases with EDR at least for EDR > 5.8.
3. The crystallite sizes, D_{200} and D_{020}, along the a- and b-axes rapidly decrease with increasing EDR at lower EDR, and then become substantially constant at EDR > 5.8. These values likely correspond to the average lateral dimensions of microfibrils.
4. The long period of the folded-chain crystals decreases on morphological transformation to the fibrous structure, but stays nearly constant at higher EDR (> 5.8).
5. The crystallite size, D_{002}, along the c-axis also decreases rapidly at lower EDR and then increases steadily with increasing EDR at > 5.8.
6. The SAXS intensity decreases with increasing EDR.
7. The structure of noncrystalline regions, studied by NMR, rapidly changes with increasing EDR and reaches a limit at higher EDR.

These results indicate that the transformation of the initial lamellar to the fibrous morphology is almost completed at an EDR of 5.8. The chain-extended crystalline component grows during further deformation via pulling out of successive folds of crystal blocks within microfibrils, consistent with the steady increases in D_{002} and density and by the reduction in SAXS intensity with EDR. As the new crystalline component improves the crystal continuity, this component detected by x-ray diffraction is referred to as crystalline tie molecules, hereafter. These features may be well explained on the basis of the Peterlin's model[4] which describes the regular changes in morphology and properties with DR.

Figure 1 shows a simplified Peterlin model for a fibrous structure. The main assumption of this new model is that the chain-extended crystalline components, connecting longitudinal crystallite at the lateral fibril surfaces, grow during the deformation by pulling out of successive folds from the crystal blocks. The model (a) in Figure 1 shows the fibrous structure formed just after transformation from the initial lamellae and contains few crystalline tie molecules (model (a) is the reference morphological state). The model (b) shows the formation of crystalline tie molecules during deformation at the expense of the chain folds of crystal blocks within microfibrils. This model is also similar to Ward's model[5] in emphasizing the growth of crystalline tie molecules. Where N is the total number of chain contained in the cross-section of a microfibril, n_1 the number of chain stems in a folded-chain crystal block, n_t the number of crystalline tie molecules in a microfibril; L_{core} is the crystalline stem length of the folded-chain crystal. The density, ρ_s, for a reference state (Figure 1-a) may be given by:

$$\rho_s = \frac{L_{core}\rho_c + (L - L_{core})\rho_a}{L}$$

As the lateral crystallite sizes, D_{200} and D_{020}, are almost constant at EDR > 5.8 (see Table 1), the total number of chains contained in the cross-section of a microfibril may also stay approximately constant; i.e., $N = n_t + n_1 = $ const.

Figure 2 shows the result of calculations for the average length, $D_{002,t}$, and the fraction, n_t/N, of the crystalline tie molecules as a function of EDR. The fraction increases rapidly with EDR reaching ~39% at EDR 30. In contrast, the average chain length is insensitive to EDR, being in the range of 470-540 Å, about 2.5 times the lamellae long period.

Ward, et al.[5] also estimated the fraction and length of the chain-extended crystalline components in HDPE films drawn at 75°C using a somewhat different model and method. Their analysis based on the modulus data, shows that the fraction increases from 5 to 40% in the DR range of 5-30, in agreement with the present results. They also estimated the crystal length distribution by applying a statistical distribution function to the weight-average crystallite length determined by the x-ray line broadening. Their estimates show that, although a few very long crystals (~1000 Å) may grow at DR 30, the crystals having length of 1~3 times the lamellar long period are predominant. Fischer, et al.[18] also analyzed the data of Ward, et al. Their results show that the fraction increases from 20 to 56% in the DR range of 9~30 with the length being about 600 Å almost independent of DR.

Thomas, et al.[9], using dark field electron microscopy determined the crystal length distribution in HDPE extrudates and found that it shifted to longer size with increasing EDR. These samples were prepared from the same HDPE (Alathon 7050) as was used here and was also extruded to a similar EDR, although the extrusion temperature was different (120 vs. 90°C). Thus, the weight-average length of all crystals (L_w) and that of longer crystalline components (L_t) at each EDR, determined from their data, are compared with these new results in Table 2. The former corresponds to the D_{002} and the latter to the $D_{002,t}$ in the present notations. For the calculation of the L_t values, the crystals longer than 200 Å were taken into account, assuming a constant lateral size. As shown in Table 2, the D_{002} increases steadily with EDR from 188 Å at EDR 5.8 to 327 Å at EDR 30. The D_{002} at each EDR is slightly smaller than the electron microscopy L_w, likely due to the higher extrusion draw temperature (120 vs. 90°C) and also the failure to detect the crystallites smaller than 50 Å in the latter study[9]. However, the $D_{002,t}$ values (470–540 Å) are ~100 Å larger than the L_t values (370–413 Å). When the lateral crystallite size varies within a sample, as was observed in the electron microscopy work, the L_t value calculated assuming the size being constant may differ significantly from the actual weight-average crystal length. Nevertheless, the trends of $D_{002,t}$ and L_t with EDR are similar. In view of these facts, it may be concluded the present estimates are in fairly good agreement with the previous electron microscopy data.

In conclusion, the growth of chain-extended crystalline tie molecules during solid-state extrusion has been evaluated. The results are in fairly good agreement with the electron microscopy work reported by Thomas, et al.[9] on similarly prepared polyethylenes. The present and previous studies suggest that this important crystalline component is formed by pulling out of folds within microfibrils during deformation of the fibrous morphology. The average length of these components were about 2.5 times the lamellar long period and insensitive to EDR. In contrast, the fraction of this component increases significantly with EDR for extrusion at 90°C, at least up to EDR 30.

Table 1: Characterization of High-Density Polyethylene Extrusion Drawn at $90^\circ C$

EDR	E (GPa)	ρ (g cm^{-3})	X_v (%)	SAXS* Intensity	L	L_{core}	D_{200}	D_{020}	D_{002}
1	2	0.960	71.4	---	306	218	187	200	---
5.8	12	0.964	74.3	---	193 263	143 195	111	127	188
12	22	0.968	77.1	8.4	186	143	102	124	193
25	38	0.975	82.1	4.1	194	143	105	121	264
30	45	0.978	84.3	2.8	195	143	103	123	327

*Intensity per unit volume, rad. count sec^{-3} cm^3, represented by the product of the peak intensity and the half width of the first order scattering.

Table 2: Average-lengths and Fractions of Crystalline Tie Molecules Estimated in This Work and From the Data of Thomas et al.[9]

EDR	D_{002} (Å)	$D_{002,t}$ (Å)	$n_t N$	EDR	L_w (Å)	L_t (Å)
5.8	188	---	0	---	---	---
12	193	490 $\pm$ 100	0.11	12	270	370
25	264	470 $\pm$ 20	0.31	24	270	320
30	327	540 $\pm$ 20	0.39	36	390	440

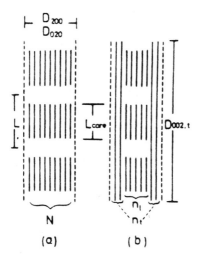

Figure 1. Schematic Representation of the Peterlin's Models for a
Fibrous Structure; (a): just after transformation to a
fibrous structure, (b): formation of crystalline tie
molecules after further deformation.

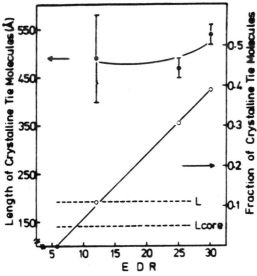

Figure 2. Growth of Crystalline Tie Molecules as a Function of
EDR. ●), average-length ($D_{002,t}$); and 0), their
fraction, (n_t/N). Broken lines show the long period
(1) and crystalline stem length (L_{core}) of folded-
chain crystals in mircofibrils.

GAS PERMEATION WITH CO_2 AS A PROBE FOR AMORPHOUS ORIENTATION

Perspective

It is well known that the molecular chains in polymers can be highly oriented in the direction of draw; the samples thus become anisotropic. Along the chain axis the atoms are covalently bonded; transverse to these covalent bonds, there are generally only weaker van der Waals interactions. The anisotropy of properties such as thermal conductivity and linear expansivity have been reported and explained by the orientation resulting from uniaxial draw[1,2].

The permeability of polymers to gases is generally interpreted by means of classical solution-diffusion equilibria and kinetics. For a polymeric film, of thickness of 1, the external (CO_2) pressure difference across the film may be considered $P_1 - P_2$. The gas flux per unit film area per unit time can thus be expressed as:

$$J = D \cdot S \frac{P_1 - P_2}{1} \tag{1}$$

Where D is the diffusivity of the gas in the polymer, S is the solubility constant, expressing the volumetric gas solubility (at 0^0C and 1 atm.) per unit volume of polymer per unit pressure. At thermodynamic equilibrium, transport for the permeability coefficient P_g ($cc[STP]cm/cm^2 \cdot sec \cdot cmHg$) is:

$$P_g = D \cdot S \tag{2}$$

The permeability is thus controlled by the diffusivity and solubility that are dependent on the morphology of polymer and its free volume. Therefore, by determination of permeability, information about the internal microscopic structure of polymer can be obtained[3,4]. The gas permeability of polymers to gas is known to be influenced by crystallinity[6-8]. It is generally found that the permeability coefficient decreases with polymer crystallinity.

Smith, et al. report that the permeability and diffusivity of gas in biaxially-oriented atactic polystyrene increases when the polystyrene was subjected to a static tensile strain, but subsequently they decreased progressively with time. They attributed these effects to a volume change[9,10] Little work has been reported on uniaxial-oriented and amorphous polystyrene.

In our hands, solid-state coextrusion has been found to be an effective technique for obtaining highly oriented polymer[11-13]. In previous studies on uniaxial draw by solid-state coextrusion, it was found that the glass transition temperature, T_g, did not change with draw, i.e. the free volume of atactic polystyrene was not a sensitively related orientation. As an alternate probe, the permeability through uniaxial orientated polystyrene has been studied. The draw ratio dependence of the transport coefficients for CO_2 in ultradrawn amorphous polystyrene (PS) and polyethylene (PE) are thus reported here. The deformation process used seems to be particularly suited to permeability studies since deformation is conducted under modest pressures reducing any tendency for voiding.

Experimental

An atactic PS, M_w of 6×10^5 and an M_w/M_n of 1.10 was obtained from Pressure Chemical Co., Pittsburgh. The powder was mold-pressed into films of about 0.6 mm thickness on being held at 180^0C and 20,000 psi for ten minutes, followed by quenching in ambient water. The films obtained were transparent.

High density PE, Marlex EHM 6033, with a M_w of 195,000 was supplied by the DuPont Co. The granules of this polymer were also molded in a similar way, but at 150^0C, followed by quenching in ice water. An opaque film of about 0.6 mm thickness was obtained.

The uniaxial draw by solid-state coextrusion was essentially the same as the technique as described[11-13]. The polymer films were inserted into split high-density PE billets and press-fitted into the barrel of an Instron rheometer. The billet assemblies were then pushed through conical brass dies of 20^0 included entrance angle at 180^0C and 90^0C for atactic PS and the PE respectively. The draw ratio was determined from the ratio of the displacements of ink markers in the films before extrusion L_1 and after extrusion, L_2, i.e. $\lambda = L_1/L_2$. The displacement also indicated that the deformation was essentially extensional rather than shear.

Permeability and Diffusion Coefficient

The permeability of CO_2 in polymer films was measured with gas chromatographically (using a Gas Chromatograph 102, Shanghai Analytic Instrument Factory). Hydrogen was used as the carrier gas. A stainless steel column with a length of 1 m and of 3 mm inner diameter was filled with 60-80 mesh GDX-401 support, polydivinylbenzene porous bead (from Chemical Reagent Co., Shanghai, China). The detector used in this measurement was a thermal conductivity cell equipped with tungsten filament. A commercial tank of CO_2 was used directly as permeant.

A small sample was tightly kept in a specially designed permeation cell, the downstream face of the cell through a triplet valve connected with the entrance to the GC column. The GC permeation measurement was carried out at 25^0C, the carrier gas flow was 20 ml/min, a detector current of 200 mA. The CO_2 pressure on the film is 0.8 Kg/cm^2. The permeation rate of CO_2 increased with time until a solution-diffusion equilibrium was reached. The volume of CO_2 permeated per unit time was measured by the peak area on the chromatogram from which the permeability of CO_2 at standard condition was calculated by:

$$P_g = \frac{T_o \cdot P \cdot V \cdot l}{P_o \cdot T \cdot A \cdot t \cdot 76} \tag{3}$$

where P_o and T_o are the pressure and temperature at standard condition, P and T are the experimental pressure and temperature. V is the volume of CO_2 permeated through the film, l is the film thickness, A is the film area and t is time of permeation.

The value of P_g was calculated at diffusion equilibrium contitions. The apparent diffusion coefficient, D_a, was obtained by the time-lag θ_L as follows:

$$D_a = \frac{l^2}{6\,\theta_L} \tag{4}$$

Birefringence

The total birefringence of drawn polymer was measured with a Zeiss polarizing microscope equipped with a Calspar rotary compensator (manufactured by Zeiss). The thickness of the samples sliced for test was taken an average of several micrometer measurements along the films.

Thermal Analysis

The glass transition temperature, T_g, of PS and the crystallinity of PE were measured in a Perkin-Elmer DSC-2. The sample weights were between 10 and 12 mg., the heating rate $20^0 C\ min^{-1}$ and the sensitivity 2 mcal/sec. Indium and lead were used as standard for calibration. The percent crystallinity of PE was calculated from the area of melting peak using the known heat of fusion of Indium and taking the enthalpy of fusion of·100% crystalline PE as 68.5 cal/gm.

Results

The birefringence of PS increased with draw (see Figure 1). The molecular orientation function can be characterized by the Hermans - Stein equation:

$$f = \frac{1}{2} (3 \cos^2\theta - 1)$$

Where f is the orientation function , θ is the average angle between chain axis and draw direction. The birefringence Δn of an oriented sample is given by:

$$\Delta n = \Delta n_o f \tag{6}$$

where Δn_o is the (intrinsic) birefringence of the fully oriented PS. For PS the intrinsic birefringence from bond polarizabilities has been calculated by Stein:

$$\Delta n_o = 0.194 - 0.51 \cdot \overline{\cos^2 \omega} \tag{7}$$

where ω is the angle between the normal to the plane of the benzene ring and the chain axis. From birefringence and infrared dichroism measurements, Milagin, et al.[18] and Jasse and Koenig[19] concluded that the average value for ω is 38^0. This value leads to a $\Delta n_o = -0.123$, which can then be substituted into Equation (6) to give the orientation function, f (see Table 1). The relationship between the permeability and draw ratio is shown in Figure 2. At pressure of 0.8 Kg/cm^2, the permeability in undrawn PS is 8.0 x 10^{-10} (cc [STP]cm/cm$^2 \cdot$ sec$\cdot$cmHg). The P_g decreased to 1.0 x 10^{-10} when draw ratio reached 5. Gas permeability is the product of diffusion and solubility. To thus explain the permeability decrease, the apparent diffusion coefficient was measured for PS at several draw ratios, see Table 1. The table indicates that the diffusion coefficient decreases with draw, whereas the solubility remains near constant. The main reason for decrease in permeability is thus likely the restriction of diffusion in drawn polymer.

The solubility of gas in a polymer is related to adsorption and to absorption in the polymer free volume. The constant solubility coefficient is consistent with a constant free volume on draw. From DSC data the glass transition temperature is 100 $\pm$ 0.5^{0}C for PS and the density 1.0485, also uninfluenced by draw. From all sides, it appears confirmed that the free volume in atactic polystyrene is essentially unchanged by this uniaxial deformation.

Consequently, the reason for the decrease in diffusion coefficient is the orientation and conformation changes of the polymer chains caused by draw. The results of small-angle neutron scattering on coextrusion oriented PS to which deuterium-labelled polymer was added show that the radius of gyration change from spherical symmetry to elliptical. At draw ratio of 5 the ratio of radii of longitudinal radius of gyration to transverse is 11[20]. The change of polymer conformation should influence the distribution of free volume in polymer, i.e. after deformation the density becomes anisotropic.

Cohen and Turnbull,[5] argued that molecular transport processes occur by the movement of molecules into voids formed by redistribution of the free volume that have a size greater than a critical value. Starting from this concept the equation for diffusion D is given by:

$$D = A \cdot \exp(-\gamma\ V^* / V_f)\qquad(8)$$

where A is transport molecule-related constant, $V^* =$ a critical volume for transport and γ is a constant of near unity. V_f is expansion volume which equals to zero at T_o; it can be expressed as:

$$V_f = \alpha V_m (T - T_o) - \beta \cdot V_p \Delta P\qquad(9)$$

where α is thermoexpansivity and V_m is a mean value of molecular volume over the relevant temperature range, β is the compressibility and V_p is a mean molecular volume for the pressure increment ΔP. For a solid, the distribution of free volume is restricted. In this case, instead of a redistribution of free volume, the dissolved molecules will diffuse through micro-voids with sizes larger than or equal to the critical volume. For this case, the second term of equation (9) can be neglected. For an anisotropic thermoexpansivity, as for uniaxially drawn polymer, as shown previously[2], the planar thermoexpansivity α parallel to the draw direction is related to the linear expansion:

$$\alpha = \alpha_{\underline{1}} + \alpha_{||}\qquad(10)$$

All this leads to the following equation:

$$D = A \exp[-\gamma\ V^* / V_m\ (\alpha_{\underline{1}} + \alpha_{||})\Delta T]\qquad(11)$$

As shown in Figure 3, in linear relationship was observed between $\ln P_g$ and $1/(\alpha_{\underline{1}} + \alpha_{||})$. The orientation dependence of permeability can be explained by the Cohen and Turnbull's equation.

Zachariades, et al. has reviewed the draw of polyethylene[21]. Generallly an increase in crystallinity is observed. In present work, a decrease in crystallinity was noted at lower draw ratio followed by an increase above draw ratio 3. The density decrease with draw has also been confirmed by Chuah, DeMicheli and Porter[22]. They found that the observed decrease in density with draw is related to the initial morphology and density, and is explained by Peterlin's model of deformation. It was also found that the smaller the initial density, the lower the draw ratio for the density minimum.

Michaels[23-25] and Peterlin[26] have studied the change in CO_2 permeability in PE on draw. At the onset, permeability in-

96

creased reaching a maximum at deformation of 35%. With further draw, permeability decreased. They both described the diffusion coefficient of small molecules in amorphous phase by following Equation:

$$D = D_a/\tau \cdot \delta \tag{12}$$

where D_a is the diffusion coefficient in the amorphous phase where τ is a geometric factor connected with the length and cross-section of the diffusion trace, δ = the mobility of chain segments associated with the crystalline linkage. They thus describe the permeability in semicrystalline polymer as:

$$P_g = S_z \chi \cdot D_z/\tau \cdot \delta \tag{13}$$

$$S = S_a \cdot \chi \tag{14}$$

where S_a is the solubility coefficient of small molecules in the amorphous phase, χ = amorphous volume fraction. Equation 13 means that permeability in a semicrystalline polymer is linked with crystallinity, free volume and orientation. At small elastic deformation, there were no marked changes in τ, δ and χ, but free volume enlarged, therefore D_a and S_a increased. This induced an increase in permeability; with plastic deformation, crystallinity developed and τ,δ also went up. Consequently, diffusion was restricted, and thus the permeability reduced. By the theory above, Kurado described quantitatively[6] the permeability of polyester films of different crystallinity. A linear relationship between log P_g and $(1 - 3\sqrt{\chi_c})^2$ was obtained where χ_c is volume fraction of crystallinity. This relationship does not hold, however, for our ultra drawn PE.

The permeability of CO_2 in ultradrawn PE is shown in Figure 5. The highest value is around draw ratio 1.5 where plastic deformation occurs and the lowest crystallinity is observed. After that, ln P_g decreases linearly with draw. Indeed a linear relationship exists between ln P_g and crystallinity in this study range. Draw ratio of 2.5 can be considered as a transition on draw from spherulitic to fibrillar.

The calculated apparent diffusion coefficient D_a and solubility constant S for the amorphous phase of PE are shown in Table 2. It can be seen that because of increased crystallinity and orientation, the trace for diffusion becomes longer and the mobility of "segments" declines. Thus D is reduced and S_a, in the range of our study, does not significantly change. This is similar to the permeability of CO_2 through drawn PS reported above, i.e. the free volume on draw is essentially unchanged. The reason for the decrease in permeability of CO_2 in ultradrawn PE is the diminution of diffusion of CO_2 in the amorphous phase.

Conclusion

The permeability to CO_2 gas for uniaxially drawn PS and PE decreases differently with draw ratio. For PS, likely due to an anisotropic distribution of free volume on orientation, the diffusion decreases with draw yet the solubility constant and thus the free volume is essentially unchanged. For PE, the decrease with draw in permeability and diffusion follows a different pattern due to changes in both orientation and fractional crystallinity on draw.

Table 1: The Permeability of CO_2 in Ultradrawn PS at $25^\circ C$.

Draw Ratio	f_x 10^2	P_g (cc[STP]cm/cm^{2}sec cmHg) x 10^{10}	D(cm^2/sec) x 10^{10}	S(cc[STP]cc cmHg)
1.0	0	8.0	24.0	0.33
1.8	1.6	5.8	19.0	0.31
3.1	4.8	2.9	7.5	0.39
4.4	12.2	1.5	---	---
5.0	1.70	1.0	4.6	0.21

Table 2: The Permeability of CO_2 in Ultradrawn PE at $25^\circ C$.

Draw Ratio	Crystal-linity (%)	P_g (cc[STP]cm/cm^{2}sec cmHg) x 10^{10}	D(cm^2/sec) x 10^9	S(cc[STP]/cc cmHg)
1.0	70.6	1.8	1.95	0.34
2.5	65.5	2.6	2.6	0.30
3.2	69.5	2.2	1.9	0.38
4.7	76.2	1.5	1.8	0.34
6.2	77.7	0.95	1.2	0.30

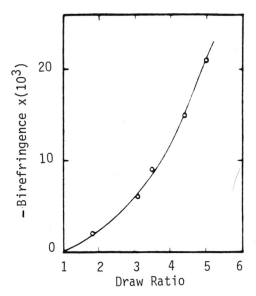

Figure 1.   Birefringence of Extruded PS as a Function of Draw Ratio.

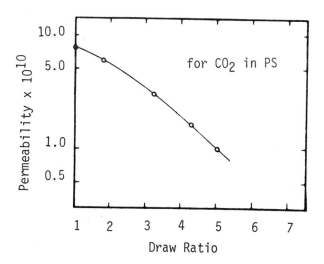

Figure 2.   Permeability of CO_2 through Extrusion-Draw PS as a Function of Draw Ratio.

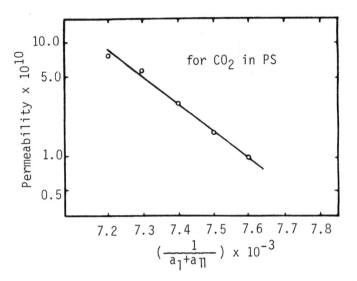

Figure 3. Permeation Flux as a Function of $[1/(\alpha_1 + \alpha_{\parallel}^{\parallel})]$ for CO_2 in Extrusion-Drawn PS.

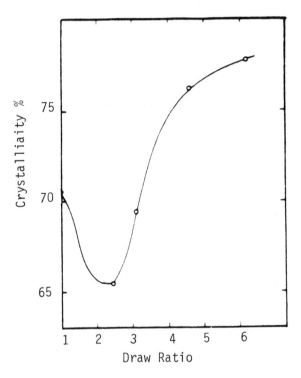

Figure 4. The Fraction of Crystallinity in PE as a Function of Draw Ratio.

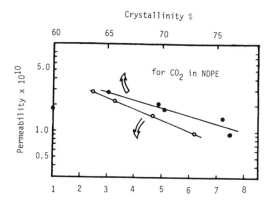

Figure 5. Permeation Flux as a Function of Crystallinity (Upper) and Draw Ratio (Lower Curve) for CO_2 in Extrusion Drawn PE.

REFERENCES

Section I

RECENT DEVELOPMENTS IN HIGH MODULUS POLYMERS

1. Polymer Science and Engineering: Challenges, Needs and Opportunities Report of the Ad Hoc Panel on Polymer Science and Engineering, National Research Council (1981).
2. Organic Polymer Characterization, NMAB 332, National Materials Advisory Board (1977).
3. Polymer Materials Basic Research Needs for Energy Applications, CONF-780643, U.S. Department of Energy (1978).
4. Workshop Proceedings: Morphology of Polyethylene and Cross-Linked Polyethylene, EPRI EL-2134-LD (1981).
5. Organic Matrix Structure Composites: Quality Assurance and Reproducibility, NMAB-365, National Materials Advisory Board (1981).
6. Materials for Lighweight Military Combat Vehicles, NMAB-396, National Materials Advisory Board (1982).
7. W. B. Black, "High Modulus/High Strength Organic Fibers," Ann. Rev. Mater. Sci., 10, 311 (1980).
8. W. B. Black and J. Preston, Editors, High Modulus Wholly Aromatic Fibers, Marcel Dekker, NY (1973).

9. H. Blades, U.S. Patent 3,869,430, "High Modulus, High Tenacity
    Poly(p-Phenylene Terephthalamide) Fiber," assigned to
    DuPont (1975).

10. T. E. Helminiak, "The Air Force Ordered Polymers Research
    Program: An Overview," A.C.S. Org. Coat. and Plast.
    Preprints, 4, 475 (1979).

Section II

PREPARATION AND CHARACTERIZATION OF DRAWN THERMOPLASTICS

1. P. B. Bowden and R. J. Young, J. Mater. Sci., 9, 2034 (1974).
2. D. M. Bigg, Polym. Eng. & Sci., 16, 725 (1976).
3. J. H. Southern and R. S. Porter, J. Appl. Polym. Sci., 14,
    3205 (1970).
4. N. E. Weeks and R. S. Porter, J. Polym. Sci., A2, 12, 635
    (1974).
5. N. J. Capiati, S. Kojima, W. G. Perkins and R. S. Porter, J.
    Mater. Sci., 13, 334 (1977).
6. W. T. Mead and R. S. Porter, J. Polym. Sci., C, in press.
7. S. Kojima and R. S. Porter, J. Polym. Sci., A-2, 16, 1729 (1978).
8. W. G. Perkins, N. J. Capiati and R. S. Porter, Polym. Eng. &
    Sci., 16 (1976).
9. A. E. Zachariades, W. T. Mead and R. S. Porter, "Ultrahigh
    Modulus Polymers," A. Ciferri and I. M. Ward, Editors,
    Applied Science, London, (1979).
10. S. Murayama, K. Imada and M. Takayanagi, International J. Polym.
    Mat., 2, 125 (1973).
11. S. Middleman, "Fundamentals of Polymer Processing," McGraw-Hill,
    NY, (1977).
12. A. G. Kolbeck and D. R. Uhlmann, J. Polym. Sci., 15, 27 (1977).
13. T. Kanamoto, A. E. Zachariades and R. S. Porter, Polym. J.,
    11, 4, 307 (1977).
14. W. T. Mead, C. R. Desper and R. S. Porter, J. Polym. Sci.,
    Polym. Phys. Ed., 17, 859 (1979).
15. C. J. Farrell and A. Keller, J. Mater. Sci., 12, 966 (1977).
16. A. S. Gibson, G. R. Davies and I. M. Ward, Polymer, 19, 683
    (1978).
17. A. Peterlin, J. Mater. Sci., 6, 490 (1971).
18. E. S. Clark and L. S. Scott, Polym. Eng. & Sci., 14, 682 (1974).
19. R. S. Porter, ACS Polym. Preprints, 122 (1971).
20. L. R. G. Treloar, Polymer, 1, 95 (1960).
21. I. Sakurad, T. Ito and K. Nakamura, J. Polym. Sci., C15, 75
    (1966).
22. A. E. Zachariades, P. D. Griswold and R. S. Porter, Polym. Eng.
    & Sci., 18, 861 (1978).
23. A. E. Zachariades, T. Shimada, M. P. C. Watts and R. S. Porter,
    ACS Polymer Preprints, 20, 2 (1979).
24. T. Shimada, A. E. Zachariades, W. T. Mead and R. S. Porter,
    J. Crystal Growth, accepted.

25. W. T. Mead, T. Shimada, A. E. Zachariades and R. S. Porter, Macromolecules, 12, 473 (1979).
26. A. E. Zachariades and R. S. Porter, J. Appl. Polym. Sci., 24, 1371 (1979).
27. A. E. Zachariades, E. S. Sherman and R. S. Porter, J. Appl. Polym. Sci., 24, 2137 (1979).
28. M. P. C. Watts, unpublished results.
29. A. E. Zachariades, B. Appelt and R. S. Porter, ACS Polymer Preprints, 20, 2 (1979).
30. W. G. Perkins, Ph.D. Thesis, University of Massachusetts, (1978).
31. Y. Sakuma and L. Rebenfeld, J. Appl. Polym. Sci., 10, 637 (1966).
32. J. S. Wang, J. R. Knox and R. S. Porter, Polym. J., 10, 619 (1978).

Section III

DEFORMATION OF LOW DENSITY POLYETHYLENES

1. H. Ulrich, "Introduction to Industrial Polymers," Hanser, NY (1981).
2. A. Buckley and H. A. Long, Polym. Eng. Sci., 9, 115 (1969).
3. J. M. Alexander and P. J. H. Wormell, Annais. C.I.R.O., XVIV, 21 (1971).
4. N. Capiati, S. Kojima, W. Perkins and R. S. Porter, J. Mater. Sci., 12, 334 (1977).
5. W. T. Mead and R. S. Porter, J. Polym. Sci., Polym. Symp., 63, 289 (1978).
6. S. Kojima and R. S. Porter, J. Appl. Polym. Sci., Appl. Polym. Symp., 33, 129 (1978).
7. W. G. Perkins, N. J. Capiati and R. S. Porter, Polym. Eng. Sci., 16, 200 (1976).
8. H. H. Chuah, R. E. DeMicheli and R. S. Porter, J. Polym. Sci., Polym. Lett., 21, 791 (1982).
9. J. H. Southern, R. S. Porter and H. E. Blair, J. Polym. Sci., A-2, 10, 1135 (1972).
10. S. Kojima, C. R. Desper and R. S. Porter, J. Polym. Sci., Polym. Phys. Ed., 16, 1721 (1978).
11. A. E. Zachariades, W. T. Mead and R. S. Porter, "Ultra-High Modulus Polymers," A. Ciferri and I. M. Ward, Editors, Applied Science, Essex, England (1979).
12. N. E. Weeks and R. S. Porter, J. Polym. Sci., Polym. Phys. Ed., 13, 2049 (1975).
13. J. H. Southern and R. S. Porter, J. Macromol. Sci.-Phys., B4, 541 (1970).
14. W. T. Mead and R. S. Porter, Intern. J. Polym. Mater., 7, 29 (1979).
15. J. H. Southern and R. S. Porter, J. Appl. Polym. Sci., 14, 2305 (1970).

16. D. J. H. Sandiford and A. H. Wilbourn, "Polyethylene," A. Renfrew and P. Morgan, Editors, Iliffe and Sons, London (1960).

17. R. J. Farris, M. S. Thesis, University of Utah (1969).

18. G. L. Wilkes, J. Macromol. Sci. Chem., C10, 149 (1974).

19. G. Cappaccio, A. G. Gibson and I. M. Ward, "Ultra High Modulus Polymers," A. Ciferri and I. M. Ward, Editors, Applied Science, Essex, England (1979).

20. M. Takayanagi, "Deformation and Fracture of High Polymers," H. H. Kausch, J. A. Hassell and R. I. Jaffee, Editors, Plenum Press (1972).

21. P. S. Hope and B. Parsons, Polym. Eng. Sci., 20, 597 (1980).

22. P. S. Hope and B. Parsons, Polym. Eng. Sci., 20, 589 (1980).

23. R. S. Porter, "Exploring the Limits of Polymer Properties: Preparation Methods and Evaluation," NSF proposal.

24. F. DeCandia, R. Russo, V. Vittoria and A. Peterlin, J. Polym. Sci., Polym. Phys. Ed., 20, 1175 (1982).

25. R. S. Porter, personal communication.

26. A. Tsuruta, T. Kanamoto, K. Tanoka and R. S. Porter, unpublished.

27. J. Shultz, "Polymer Materials Science," p. 210, Prentice-Hall, NJ (1974).

28. J. Brandrup and E. H. Immergut, "Polymer Handbook," p. VI 4H, Interscience Publishers, NY (1967).

29. F. DeCandia, R. Russo, V. Vittoria and A. Peterlin, J. Polym. Sci., Polym. Phys. Ed., 20, 269 (1982).

30. W. Glenz, N. Morosoff and A. Peterlin, J. Polym. Sci., Polym. Lett., 9, 211 (1971).

31. P. Meares, "Polymers: Structure and Bulk Properties," Chapter 5, Van Nostrand Reinhold, London (1965).

32. S. M. Aharoni and J. P. Sibilia, J. Appl. Polym. Sci., 23, 133 (1979).

33. P. S. Hope, A. G. Gibson and I. M. Ward, J. Polym. Sci., Polym. Phys. Ed., 18, 1243 (1980).

34. J. Clements, G. Capaccio and I. M. Ward, J. Polym. Sci., Polym. Phys. Ed., 7, 693 (1979).

35. W. T. Mead and R. S. Porter, J. Appl. Phys., 47, 4278 (1976).

36. D. P. Pope, J. Polym. Sci., Polym. Phys. Ed., 14, 811 (1976).

37. L. Koski, J. Thermal Anal., 13, 467 (1978).

38. J. P. Bell and J. H. Dumbleton, J. Polym. Sci., A-2, 7, 1033 (1969).

39. G. E. Sweet and J. P. Bell, J. Polym. Sci., A-2, 10, 1273 (1972).

40. F. DeCandia, R. Russo, V. Vittoria and A. Peterlin, J. Polym. Sci., Polym. Phys. Ed., 20, 269 (1982).

41. F. DeCandia, A. Perullo, V. Vittoria and A. Peterlin, J. Appl. Polym. Sci., 28, 1815 (1983).

42. I. M. Ward, "Mechanical Properties of Solid Polymers," Chapter 10, Wiley, London (1971).

43. V. B. Gupta, A. Keller and I. M. Ward, J. Macromol. Sci.-
    Phys., B2, 139 (1968).
44. D. W. Hadley, P. R. Pinnock and I. M. Ward, J. Mater. Sci.,
    4, 152 (1969).
45. F. C. Frank, V. B. Gupta and I. M. Ward, Phil. Mag., 21,
    1127 (1970).
46. V. B. Gupta and I. M. Ward, J. Macromol. Sci.-Phys., B2, 89
    (1968).
47. V. B. Gupta and I. M. Ward, J. Macromol. Sci.-Phys., B1, 373
    (1967).
48. I. M. Ward, J. Polym. Sci., Polym. Symp., 58, 1 (1977).
49. S. Hoshino, J. Powers, D. G. Legrand, H. Kawai and R. S. Stein,
    J. Polym. Sci., 58, 185 (1962).
50. R. S. Stein and F. H. Norris, J. Polym. Sci., 21, 381 (1956).

Section IV

DEFORMATION OF HIGH DENSITY POLYETHLENE

1. A. Cifferi and I. M. Ward, "Ultra-High Modulus Polymers," Appl.
   Sci. Publ., London, (1979).
2. A. E. Zachariades and R. S. Porter, Editors, "Strength and
   Stiffness of Polymers," Plastic Engineering Series, Vol.
   4, Marcel Dekker, (1983).
3. T. Kanamoto and K. Tanaka, Plastics, 34, 10, 25 (1983).
4. A. Peterlin, Polym. Eng. Sci., 18, 488 (1978), and A. Peterlin,
   J. Mater. Sci., 6, 490 (1971).
5. A. G. Gibson and I. M. Ward, Polymer, 19, 683 (1978).
6. M. Takayanagi, K. Imada and T. Kajiyama, J. Polym. Sci., Part
   C, 15, 263 (1966).
7. D. S. Prevorsek, P. J. Harget, R. K. Sharma and A. C. Reim-
   schuessel, J. Macromol. Sci., B-8, 127 (1973).
8. T. Kanamoto, S. Fujimatsu, A. Tsuruta, K. Tanaka and R. S.
   Porter, Repts. Progr. Polym. Phys., Japan, 24, 185 (1981).
9. E. S. Sherman, R. S. Porter and E. L. Thomas, Polymer, 23,
   1069 (1982).
10. N. E. Weeks, S. Mori and R. S. Porter, J. Polym. Sci., Polym.
    Phys. Ed., 13, 2031 (1975).
11. G. Capaccio and I. M. Ward, J. Polym. Sci., Polym. Phys. Ed.,
    19, 667 (1981).
12. R. G. Snyder and J. R. Scherer, J. Polym. Sci., Polym. Phys.
    Ed., 18, 421 (1980).
13. Y. K. Wang, D. A. Waldman, R. S. Stein and S. L. Hsu, J. Appl.
    Phys., 53, 6591 (1982).
14. A. Tsuruta, T. Kanamoto, K. Tanaka and R. S. Porter, Polym.
    Eng. Sci., 23, 521 (1983).
15. T. Kanamoto, A. Tsuruta, K. Tanaka, M. Takeda and R. S. Porter,
    Polym. J., 15, 327 (1983).
16. P. D. Griswold, A. E. Zachariades and R. S. Porter, Polym. Eng.
    Sci., 18, 861 (1981).

17. M. Ito, T. Kanamoto, K. Tanaka and R. S. Porter, Macro-molecules, 14, 1779 (1981).
18. L. Fischer, R. Haschberger, A. Ziegeldorf and W. Ruland, Colloid Polym. Sci., 260, 174 (1982).

Section V

GAS PERMEATION WITH CO_2 AS A PROBE FOR AMORPHOUS ORIENTATION

1. C. . Choy, Polymer, 18, 984 (1977).
2. L. H. Wang, C. L. Choy and R. S. Porter, J. Polym. Sci., A-2, 20, 633 (1982).
3. N. E. Droili and C. Migliaresi, J. Appl. Polym. Sci., 19, 1999 (1975).
4. W. Vieth and W. F. Wuergh, J. Appl. Polym. Sci., 13, 1164 (1959).
5. M. H. Cohen and D. Turnbull, J. Chem. Phys., 31, 1164 (1959).
6. N. Kurosa, San-I Gakkaishi, 35 (10), T-413 (1979).
7. W. J. Koros and D. R. Paul, J. Polym. Sci., A-2, 16, 2171 (1978).
8. S. W. Lasoski, Jr., and W. H. Cobbs, J. Polym. Sci., 36, 21 (1959).
9. G. Levita and T. L. Smith, Polym. Eng. Sci., 21, 936 (1981).
10. T. L. Smith, W. O. Oppermann, A. H. Chan and G. Levita, Amer. Chem. Soc., Polym. Preprints, 24 (1), 83 (1983).
11. P. D. Griswold, A. E. Zachariades and R. S. Porter, in "Flow Induced Crystallization in Polymer Systems," R. L. Miller, Editor, Gordon and Breach, NY, pp. 205-211 (1979).
12. A. E. Zachariades, E. S. Sherman and R. S. Porter, J. Appl. Polym. Sci., 24, 2137 (1979).
13. A. E. Zachariades, E. S. Sherman and R. S. Porter, J. Polym. Sci., Polym. Letters Ed., 17, 255 (1979).
14. B. Appelt, L. H. Wang and R. S. Porter, J. Mater. Sci., 16, 1763 (1981).
15. T. L. Caskey, Modern Plast., 45 (4), 148 (1976).
16. J. Brandrup and E. H. Immergut, "Polymer Handbook," John Wiley & Son, NY, (1975).
17. R. S. Stein, J. Appl. Phys., 32, 1280 (1961).
18. M. F. Milagin, A. D. Gabarayvea and J. T. Shishkin, Polym. Sci. USSR, A12, 577 (1970).
19. B. Jasse and J. L. Koenig, J. Polym. Sci., Polym. Phys. Ed., 17, 799 (1979).
20. G. Hadziioannou, L. H. Wang, R. S. Stein and R. S. Porter, Macromolecules, 15, 880 (1981).
21. A. E. Zachariades, W. T. Mead and R. S. Porter, Chem. Rev., 80, 351 (1980).

22. H. H. Chuah, R. E. DeMicheli and R. S. Porter, J. Polym. Sci., Polym. Letters Ed., accepted (1983).
23. A. S. Michaels, W. R. Vieth and H. J. Bixler, J. Appl. Polym. Sci., 3, 2735 (1964).
24. A. S. Michaels and R. E. Parker, Jr., J. Polym. Sci., 41, 53 (1959).
25. A. S. Michaels and H. J. Bixler, ibid, 50, 413 (1961).
26. H. Yasuda and A. Peterlin, J. Appl. Polym. Sci., 18, 531 (1974).

STRUCTURE/PROPERTY RELATIONSHIPS IN HIGH-STRENGTH STEELS

G. B. Olson

Massachusetts Institute of Technology
Cambridge, MA
Consultant to AMMRC, Watertown, MA

Cyril Smith[1] has called attention to the importance of a structural hierarchy in materials. Although the structural elements of concern to materials science span finite size limits ranging from atomic dimensions to the scale of a manufactured component, the structural hierarchy within these limits is of infinite complexity, so that we can never expect to fully characterize the structure of any real material. This may make the task of understanding structure/property relationships seem impossible, but the situation is not quite so hopeless if we accept the dual nature of structure/property relationships emphasized by Morris Cohen[2]. While we tend to think of properties as controlled by structure, it is the nature of the intellectual exercise of materials science that we can equally well consider structure to be controlled by properties. That is, what we see when we look in a microscope is a function of what we are trying to explain.

We here discuss the structure of steels from the viewpoints of strength, toughness, and hydrogen resistance. Results are drawn primarily from current research at MIT and AMMRC. Focussing on the highest strength levels of technological interest, we emphasize martensitic steels.

STRENGTH

The structural basis of strength in the highest strength materials generally involves a size scale which has only recently become resolvable by modern microscopy and microanalysis. Despite decades of research on carbon clustering phenomena in ferrous martensites using indirect experimental techniques, the modulated structures which develop on room temperature aging have only become clearly evident with the advent of the JEOL 200CX transmission

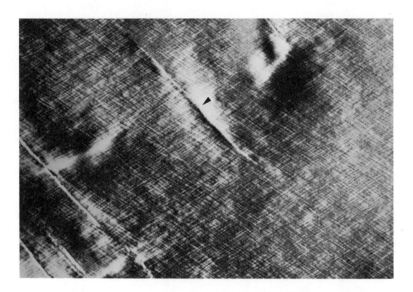

Figure 1. Bright-field Transmission Electron Micrograph of Fe-15Ni-1C Martensite Aged 5hr. at 300 K. Modulations on (20$\underline{3}$) Planes are Viewed Along [010] with Tetragonal [001] Axis Vertical. Arrow Indicates Lattice Dislocation[3].

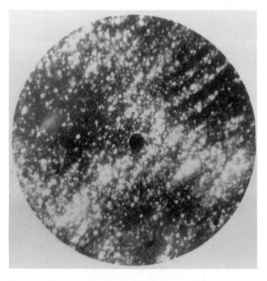

Figure 2. Field-ion Micrograph of Alloy of Figure 1 Aged 1440 hr.; Dark Bands are High-carbon Regions. 10kV Ne Image[4].

electron microscope. Figure 1 shows the ~20Å wavelength modulations that develop after 5 hr. aging of an Fe-15Ni-1C alloy which was transformed to martensite at sufficiently low temperatures to preclude carbon redistribution during the martensitic transformation[3]. The modulations lie along (203) planes of the bct martensite and form homogeneously throughout the material with little influence of substructural defects such as the dislocation indicated by the arrow. The wavelength coarsens gradually with further aging. An electrical resistivity peak is associated with an early stage of the process.

Microanalysis on the fine scale of these modulations has been made possible by the development of the atom-probe field-ion microscope (AP/FIM). Figure 2 shows a Ne field-ion image of the same modulated structure where high carbon regions image darkly[4]. Atom-by-atom analysis at a (101) pole near the (203) plane of the modulations gives the periodic variation in carbon content depicted in Figure 3 as a function of number of ions collected[4]. A gradual increase in the amplitude of the composition modulations to the final values attained in Figure 3 after 66 days aging substantiates a spinodal decomposition mechanism for the formation of this microstructure. The compositions attained in Figure 3 indicate a coherent miscibility gap with end compositions of approximately 0.2 and 11 at.pct.C at 300K. Based on Mössbauer spectroscopy[5,6] and some electron diffraction evidence[3], the high carbon regions appear to possess order of the γ'-Fe_4N type. Using a phenomenological free-energy function for ordered phases fit to fcc-bcc phase equilibrium information, a metastable phase diagram has been constructed giving a miscibility gap consistent with these compositions, with a critical temperature near 600K[3]. The modulated decomposition product not only plays an important role in strengthening, but the severe elastic distortions associated with coherent decomposition account for the inherent brittleness of "untempered" martensite[7]. Modern instrumentation has thus provided a plausible basis for technologically important phenomena which have eluded satisfactory explanation for 3000 years.

On further aging, ε-carbide is observed to form heterogeneously at high carbon regions, adopting the habit of the modulated structure. Peak hardness is associated with a mixture of fine-scale carbide precipitation and the modulations. A later stage at higher temperatures corresponding to the traditional "first stage" of tempering consists of the coarsening of a semicoherent internally-faulted plate-shaped form of the ε-carbide. The crystallography, morphology, and substructure of the carbide are consistent with an invariant-plane-strain transformation in which iron diffusion is unnecessary[3]. The shape strain accounts for an observed increase of matrix dislocation density and important stress relief phenomena observed during early tempering[8].

Although dislocation substructure evidently plays no significant

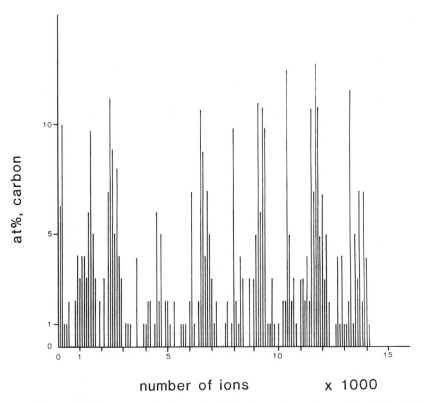

Figure 3. AP/FIM Composition Profile Probing Through (101) Pole
With Each Data Point Averaged Over 100 ions. Fe-15Ni-1C
Martensite Aged 1580 hr.[4].

Figure 4. Bright-field Transmission Electron Micrograph of 14Co-
10Ni Secondary Hardening Steel Tempered 5 hr. at 780 K.
Arrows Indicate Fine M_2C Alloy Carbide Particles in High
Dislocation Density Martensitic Substructure[10].

role in the peak-hardness microstructure of carbon martensites, its role appears to be quite important in secondary hardening steels which are strengthened by alloy carbide precipitation at higher temperatures. Retardation of dislocation recovery during tempering by alloying with Co has provided fine heterogeneously nucleated M_2C alloy carbide dispersions which give very efficient strengthening per unit of alloy carbon content[9]. The dispersions have sufficient overaging resistance that long time tempering treatments can be employed to fully dissolve the less stable Fe_3C carbides which normally limit fracture toughness in secondary hardening steels. Figure 4 shows the unusual high dislocation density and fine dispersion of M_2C carbides obtained in a 14Co-10Ni-2Cr-1Mo-0.15C steel after tempering 5 hr. at 780K[10]. AP/FIM microanalysis of M_2C carbides in the peak-hardness microstructure of typical secondary hardening steels reveals complex compositions containing Mo, Cr, W, Fe and V[11] and similar results have been recently obtained on the steel of Figure 4[12]. The complexity of the carbides is believed to play an important role in their coarsening resistance[13]. Exploitatio of these phenomena has formed the basis of the highest-performance steels developed to date.

TOUGHNESS

Substantial improvements in the cleanliness of high-strength steel have been motivated by models of ductile fracture via coalescence of inclusion-initiated voids. It now appears sufficient cleanliness has been achieved that ductile fracture of the highest-toughness high-strength steels is no longer inclusion-controlled, but occurs instead by a plastic shear instability producing a "zig-zag" fracture path at a mode I crack tip[14]. The process of plastic shear instability can be isolated in pure shear deformation experiments such as thin-walled torsion tests. Figure 5 shows strain profiles across the gage section measured from grid patterns on such a specimen of R_c55 hardness 4340 steel tested to failure at an isothermal strain rate of $\dot{\gamma} = 10^{-3}s^{-1}$[15]. Profiles from three different grid lines are shown. After uniform straining to the level γ_i, indicated by the dashed line, sharp strain peaks develop in the form of shear bands, some of which ultimately lead to shear fracture. The flow stress is observed to pass through a maximum at the instability strain γ_i.

The dependence of γ_i on hardness level is shown by the lower curve in Figure 6 for the hardness range of R_c30 to 56 covering the classical three stages of tempering. Data are shown for both torsion tests and a "linear-shear" specimen geometry[16]. The instability strain generally diminishes with increasing hardness, but shows a local minimum near $R_c 50$ corresponding to stage II tempering where retained austenite decomposes to form interfacial cementite films promoting "tempered martensite embrittlement". An increase in γ_i by about a factor of four is obtained by testing with a compressive

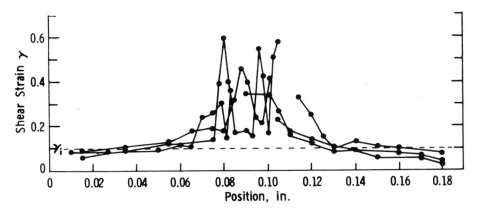

Figure 5. Shear Strain Profiles Measured Across Gage Section of Gridded Thin-Wall Torsion Specimen of ESR 4340 Steel Tempered 1 hr. at 430 K (stage I) to $R_c 55$ Hardness. Specimen Tested Isothermally at $\dot{\gamma} = 10^{-3}\ s^{-1}$. Dashed Line Identifies Instability Strain, γ_i[15].

stress of one-third the yield strength applied normal to the shear plane, as depicted by the upper curve in Figure 6. Such a pressure-dependent behavior is consistent with a void-softening mechanism of shear instability. Scanning electron microscopy of R_c30 material

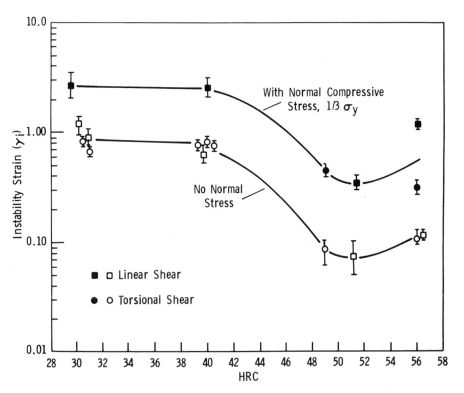

Figure 6. Measured Shear Instability Strain vs. Hardness Level for Air-Melted 4340 Steel Tested with Thin-Wall Torsion and Linear-Shear Specimen Geometries. Open Points Represent Pure Shear Tests; Solid Points Depict Effect of Superimposed Normal Compressive Stress of One-Third the Yeild Strength,
$\dot{\gamma} = 10^{-3} \, s^{-1}$ [16].

(stage III tempering) deformed just into the instability region shows void formation on cementite particles. Transmission electron microscopy of the R_c56 material tempered in stage I, corresponding to fine-scale ε-carbide precipitation, shows void formation on alloy carbides undissolved during austenitizing treatment. When the shear tests are conducted under adiabatic conditions at strain rates of 10^2 to $10^4 s^{-1}$ the instability strain is reduced at lower hardness levels, consistent with the additional effect of thermal softening. However, a significant reduction is not observed at the highest hardness levels, indicating a dominant role of the microstructure-based void softening, with thermal softening exerting only a minor influence.

The important role of undissolved alloy carbides in the high-hardness stage I tempered steel suggests a strong effect of solution treatment on shear instability resistance. This is supported by observed increases in the K_{IC} toughness of 4340 steels with higher austenitizing temperatures[17,18]. The benefit of high austenitizing vanishes after tempering beyond stage I, presumably due to a dominance of void softening by precipitated cementite. There is also evidence of a correlation between alloy carbide content and the ballistic performance of stage I tempered 4340 steels under conditions of shear-instability-controlled plugging[19].

Although the sharp-crack K_{IC} toughness of 4340 steels is raised by high austenitizing treatments, this is often attended by reduction in the blunt-notch toughness (e.g. Charpy C_V energy) and tensile ductility[18]. This appears to be related to the grain coarsening that normally accompanies high austenitizing, promoting more brittle failure modes under blunt-notch and smooth-bar test conditions. Recent research on rapid solidification processing (RSP) of steels has provided the means to avoid the excessive grain growth via the grain boundary pinning effect of finely dispersed stable inclusion phases which resist coarsening at high austenitizing temperatures[20,21]. Figure 7 summarizes the influence of austenitizing temperature T_A on the room temperature fracture properties of both RSP and conventionally processed 4Mo-0.25C martensitic steels tempered in stage I at $150°C$[22]. The shear-instability-controlled sharp-crack toughness K_Q of both the RSP and conventional steel increases substantially on raising the austenitizing temperature from 1025 to $1150°C$ to dissolve alloy carbides. In this lower-carbon steel the room temperature Charpy C_V energy also increases with austenitizing temperature for both steels, but the RSP steel shows a distinct advantage in smooth-bar fracture ductility (% RA) after austenitizing at $1150°C$ where its grain size is an order of magnitude smaller than that of the conventionally processed steel. Although the room temperature C_V energy of the two steels is

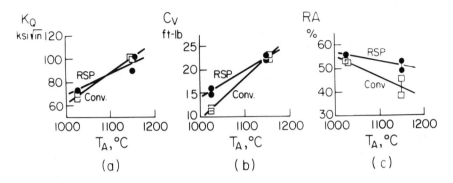

Figure 7. Influence of Austenitizing Temperature, T_A, on Room-Temperature Fracture Properties of Rapidly Solidified (RSP) and Conventionally Processed (Conv.) 4 Mo-0.25c Steels Tempered 1 hr. at 150C; (a) Sharp-Crack K_Q Toughness, (b) Blunt-Notch Charpy Impact Energy, (c) Tensile Reduction in Area[22].

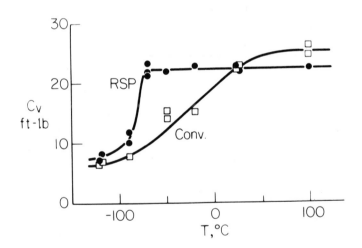

Figure 8. Charpy Ductile-Brittle Transition Curves for RSP and Conventional 4Mo-0.25C Steel After 1150°C Austenitizing, 150°C Temper[22].

identical after 1150^0C austenitizing, the advantage of the RSP steel's finer grain size is apparent in the sharply reduced ductile-brittle transition behavior shown in Figure 8. The behavior of the 4Mo-0.25C steel demonstrates that the benefits of high-temperature solution treatment can be practically applied if stable grain-refining dispersions can be maintained at a sufficiently fine size to minimize their own contribution to void formation.

For a given array of void-forming particles, resistance to shear instability can be further increased by control of the material plastic flow properties to enhance flow stability. A particularly effective mechanism for this is transformation plasticity, i.e. the operation of a martensitic transformation as a deformation mechanism. Unusual constitutive flow relations can be predicted in this case directly from transformation kinetic theory[23]. In addition to the dilatant plasticity behavior associated with the transformation volume change, the transformation can impart an upward curvature to the stress-strain curve which is especially effective in stabilizing plastic flow. The influence on fracture toughness is summarized in Figure 9 for a series of precipitation-hardened metastable austenitic steels with a 180 ksi yield strength[24]. The incremental increase in J-integral toughness relative to that of stable austenite is plotted vs. a temperature parameter which is normalized with respect to M_d, the highest temperature at which martensite forms with deformation, and M_s^σ, the temperature at which martensite forms at yielding; the latter temperatures are determined for the crack-tip stress state. Solid points indicate shear-instability-controlled fracture, while open points depict a more brittle flat mode associated with excessive formation of plate martensite at low temperatures. The overall behavior is consistent with the stress-strain-curve-shaping effect of transformation plasticity, indicating an optimum degree of crack-tip transformation near the M_s^σ temperature where fine strain-induced martensite forms in the fracture process zone. The two curves labelled HV represent alloys with a 4% transformation volume change while LV denotes a lower volume change of 2.5%. The greater toughening effect of the larger volume change is attributed to a favorable change of crack-tip stress state which inhibits void softening in accordance with models of void-induced shear insta-bility[25].

The maximum observed toughening increment of 60% corresponds to a total fracture toughness of K_{IC} = 230 Ksi $\sqrt{in}$ which appears to be the highest toughness ever achieved at this strength level. The possibility that retained austenite dispersions of controlled stability could provide similar toughening effects in martensitic steels is under investigation.

The "stress corrosion" cracking of high strength martensitic steels is generally believed to occur by a hydrogen embrittlement mechanism via cathodic charging at the crack tip[26]. The fracture

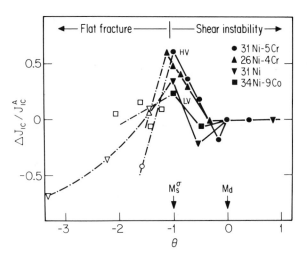

Figure 9. Incremental Change of J_{IC} Toughness Relative to Stable Austenite Thermodynamic Stability Defined by Normalized Temperature Parameter $\theta = (T-M_d)/M_d - M_s\sigma)$. Solid Points Represent Shear-Instability-Controlled Fracture, While Open Points Depict Flat Tensile Fracture Mode. HV Identifi Alloys with 4% Volume Change, LV 25% VOlume Change[24].

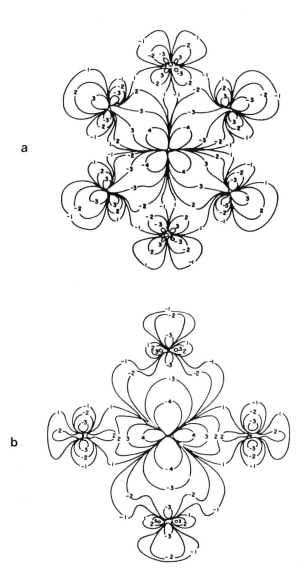

Figure 10. Wave-Function Contour Maps for Majority-Spin Bonding d
Orbitals of 15-Atom bcc Fe Cluster; (a) 1 $t_{2g}\uparrow$ Orbital
Plotted on (110) Plane Depicting Near-Neighbor Bonding
Interaction Along <111> Directions, (b) 1 $e_{g}\uparrow$ Orbital
Plotted on (100) Plane Showing Second-Neighbor Bonding
Interaction Along <001> Directions[29].

follows an intergranular path along boundaries enriched in metalloid impurity segregants. These impurities appear to directly reduce boundary cohesive strength as well as increase the sensitivity to hydrogen. Extensive studies of impurity segregation phenomena have provided some improvement in intergranular hydrogen resistance through control of alloy cosegregation effects[27], but the actual mechanism by which metalloids and hydrogen affect boundary cohesion has received attention only very recently.

While toughness is determined at the scale of particle dispersions and strength is controlled at near-atomic dimensions, the structural basis of boundary cohesive properties lies at the electronic level which is beyond the realm of traditional metallurgy; this requires the interdisciplinary approach of materials science. Fairly recently, quantum mechanical methods of computing the complete electronic structure of small atom clusters have been developed which are especially useful for the localized defect phenomena of importance to materials[28]. An example of an application particularly appropriate to the cohesive properties of steels is the computed electronic structure of a fifteen atom bcc Fe cluster summarized by Figure 10[29]. Wave function contours corresponding to the two highest occupied molecular orbitals are plotted for sections in the (110) and (100) planes. These orbitals, which are of special importance to the bonding behavior, depict the basis of directional characteristics of the bonding, Figure 10a showing 1st neighbor [111] bonding interaction and Figure 10b 2nd neighbor [001] bonding. When a uniaxial strain is applied in the [001] direction, the electronic state corresponding to Figure 10b changes from a bonding interaction to a localized magnetic state. Such a relaxation at an (001) free surface can account for an observed increase of local magnetic moment for such surfaces, and an attendent reduction of unsatisfied bond interactions can then provide a rationale for the free-surface-energy anisotropy which underlies the cleavage behavior of iron[30].

Application of the cluster methods to grain boundaries has been made possible by the development of structural unit models which describe boundary structures in terms of basic packing polyhedra[31]. Figure 11 depicts a representation of a tilt boundary core in terms of capped trigonal prisms, a particularly common unit in grain boundary structures. This unit is of special significance to the intergranular decohesion problem as it represents the nine-fold coordination environment of metalloid atoms in many metal/metalloid compounds such as Fe_3P. It is thus possible to fairly precisely specify the coordination geometry of preferred P sites in an Fe

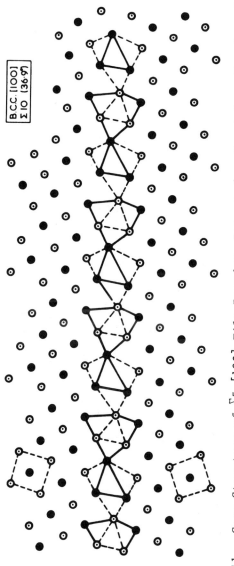

B.C.C. [100]
Σ 10 [36.9°]

Figure 11. Core Structure of Σ5 [100] Tilt Boundary Between bcc Crystals. Boundary Contains Stacks σ of Two Types of 9-Atom Capped Trigonal Prism, Differing in Their Degree of Distortion, and Staggered by Half the Lattice Parameter in a Direction Normal to the Plane of the Figure. The Centers of These Polyhedral Units are the Preferred Sites of P Atoms in Fe Grain Boundaries[31].

grain boundary as a basis for cluster calculations of local electronic structure. Calculations on similar clusters have already demonstrated a possible metalloid embrittlement mechanism whereby charge transfer to the P atom reduces Fe-Fe bonding and creates an easy fracture path at the boundary[32]. The electronic effect is consistent with the relative embrittling potency of different metalloids and the influence of cosegregating alloy elements such as Mn and Cr on embrittlement[32]. These results set the stage for calculations of the manner in which hydrogen may catalyze the decohesion of metalloid-enriched boundaries[30]. A fundamental understanding of the electronic basis of intergranular cohesion will very likely lead to the prediction of cohesion enhancing boundary segregants that will mitigate the embrittling effect of those impurities which are too costly to entirely eliminate from steels.

CONCLUSIONS

Rather than discussing quantitative structure-property relations which have been well established in ferrous metallurgy, we have here emphasized an exciting new frontier of understanding which is at an early qualitative stage. Steady progress in experimental and theoretical research is catalyzed by a maturing of instrumentation and calculational techniques which have only now reached the level of structure that controls those properties which have always been of greatest technological importance. Combined with the vast database of experience which is unique to steels, this has allowed a new depth of understanding in what has long been the most sophisticated class of materials. Despite greatly increased research efforts in other "new" materials classes over the past decade, often at the expense of ferrous metallurgy, the cutting edge of materials science is still made of steel.

ACKNOWLEDGEMENTS

Research at M.I.T. on structure/property relationships, martensitic transformations, and transformation plasticity is sponsored by ONR, NSF, and DOE. Research at AMMRC on shear instability and strengthening mechanisms is supported by the Metals Research Division of the Metals and Ceramics Laboratory. The author is grateful for helpful discussions with M. Cohen, K. A. Taylor, M. E. Eberhart, and K. H. Johnson of M.I.T.; M. Azrin, J. G. Cowie, E. B. Kula, and R. J. Harrison of AMMRC; and G. D. W. Smith of Oxford University.

REFERENCES

1. C. S. Smith, A Search for Structure, MIT Press, (1978).
2. M. Cohen, "Unknowables in the Essence of Materials Science and Engineering," J. Mater. Sci. and Eng. 25, p.3 (1976).
3. K. A. Taylor, "The Aging and First Stage of Tempering of Fe-Ni-C Martensites," Sc.D. Thesis, MIT, June (1985).
4. L. Chang, A. Cerezo, G. D. W. Smith, M. K. Miller, M. G. Burke, S. S. Brenner, K. A. Taylor, T. Abae, and G. B. Olson, J. de Physique 45, C9-409 (1984).
5. W. K. Choo and R. Kaplow, Acta Metall. 21, p.725 (1973).
6. H. Ino, T. Ito, S. Nasu, and U. Gosner, Acta Metall. 30, p. 9 (1982).
7. G. B. Olson and M. Cohen, Metall. Trans. A 14A, p.1057 (1983).
8. R. L. Brown, H. J. Rack, and M. Cohen, Mater. Sci. Eng. 21, p.25 (1975).
9. G. R. Speich, D. S. Dabkowski and L. F. Porter, Metall. Trans. 4, p.303 (1973).
10. J. E. Krzanowski, AMMRC unpublished research.
11. K. Stiller, L. E. Svensson, P. R. Howell, W. Rong, H. O. Andren, and G. L. Dunlop, Acta Metall. 32, p.1457 (1984).
12. L. Chang and G. D. W. Smith, Oxford University, unpublished research.
13. G. L. Dunlop and R. W. K. Honeycombe, Metal. Sci. J. 12, p.367 (1978).
14. J. Vander Avyle, ScD. Thesis, MIT, (1974).
15. G. B. Olson, M. Azrin, and N. J. Tsangarakis, Proc. 29th Sagamore Army Materials Conference - Part II AMMRC MS84-2, p.139 (1984).
16. J. G. Cowie, M. Azrin, and G. B. Olson, AMMRC, unpublished research.
17. G. Y. Lai, W. E. Wood, R. A. Clark, V. F. Zackay, and E. R. Parker, Metall. Trans. 5, p.1663 (1974).
18. R. O. Ritchie, B. Francis, and W. L. Server, Metall. Trans. A 7A, p.831 (1976).
19. L. Soffa, and A. Hirko, Hughes Helicopters, Inc., private communication.
20. M. Suga, J. L. Goss, G. B. Olson, and J. B. Vander Sande, in Proc. 2nd Intl. Conf. Rapid Solidification Processing: Principles and Technologies ed. R. Mehrabian, B. H. Kear, and M. Cohen (Claitor's, Baton Rouge), p. 364 (1980).
21. C. Y. Hsu, "Grain-Growth Mechanisms in Rapidly Solidified Steels," Sc.D. Thesis, MIT, February (1984).
22. P. M. Fleyshman, "Fracture Toughness of Rapidly Solidified High Strength Steels," S.M. Thesis, MIT, September (1982).
23. G. M. Olson, in Deformation, Processing, and Structure, ed. G. Krauss (ASM, Metals Park), p.391 (1983).
24. R. H. Léal, "Transformation Toughening of Metastable Austenitic Steels," Sc.D. Thesis, MIT, June (1984).

25. A. Needleman and J. R. Rice, in Mechanics of Metal Sheet
    Forming, ed. D. P. Koistinen and N. M. Wang (Plenum, NY),
    p.237 (1978).
26. G. Sandoz, Metall. Trans. 3, p.1169 (1972).
27. S. K. Banerjee, C. J. McMahon, Jr., and H. C. Feng, Metall.
    Trans. A 9A, p.237 (1978).
28. K. H. Johnson, Adv. Quantum Chem. 7, p.143 (1973)
29. C. Y. Yang, Sc.D. Thesis, MIT, (1977).
30. M. E. Eberhart and K. H. Johnson, MIT, private communication.
31. M. F. Ashby, F. Spaepen, and S. Williams, Acta Metall. 26,
    p.133 (1978).
32. R. P. Messmer and C. L. Briant, Acta Metall. 30, p.457 (1982).

CHEMICAL PROPERTIES OF REAL AND IDEAL GLASS SURFACES

Carlo G. Pantano

Department of Materials Science and Engineering
Pennsylvania State University
University Park, PA 16802

ABSTRACT

In general, differences in composition and structure may exist
between the bulk and 'real' surface, or near surface region, of a
multicomponent glass. These differences are due to high
temperature surface chemical phenomena which occur during the
formation and cooling of a melt glass surface, as well as to
surface chemical reactions under ambient conditions. This is in
contrast to an 'ideal' glass surface whose composition and structure are identical to the bulk. It is likely that ideal glass
surfaces can be created only by fracturing bulk, homogeneous, microstructure-free glass in an ultra-high environment. These ideal
glass surfaces are well-defined and reproducible; therefore, they
are excellent precursors for fundamental glass surface studies.
They can also be used as calibration standards and control specimens
for the interpretation of 'real' glass surface analyses. The
chemical properties of some 'real' and 'ideal' glass surfaces will
be discussed and related to macroscopic properties including dynamic
fatigue and the peel strength of laminated composites.

INTRODUCTION

The surface composition and structure of multicomponent glasses
are influenced by most glass manufacturing processes. During the
formation and cooling of a melt glass surface, for example, re-
orientation of the silicate structure, segregation of impurities,
volatization of modifiers, water and other impurities, and
chemisorption phenomena will modify and determine the final characteristics of the glass surface. The extent of this chemical and
structural modification of the surface is, of course, a complex

127

function of the bulk glass composition and the thermal and atmospheric history of the glass surface. Obviously, any high temperature alteration of the surface will influence subsequent adsorption, hydration or other surface chemical behavior in a manner which cannot be predicted simply from a knowledge of the bulk composition. Thus, it is generally observed that melt surfaces have better hydration resistance and lower chemical reactivity than ground, polished or fracture surfaces. The high temperature modification of the surface due to volatization of water, impurities and alkali, and the more favorable molecular orientation of the silicate network, are possible reasons for this. The presence of residual stress at the melt surface due to non-uniform cooling or composition gradients can also influence the chemical reactivity. Clearly, the ability to characterize these 'real' glass surfaces in relation to their processing and properties is extremely important. This is especially evident when one recalls that the reliability of a high-strength glass fiber is determined in its outermost 5-50 nm or that the adhesive strength of a glass/polymer interface is controlled over a region which extends only .5-5 nm.

A 'real' glass surface can be defined as the region in which the composition and structure are clearly distinguished from the bulk. This is in contrast to an 'ideal' glass surface whose composition and structure are identical to the bulk (except for the presence of dangling bonds). It is likely that an ideal glass surface can be created only by fracturing bulk, homogeneous, microstructure-free glass in an ultra high vacuum environment.

The study and analysis of 'ideal glass' surfaces can have many objectives. On the one hand, these 'ideal' surfaces are excellent calibration standards and controls for the interpretation of 'real' glass surface analyses because the surface composition and in-depth profile are known with certainty. Primarily though, they provide reproducible model surfaces which can be used to understand the nature of dangling bonds and the relative effects of surface reconstruction, adsorption and volatization as a function of temperature and environment. In a sense, a more fundamental understanding of 'ideal' surfaces is required to fully define the reaction mechanisms responsible for the behavior and properties of 'real' surfaces during processing in, or exposure to, particular atmospheric and thermal conditions. Figure 1 shows, for example, the microstructure observed one hour after exposing the fresh fracture surface of a complex multicomponent glass to air. Clearly, this type of surface modification will influence the glass strength, its adhesive strength in seals and composites, its optical properties, and its coefficient of friction. Of course, other environmental conditions and/or elevated temperatures would modify this surface in a different fashion, and therefore, one can study the surface chemistry by creating those 'ideal' fracture surfaces under varying, but controlled conditions. The important point is that the initial condition of the 'ideal' surface is well defined and can be reproduced. In this paper we simply use the surface analyses of 'ideal' glass surfaces to calibrate and interpret the corresponding analyses of some 'real' glass surfaces.

Figure 1.   Electron Micrograph of a Glass Fracture Surface (bar = 1 μm); this Micrograph was Obtained a Few Hours After Fracture Although the Smaller Features were Visible within One Hour; This is a Multicomponent Glass Composition Used for Television Picture Tubes (from Philips Technical Rev., 26(3), 85 (1965)).

SURFACE ANALYSIS

A wide variety of instrumental methods have been developed for the chemical and compositional analysis of surfaces. These analyses are carried out in an ultra high vacuum environment utilizing a beam of energetic electrons, ions, or x-rays to probe the solid surface. All of these input probes result in the emission or scattering of particles which carry information about the surface. The surface sensitivity, which is typically .5-10.0 nm, is due to the limited escape depth of the emitted particle. Most of these methods utilize, or can be combined with, ion-beam etching to provide in-depth compositional profiles.

Some of the more common surface analysis techniques include Auger electron spectroscopy (AES), x-ray photoelectron spectroscopy (XPS), secondary-ion mass spectroscopy (SIMS), sputter-induced photon spectroscopy (SIPS), ion scattering spectroscopy (ISS), Rutherford backscattering (RBS), and nuclear reaction analyses (NRA). The major distinctions among these techniques are the type of information provided (e.g., stoichiometry, trace element analysis, oxidation state, hydrogen detection, depth-profile, monolayer sensitivity, etc.), the spatial resolution (e.g., 50 nm for AES vs 100 μm for XPS), and the quality of the analysis (e.g., precision and accuracy, surface sensitivity, sample damage, charging, sputtering artefacts, etc.). The theory, instrumentation and applicability of these methods have already been presented in other volumes of this conference series[1-3], as well as in many books[4,5] and journals[6-9]. Thus, there is little need to present a discussion of that kind here. Since the routine application of these methods is now about fifteen years old, it is perhaps of more value to assess their ability to characterize real glass surfaces.

The primary objective of most glass surface analyses is to measure the surface composition. Unfortunately, the quantitative analysis of glass surface composition is by no means a routine matter. The use of elemental sensitivity factors to correct the measured signals is a common procedure, but the accuracy is limited to <30-50% even in simple binary systems. In complex multicomponent glasses, the accuracy is further degraded. In general, then, it is necessary to prepare glass standards for signal calibration. The 'ideal' surface of the glass is the only meaningful standard, and therefore, it is necessary to fracture these glass standards in vacuum. Figure 2 shows, for example, the observed relationship between the tin to oxygen Auger signal ratios and the bulk tin to oxygen atomic ratio (the bulk glass composition, and hence the composition of the 'ideal' surface, are obtained independently using standard spectrochemical methods). Of course, a similar curve can be obtained for any species, in any glass compositional system, and for methods other than AES[7]. The slope of the line in Figure 1 is essentially the Auger sensitivity factor for tin, but

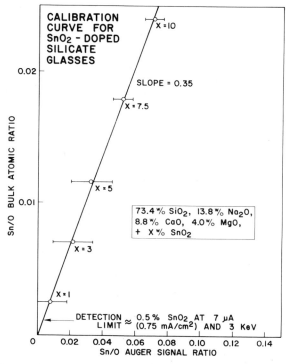

Figure 2. The Relationship Between the Atomic Ratio of Sn to O and the Auger Signal Ratio of $Sn(M_4NN)$ to $O(KLL)$; Auger Signals Obtained from Standard Glass Fracture Surfaces Created in Air and in Vacuum.

131

in contrast to elemental sensitivity factors, it accounts for the complexity of this multicomponent glass composition. This sensitivity factor can be used to quantitatively determine the relative tin content from the Auger analysis of real glass surfaces (in this composition range). If at least one of these 'ideal' surfaces is analyzed during the course of a 'real' glass surface analysis, it is possible to obtain the absolute tin concentration. One might argue that a comparable relative or absolute concentration calibration can be achieved more easily by sputter-cleaning the glass standards after fracture or polishing in air (or even by sputter-etching the 'real' glass surface to expose the bulk glass). It is shown below, however, that multicomponent oxide glasses are susceptible to sputtering artefacts which prevent the creation of an 'ideal' surface by sputter etching.

Of course, real glass surfaces are often covered to varying degrees with adsorbed water and carbonaceous species. Most of these techniques are insensitive to the presence of hydrogen species, and therefore, their unknown presence can influence the accuracy of a quantitative analysis. The SIMS, SIPS and NRA techniques can detect the presence of hydrogen, but because their surface sensitivity is limited to ~5-10 nm, it is difficult to measure adsorbed hydrogen species in the surface monolayer. The only methods available for detecting adsorbed hydrogen species are electron or photon stimulated desorption (ESD-PSD)[10] or statical secondary ion mass spectroscopy (SSIMS)[11], neither of which are well-developed techniques. On the other hand, adsorbed carbon species are very easily detected with AES & XPS whose surface sensitivity is ~.5-5.0 nm. In either case, though, the standards needed for their accurate quantitative analysis requires the deposition of calibrated amounts of the hydrogen or carbon species of interest upon 'ideal' glass surfaces. This is by no means a routine analytical procedure.

In the case of carbonaceous species, it is perhaps of more importance to define the chemical state than the quantity. XPS is really the only method which can provide this information in a direct fashion. But while it is relatively straightforward to verify the presence of carbonates, silicones, carbides, etc., it is virtually impossible to distinguish between hydrocarbons and elemental carbon. Unfortunately, most, if not all, of the carbon observed on real glass surfaces is one or both of the latter. Figure 3 shows the high-resolution carbon 1s photoelectron spectra for various glass surfaces. The SiO_2, commercial soda-lime-silicate, and 30% Na O -20% CaO - 50% SiO_2 surfaces were prepared in vacuum and then exposed to the laboratory air during which time these carbonaceous species were adsorbed. Since it is unlikely that there was any appreciable concentration of elemental carbon soot in the laboratory environment, the main peak at 285 eV is probably due to a hydrocarbon species in those cases. The treated commercial soda-lime-silicate, on the other hand, is the real surface of a glass

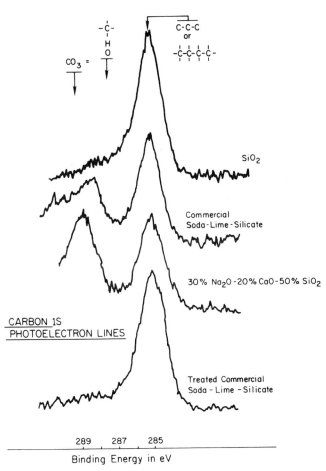

Figure 3. The High Resolution Carbon 1s XPS Spectra for Glass Surfaces. The Surfaces of the Upper Three Glasses were Prepared in Vacuum; These Carbonaceous Species were Adsorbed Onto the Clean Surfaces During a One-Hour Exposure to the Laboratory Atmosphere.

bottle. It is likely that the hot-end of a glass container plant, as well as the high temperature fluorocarbon surface treatment, does provide a source of elemental carbon soot. Here, the ability to detect hydrogen with XPS would facilitate a positive identification. These spectra also show the presence of a carbonate species on the untreated commercial glass and high-soda lime glass surfaces. Its absence on the fluorine-treated surface correlates with the corresponding depletion of sodium due to dealkalization by the fluorocarbon gas and is consistent with the spectra observed for the alkali-free SiO_2 and high-alkali glass. These carbonates may well correspond to the features observed in Figure 1.

It is quite common to use ion-beam sputtering to provide an in-depth compositional analysis of real glass surfaces. In those instances where a coating or distinct surface layer (e.g., due to leaching or hydration) is present, the qualitative profile (signal versus sputtering time), in itself, has proven to be extremely useful. And with effort, of course, these profiles can be quantified to better characterize the surface. However, in the absence of a distinct coating or surface layer, these sputter profiles can be distorted and must be interpreted with great caution (7). This is due to the differences in sputtering yield which may exist for the various constituents of a multicomponent glass. That is, the surface composition can change even in the absence of any real in-depth compositional gradient. Figure 4 shows, for example, the Auger sputter profile of tin obtained for the 'ideal' surface of a glass whose bulk composition is $66SiO_2$, $13Na_2O$, $8CaO$, $4MgO$ and $9SnO_2$. The measured tin profile is clearly a gross distortion of the true profile which is flat for this ideal surface. This sputtering effect can generate a number of problems. Very often, a real glass surface is analyzed (without the use of an ideal surface 'control') by sputter profiling into the bulk or by sputter-cleaning a glass standard prepared in air. Any differences in the signal for the unsputtered real surface, and the sputtered surface, are attributed to differences in composition between the surface and bulk. Obviously, the sputtering effect noted above can lead one to erroneous conclusions about the characteristics of the real glass surface. This is our primary rationale for routinely examining 'ideal' glass surfaces during the course of the real glass surface analysis.

Moreover, in those instances where a true in-depth profile does exist, the profile is distorted, and thereby, not ideally suited to a quantitative determination of, for example, a diffusion coefficient. Figure 5a presents the SIMS depth profiles for the

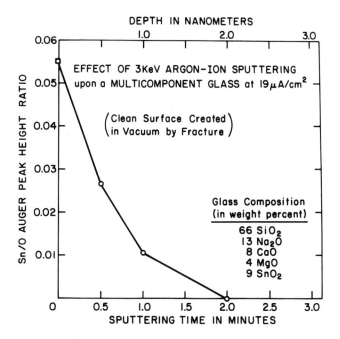

Figure 4. The Dependence of the Sn(M₄NN) to O(KLL) Auger Signal
Ratio Upon Sputter Etching; this 'Apparent' Concentra-
tion Gradient is a Consequence of the Sputtering
Process.

ideal surface of a glass with the same relative concentrations of
Na_2O, CaO and SiO_2 as float glass but doped with ~7.5 weight percent
SnO_2. Although this fracture surface was created in air, and
therefore is not quite as clean as the ideal surfaces created in
vacuum for the AES studies, it is doubtful that the surface preparation in air has influenced the glass to the depths suggested by
the $^{120}Sn^+$ and $^{42}Ca^+$ profiles. Thus, it is certain that the
'profiles' for $^{120}Sn^+$, $^{42}Ca^+$, and to some extend, $^{26}Mg^+$, $^{16}O^+$ and

$^{54}Fe^+$, are due to sputtering effects. Of course, the distortion of the expected flat concentration profiles appear less dramatic in these SIMS profiles than it did in the corresponding Auger sputter profiles. This is to be expected since AES analyzes the sputter modified surface layers, whereas SIMS analyzes the flux of secondary ions ejected from the surface. Nonetheless, it must be emphasized that these sputtering effects will have a significant influence upon the quantitative interpretation of the tin diffusion profile in the 'real' glass surface of Figure 5b. Here, a glass with the same relative concentrations of Na_2O, CaO and SiO_2 as float glass was exposed to molten tin under controlled conditions (12). Although the original objective of this analysis was to use the measured tin profile to obtain a tin diffusion coefficient, it is now obvious that this profile has been distorted by the sputtering effect.

In perhaps a more practical sense, the quality of a glass surface analysis is greatly influenced by charging, ion-migration and beam damage (13-14). Due to the insulating nature of most glasses, and the exclusive use of charged particle beam probes and/ or charged particle detection, the surfaces can accumulate charge during the course of an analysis. Very often, the charging phenomena is unstable, and thereby, prevents the acquisition of a spectra. However, there are procedures (usually based upon patience and experience) for stabilizing the surface charge so that spectra can be obtained, but seldom is the surface charge zero. This 'net' surface charge generates an electric field perpendicular to the surface which can drive the migration of mobile ionic species away from or towards the surface. Thus, the surface composition can be perturbed by beam induced migration during the analysis. The most reliable remedy for this is to cool the sample to liquid nitrogen temperature during the analysis, and thereby reduce the rate at which the surface composition is perturbed. In some cases, it is possible to record the spectra at room temperature before any measurable migration has occurred. The surface composition can also be perturbed during the analysis due to 'beam damage'. The origin of this beam damage can be one of a number of phenomena including electron or photon stimulated adsorption or desorption, electron or ion induced dissociation (e.g., reduction of oxides or decomposition of (in)organic compounds), or beam-induced heating. These effects are difficult to eliminate but they can be attenuated, in general, by reducing the incident electron or ion beam current density, by reducing the ion beam energy, or by using XPS where the photon flux is exceedingly low. Most importantly, though, one must recognize the existence of these effects in order to provide a meaningful interpretation of the data.

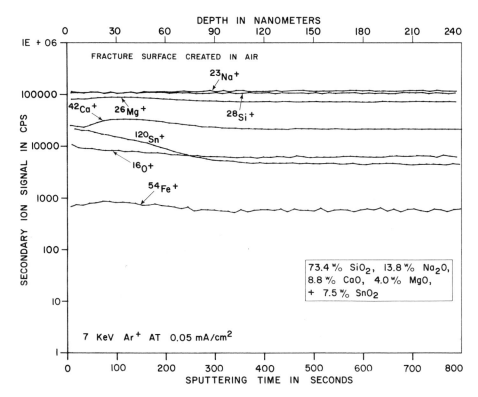

Figure 5a.   The Dependence of Secondary Ion Intensities Upon Sputter
Etching Time for: A Multicomponent Silicate Glass
Fracture Surface; again, the 'Apparent' Sn and Ca
Profiles are a Consequence of the Sputtering Process.

It is worth noting, finally, that since the RBS and NRA techniques can provide in-depth profiles without the use of ion beam
etching, they are not subject to the sputtering artefacts which

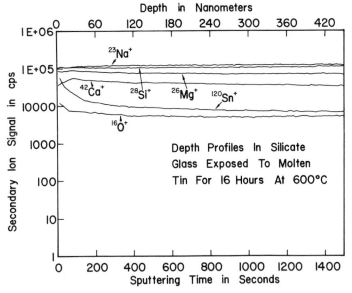

Figure 5b.   A Multicomponent 'Sn-free' Silicate Glass after Exposure
            to Molten Tin; Although One Expects a Real Sn Concentra-
            tion Gradient in This Case, the Contribution of the
            'Sputtering Effect' Seen in (a) To The Overall Measured
            Profile is Not Easily Determined.

accompany the sputter-profiling methods. Moreover, quantitative analysis can be accomplished from first principles without the use of sensitivity factors or glass standards. Unfortunately, the surface sensitivity of these methods is of the order 10 nm, and therefore, have limited applicability for 'surface' analysis. However, these are clearly the preferred methods for quantitative depth-profiling of coatings, surface layers, or diffusion profiles where $\geq$ 10 nm depth resolution is sufficient.

SURFACE ANALYSIS FOR PROCESS OR PROPERTY CONTROL

The ultimate objective of most surface analyses is to relate the surface composition or in-depth profile to the method of preparation, surface-dependent properties, or performance. Figure 6 presents the tin concentration-depth profiles obtained with SIMS for three different samples of commercial float glass (15). Although the profiles vary somewhat depending upon the thickness and/or manufacturer, the 'normal' profile is a meaningful representation of most float glasses. The other two profiles are clearly 'anomalous' relative to the normal profile, and these influenced the performance of the materials. Problem glass A, for example, exhibited an almost instantaneous weathering (or hazing) upon exposure to the ambient atmosphere after exiting the annealing lehr. The reduced concentration of tin (oxide) in the outermost 20 nm of this glass surface is clearly evident (see Figure 6). The SIMS profiles for the other elements in the glass showed that this was compensated for by the accumulation of calcium (oxide). Thus, moisture and carbon dioxide in the ambient atmosphere were readily adsorbed onto this calcia (CaO) rich surface where the formation of calcium hydroxides and carbonates were favored. These surface reaction products not only contributed to 'hazing' of the surface themselves, but enhanced the adhesion of airborne dust and other particulates. This further degraded the appearance and cleanliness of the glass surface. The unexpected generation of this 'reactive' glass surface was attributed to organometallic contamination in the back-end of the tin-bath which upset the redox equilibrium, and thereby, influenced the float glass surface chemistry. Problem glass B is actually a tempered float glass utilized for the rear window of an automobile. The addition of a rear window defroster requires the deposition (by silk-screen patterning) of a metal/glass-frit slurry. Subsequently, a heat treatment causes the glass frit to melt, flow and bond the conductive defroster pattern to the glass window. This particular sampling of float glass, however, exhibited localized 'spalling' of the glass - near the conductive metal strips - during the heat treatment step. It was concluded that the high tin concentration in the surface of this batch of glass (see Figure 6) increased the local thermal expansion coefficient such

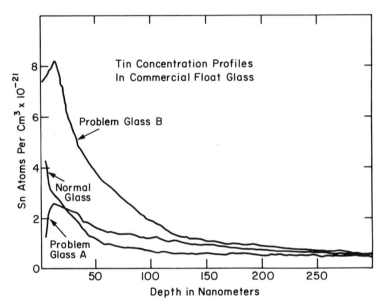

Figure 6. SIMS Depth Profiles for Tin on the 'Tin-Bath Side' of some Production Float Glasses.

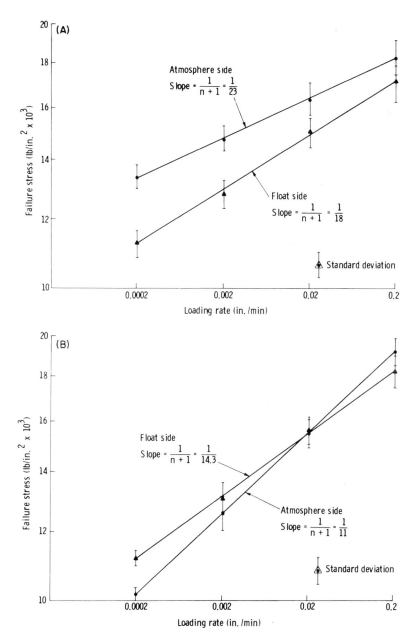

Figure 7.   Dynamic Fatigue Results for As-Received Float Glass
(Upper) and Annealed Float Glass (Lower); from 16.

that non-uniform stresses at the periphery of the metal strips initiated failure. The high tin surface concentration in this batch of float glass could also be attributed to poor process control in the tin bath; that is, the tin penetration into these surfaces is proportional to the oxygen activity in the tin bath which, again, depends upon the redox equilibria in the tin bath.

The strength and fatigue behavior of glasses is very much dependent upon surface composition and structure. Figure 7a presents some dynamic fatigue data for the opposing surfaces of commercial float glass (16). In the 'as-received' condition, one observes that the 'atmosphere' side is stronger than the 'float' side at all loading rates. This is due to the presence of a compressive residual stress on the 'atmosphere' side. The 'atmosphere' is actually the inert gas which blankets the tin bath, and thereby, enhances the volatilization of modifier species during processing. The depletion of modifier species on the 'atmosphere' side surface lowers the thermal expansion coefficient, whereas any loss of modifier species on the 'tin' side surface are compensated by the penetration of tin species. Thus, the lower thermal expansion coefficient on the 'atmosphere' side generates a residual compressive surface stress during cooling. After annealing-out this residual stress, the dependence of strength upon loading rate reveals the effects of stress corrosion. The data in Figure 7b shows that the strength of the 'atmosphere' side surface is more sensitive to loading rate than the 'float' side surface. This implies that the 'atmosphere' side surface is more reactive – that is, more susceptible to stress corrosion – than the 'float' side surface. The corrosion and surface analysis data in Figures 8 and 9 provide for the interpretation of this dynamic fatigue behavior. Figure 8 shows that the 'atmosphere' side surface releases more sodium (even though it is somewhat depleted in alkali) than the 'float' side surface during exposure to deionized water at 121.5^0C in an autoclave. The leaching of sodium is a direct consequence of the in-depth penetration of water and subsequent hydration of the glass surface. Figure 9 not only verifies the enhanced reactivity of the 'atmosphere' side surface – in this case during exposure to 98% relative humidity – but shows more directly the response of these two surfaces to hydration. Most obvious is the enhanced depletion of sodium – and corresponding incorporation of hydrogen – on the atmosphere side of the float. Thus, the hydrated surface film on the 'air' side surface is much thicker than that on the 'tin' side surface after an identical exposure to a controlled humidity environment (this result was verified with a complementary set of hydrogen analyses by NRA). It is also apparent that the hydrated layer on the tin-side is thin and distinct whereas it is diffuse and barely distinguishable on the air-side. It is believed that

'molecular' water was liberated from the air-side surface in the
analytical vacuum chamber due to an extensively hydrated microporous
surface film; optical spectra verified the presence of an anti-

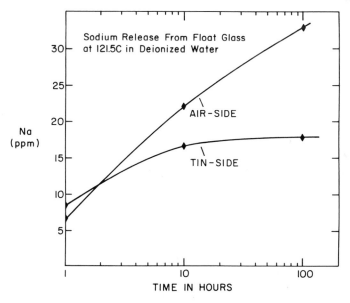

Figure 8.   The Concentration of Sodium Released from the Air-Side
and Tin-Side Surfaces of Float Glass in Deionized Water
at 121.5°C.

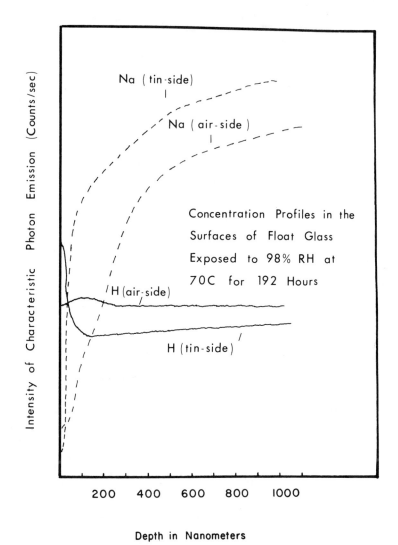

Figure 9. SIPS Depth Profiles for Sodium and Hydrogen on the Air-Side and Tin-Side Surfaces of Float Glass After Exposure to 98% Relative Humidity.

reflective layer on this surface. In contrast, the tin-side surface probably contains only 'structural' water which is stable under these vacuum conditions. Overall, these analyses show that the corrosion, surface film characteristics and reaction kinetics on the air and tin side surfaces are very different. Thus, it is not surprising that their stress corrosion susceptibilities are intrinsically different. And in the unannealed glasses, the residual stress effects due to the alkali depletion are superimposed.

Although the interfaces which comprise composite materials are not directly accessible for surface analysis, it is possible in some cases to perform a useful analysis. Here, a simple example is presented for laminates of float glass and plasticized polyvinyl butyral (PVB) (17); this is the standard safety glass design used for windshields by the automobile industry, as well as, for glazing in architectural applications. It is common practice to prepare specific PVB formulations which provide an adhesive strength suitable for the desired application. On the other hand, the adhesive strength can be adversely affected - independent of the PVB formulation - if the glass surface is not cleaned and prepared in the proper fashion. In this study, two sets of laminated glasses were prepared for evaluation. The adhesion between the glass and PVB was varied by using a different glass surface preparation for one set of laminates versus the other set, as well as, through the use of controlled additives in the PVB. These two sets of one-sided laminates were subjected to a 90^0 peel test, and then the exposed surfaces were subjected to a surface analysis by x-ray photoelectron spectroscopy (XPS). It was found that a thin layer of the polymer was left on the glass surface. This indicated that the laminates failed 'cohesively' in the PVB but near the glass/PVB interface. The high resolution XPS spectra verified that this carbonaceous polymer layer was a remnant of the PVB. Moreover, the layers were sufficiently thin that their thickness could be determined non-destructively (i.e., without sputtering) by measuring and calibrating the Si2p signal from the underlying glass substrate. Figure 10 shows that a direct correlation exists between the measured peel strength and the thickness of the polymeric 'interphase' layer left on the glass surface. It is also apparent that the method of glass surface preparation (sample set 1 vs sample set 2) will influence the nature of the interphase layer. Clearly, the amount of PVB left on the glass surface increases in proportion to adhesion in a given laminate. It remains to be shown how the chemistry of the PVB - and the chemistry of the glass surface - combine to determine the thickness and nature of this interphase layer.

SUMMARY

The capabilities of some modern methods of surface analysis were reviewed with regard to the characterization of real glass surfaces. It was shown that quantitative analysis is not yet a

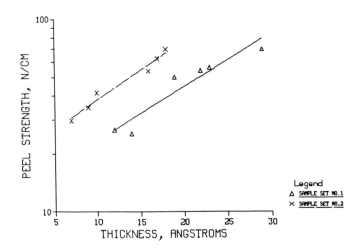

Figure 10. The Relationship Between the 90^0 Peel Strength and the Polymeric 'Interphase' Layer Thickness (Obtained by XPS) Left on the Glass Surface; from 17.

146

routine undertaking, but requires the preparation of 'ideal' glass surfaces for signal calibration. These 'ideal' glass surfaces can also be used to investigate the extent of sputtering artefacts and beam damage. In spite of these apparent limitations, though, surface analysis techniques can provide a significant amount of information about real glass surfaces. It requires only that the investigator be familiar with these limitations in his data inter- pretation, and in some instances, use the results of complementary analyses.

REFERENCES

1. A. U. MacRae, "Techniques for Studying Clean Surfaces" in Surfaces and Interfaces I, Proc. 13th Sagamore Army Materials Research Conf., J. J. Burke, N. L. Reed and V. Weiss, Editors, Syracuse University Press, (1967), pp. 29-52.
2. K. F. J. Henrich, H. Yakowitz and D. E. Newburg, "New Techniques for the Surface Analysis of Nonmetallic Solids," in Ceramics and Polymers, Proc. 20th Sagamore Army Materials Research Conf., J. J. Burke and V. Weiss, Editors, Syracuse University Press, (1975), pp. 73-102.
3. K. Vedam, "Characterization of Surfaces" in Ceramics and Polymers, Proc. 20th Sagamore Army Materials Research Conf., J. J. Burke and V. Weiss, Editors, Syracuse University Press, (1975), pp. 503-538.
4. Methods of Surface Analysis (Methods and Phenomena Series, Vol. 1), A. W. Czanderna, Editor, Elsevier, Amsterdam (1975).
5. Material Characterization Using Ion Beam, (NATO Advanced Study Institutes, Series B, Vol. 28), J. P. Thomas and A. Cachard, Editors, Plenum, NY, (1978).
6. C. G. Pantano, "Surface and In-Depth Analysis of Glass and Ceramic Materials," Bull. Am. Ceram. Soc., 60(11), 1154 (1981).
7. C. G. Pantano, J. F. Kelso and M. J. Suscavage, "Surface Studies of Multicomponent Glasses: Quantitative Analysis, Sputtering Effects and the Atomic Arrangement" in Advances in Materials Characterization, D. R. Rossington, R. A. Condrate and R. L. Snyder, Editors, Plenum, NY, (1983), pp. 1-38.
8. H. Bach, "Investigation of Glasses Using Surface Profiling by Spectrochemical Analysis of Sputter-Induced Radiation," J. Am. Ceram. Soc., 65(11), 527 (1982).
9. R. G. Gossink, "Application of Secondary Ion Mass Spectrometry to Glass Surface Problems," Glass Techn., 2(13), 125 (1980).

10. M. L. Knotek, "Stimulated Desorption From Surfaces," Physics Today, 37(9), 24 (1984).
11. A. Benninghoven, "Surface Investigation of Solids by the Statistical Method of Secondary Ion Mass Spectroscopy (SIMS)," Surface Sci., 35, 427 (1973).
12. M. J. Suscavage and C. G. Pantano, "Tin Penetration into a Soda-Lime-Magnesia-Silica Glass," Glass Techn. Ber., 56K, 498 (1983).
13. C. G. Pantano and T. E. Madey, "Electron Beam Damage in Auger Electron Spectroscopy," Appl. Surf. Sci., 6, 115 (1981).
14. B. M. J. Smets and T. P. A. Lommen, "Ion Beam Effects on Glass Surfaces," J. Am. Ceram. Soc., 65(6) C80 (1982).
15. L. Colombin, H. Charlier, A. Jelli, G. Debras and J. Verbist, "Penetration of Tin in the Bottom Surface of Float Glass: A Synthesis," J. Non-Crystal. Sol., 38-39, 551 (1980).
16. R. R. Tummala and B. J. Foster, "Strength and Dynamic Fatigue of Float Glass Surfaces," J. Am. Ceram. Soc., 58, 156 (1975).
17. D. J. David and T. N. Wittberg, "ESCA Studies of Laminated Safety Glass and Correlations with Measured Adhesive Forces," J. Adhesion (to appear).

THE RELATIONSHIPS BETWEEN POWDER PROPERTIES, SINTERED MICRO -

STRUCTURES AND OPTICAL PROPERTIES OF TRANSLUCENT YTTRIA

Frank C. Palilla, William H. Rhodes and Caryl S. Pitt

GTE Laboratories Incorporated

Waltham, MA 02254

ABSTRACT

This report describes the effect of physical and chemical powder features in sintering translucent Al_2O_3-doped Y_2O_3. In addition, the interrelationships of powder properties and several process parameters are described. The crucial features which lead to > 99.5% of the theoretical density in a highly translucent body have been identified as: (1) surface area, (2) agglomerate size, (3) a chemical composition free of multivalent and/or light-absorbing elements, and (4) a hydrogen fill of the pressed poly-crystalline form as a first step in the sintering cycle. For densities still closer to the theoretical upper limits, adjustments to the alumina concentration and peak sintering temperatures have been incorporated. A presintering step in air has no effect in final properties. The correlation of these relationships is de-scribed principally with regard to the sintered densities obtained. However, the correlation between sintered density and optical properties is unequivocally demonstrated.

INTRODUCTION

Yttria has been sintered and hot pressed to near theoretical density, resulting in consolidated bodies which transmit light[1]. Because of the refractoriness (MP 2464^0C), chemical stability, and long infrared wavelength cutoff (9 μm), this material has potential applications as high intensity discharge lamp envelopes, windows and infrared domes. The present investigation was aimed toward the lamp application where high translucency is required. The Al_2O_3 sintering aid was identified as a suitable additive to achieve the desired properties[2]. MgO, $MgAl_2O_4$ and La_2O_3 were also found to produce translucent or transparent (La_2O_3 doped) material[3], but studies on these materials have been published elsewhere and will not be discussed further[4].

Early in the study of Y_2O_3: (Al_2O_3) it was observed that variable results were obtained that could be attributed to chemical and physical variations among the incoming yttria powders. Consequently, a study of the interrelationship between powder properties and sintered properties was undertaken. The specifications established by this investigation relate to the physical requirements of the yttria powder, the chemical requirements for optimal transparency, and sintering parameters required to give high transmittance while maintaining a microstructure of sufficiently small grain size so that the lamp envelope remains structurally sound.

Throughout this parametric study, the attained sintered density of Y_2O_3: (Al_2O_3) compared to its theoretical density has been used as the criterion for correlation with the various parameters. The correspondence between the improvements realized in the sintered density and the ultimate objective of improved optical properties is firmly established. In addition, ceramographic examinations of the microstructural features have been used to supplement and confirm the property measurements.

EXPERIMENTAL

Sources of Materials

The incoming Y_2O_3 powder was the major variable in this study. Powders were obtained from four primary vendors: Molycorp, Megon, Rhodia (now Rhone-Poulenc) and Kolon. Although it is suspected that all powders were derived from oxalates, this study provided some variability in process history. A second aspect of this study was to investigate a number of powder lots having widely varying physical parameters. In this case, consistency of precursor salt and impurity content, and using powder from one vendor, assured powder physical parameters to be the major variables.

Physical Characteristics Investigated

The physical characteristics examined were:

(a) Surface area in square meters per gram (m^2/g) measured by a three-point Quantasorb BET unit.

(b) Mean or average agglomerate size in microns (μm) measured sedigraphically and/or by a Fisher Sub-Sieve Size Unit, respectively. [These measurements are assumed to represent agglomerates since they are an order of magnitude larger than the equivalent crystallite sizes as determined by item (c).]

(c) Equivalent BET diameter which is the average individual crystallite diameter as calculated from the measured BET surface area, assuming that all particles are nonporous spheres, using the formula:

$$\text{Equiv. BET diam. } (\mu m) = \frac{6}{\varrho \text{ (density) } g/cm^3 \times SA \ m^2/g \times 10^4 \ cm^2/m^2 \times 10^{-4} \ cm/\mu m}$$

$$= \frac{6}{\varrho \ (g/cm^3) \ SA \ (m^2/g)} \ (\mu m)$$

(d) Agglomeration factor which represents the ratio of mean or average agglomerate size and the BET-determined crystallite size.

(e) Bulk density in g/cm^3 measured by a Scott Paint Volumeter according to ASTM procedures.

(f) Tap density in g/cm^3 measured after the bulk density by continued addition of powder while tapping (~200X) to the full capacity of the 1-in. cubic cup of the volumeter.

(g) Hausner ratio which represents the ratio of the bulk density to the tap density.

(h) Mean pore diameter in microns (μm) of green discs, cut to ~1/8 in. pieces, measured by the mercury intrusion method using an Aminco porosimeter.

(i) Green density in g/cm^3 obtained by the weight and by the micrometric measurement of the diameter and thickness of the unsintered compacted ½ in. diameter discs.

(j) Compaction ratio which represents the ratio of green density to either the bulk density or the tap density, as specified.

(k) Sintered shrinkage in percent obtained by micrometric measurements of the diameter of the green discs and of those after sintering.

(1) Sintered density in g/cm^3 obtained using Archimedes principle but utilizing toluene as the medium for volume displacement.

(m) Theoretical density in percent represents the ratio of sintered density obtained to the theoretical density of yttria (5.0316 g/cm^3) and appropriately corrected for those cases containing the 0.31 m/o and 0.51 m/o alumina (5.0277 and 5.0249 g/cm^3, respectively). No correction is made for loss of alumina during the sintering process which is in the range of 40% to 60%[2].

(n) Average grain size in microns (μm) obtained by the linear intercept method[5].

(o) Total diffuse transmittance in percent as measured using a Cary double beam spectrophotometer over the visible range from 350 nm to 800 nm with the use of an integrating sphere. (Reflection losses were calculated to be 18.7%.)

(p) Specular or in-line transmittance in percent as measured with the above Cary spectrophotometer over the same wavelength range but without an integrating sphere, i.e., the light passing directly through the sintered disc and instrument aperture is measured.

(q) Powder morphology was examined by scanning electron microscopy and electron microscopy replication.

Chemical Purity Requirements

After establishment of the physical properties which yield acceptable sintered densities of $Y_2O_3:(Al_2O_3)$, the cationic chemical specifications were established on the basis of emission spectrographic analyses. The impurity levels of the common elements and rare-earth elements were first averaged for all the yttria samples which led to sintered densities of alumina-doped yttria greater than 99% of the theoretical value. Low limits were then set for elements known to impart body colors or visible absorption bands, as well as for multivalent elements whose complex chemistry could lead to thermal instabilities. Finally, higher limits were set for elements expected to be optically inert because of their electronic configurations.

Selected lots were analyzed by spark source mass spectroscopy for comparison with the emission spectroscopic analyses and for revealing potential anionic contaminants.

Process Parameters

The studies on undoped yttria were performed without further powder processing. Discs 1.25 cm diameter by 0.3 cm thick were uniaxially pressed at 350 MPa. The samples were sintered in a

W-mesh resistance furnace in controlled atmosphere beginning with 10^{-5} Torr vacuum shifting to Ar, then dry H_2 at 23^0C, and finishing with 23^0C dew point H_2 at the 2100^0C sintering temperature. The heating and cooling rates were held constant, as well as the 2-hr isothermal hold at 2100^0C.

The $Y_2O_3:(Al_2O_3)$ samples were prepared by one basic process with two sintering schedules labeled cycles A and B for reference. The common steps were:

Weigh 99.99% Y_2O_3 + $Al(NO_3)_3 \cdot 9H_2O$ to give 0.31 or 0.51 m/o Al_2O_3
↓
Wet Mix/Ball Mix
↓
Dry and Calcine 1000^0C for 1 hr
↓
Air Impact Mill
↓
Wet Mix – Polyvinyl Alcohol Binder
↓
Dry and Sieve Through 100 Mesh
↓
Uniaxially Cold Press at 350 MPa
↓
Sinter by Cycle A or Cycle B

The initial difference between cycles A and B was in back-filling the furnace chamber after the 10^{-5} Torr evacuation. In cycle A, the furnace was backfilled with H_2, while in cycle B, Ar was introduced prior to flushing the chamber with H_2 before heating began. The first hold, in the sintering cycle, was designed to incorporate a 2-hr solid state sintering cycle below the 1940^0C eutectic between Y_2O_3 and $Y_4Al_2O_9$. The temperatures were 1900^0C for cycle A and 1850^0C for cycle B. Liquid phase sintering is thought to occur at the peak sintering temperature which was 2150^0C and 2100^0C for cycles A and B, respectively. The isothermal hold was held constant at 4 hr. Closed porosity was reached prior to reaching the peak temperature, so an Ar-8% H_2 atmosphere was introduced to conserve power. The temperature was then lowered to 1940^0C for 2 hr and atmosphere shifted to 23^0C dew point H_2 to drive as much of the Al_2O_3 as possible back into solid solution (the solid solubility field is widest at the eutectic temperature) and ensure a stoichiometric specimen with the O^{-2} source in the wet H_2. This last sintering step was identical for both cycles.

Two Al_2O_3 concentrations were employed, 0.31 m/o and 0.51 m/o. Both levels were above the solid solubility limits, thus resulting in a liquid phase sintered body with varying second phase grain boundary volumes. The second phase was thought to be $Y_4Al_2O_9$, although there was too little to identify by x-ray diffraction.

Microprobe studies proved the presence of both Y and Al.

Physical Specifications

To establish the physical powder parameters for optimal translucency of sintered $Y_2O_3:(Al_2O_3)$, the sintering procedure of cycle B was maintained rigorously constant throughout this phase of the study for the Y_2O_3 containing 0.31 m/o Al_2O_3. Initially, the selection of yttria powders included those originating from different sources and/or processes. Subsequently, the same correlations were found to apply, with less scatter in the data points, when the processing histories of the powders were considered essentially the same (one vendor).

The fabrication process given above for the nonalumina containing discs was also maintained constant during this phase of the study. Although sintered yttria of itself is not transparent, this study did reveal a correspondence in the physical requirements of the "as received" powders to the Al_2O_3-doped processed powders. Because of this relationship, it may be possible to expedite the evaluation of a powder by applying the same criteria to sintered yttria while circumventing the powder processing of the alumina-doped sintered yttria.

Plotting Criteria

All of the straight line plots were generated by a computer fit using least squares criteria.

RESULTS AND DISCUSSION

As a result of the composite parametric studies which incorporated all the elements discussed in this section, it was possible to identify the following as crucial parameters: two physical properties (surface area and agglomerate size), a chemical composition free of complicated multivalent ions, and a hydrogen fill of the evacuated pressed $Y_2O_3:(Al_2O_3)$ composition as an initial step in the sintering cycle. An alumina concentration of 0.51 m/o and a peak sintering temperature of 2150^0C were identified as less critical, but nonetheless positive parameters. Some parameters, such as presintering, were found to have no obvious effect. In addition, some correlation between certain parameters and sinterability of yttria have been uncovered which are of scientific interest in themselves and possibly applicable in a general way to the overall field of sintering.

Sintered Yttria-Alumina

Figures 1 and 2 relate the physical features of agglomerate size and surface area of the "as received" yttria powders to the sintered density of the Y_2O_3:0.31 m/o Al_2O_3 discs derived from them. Admittedly, the scatter appears to be quite large, but nonetheless, the trend is clear. These same specific physical features for processed powders were also plotted for correlation with the sintered density of Y_2O_3:0.31 m/o Al_2O_3 after processing. The resultant relationships are plotted in Figures 3 and 4. It is obvious that the relationships hold, and, with less scatter than with the physical properties of the "as received" powders. As will be shown later, this change in degree of scatter has required that the initial specifications set for the commitment of a specific powder lot from a supplier must be followed by contingent specifications of the physical features after processing in order for the yttria lot to be accepted.

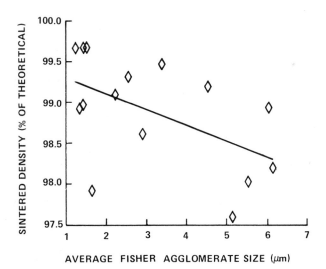

Figure 1. Sintered Density of Y_2O_3:0.31 m/o Al_2O_3 as percent of theoretical density vs average Fisher agglomerate size of "as received" yttria in μm.

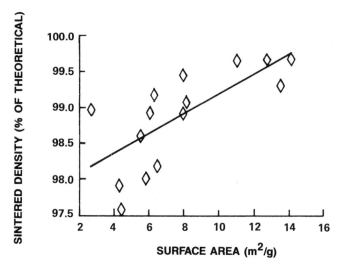

Figure 2   Sintered density of Y_2O_3: 0. 31 m/o Al_2O_3 as percent of theoretical density vs surface area of "as received" yttria in m^2/g.

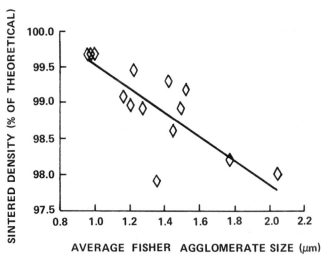

Figure 3   Processed Y_2O_3: 0.31 m/o Al_2O_3 – sintered density as percent of theoretical density vs average Fisher agglomerate size in µm

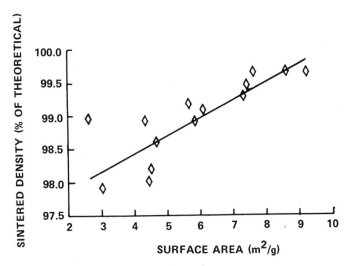

Figure 4. Processed $Y_2O_3:0.31$ m/o Al_2O_3 – sintered density as percent of theoretical density vs surface area in m^2/g.

It should be noted that dynamic physical changes occur during processing, i.e., the range of average agglomerate size narrows and the average size decreases; the surface area range also narrows and the average surface area decreases. The decrease in both physical parameters is not inconsistent. The crystallites are an order of magnitude smaller than the agglomerates, and the smallest of these crystallites grow preferentially during the powder calcining step; this tends to decrease the surface area. On the other hand, the changes in agglomerate size demonstrate that the initial large agglomerates are weakly bonded and therefore are readily commutated by the subsequent air-impact milling.

The mean agglomerate size as determined sedigraphically follows the same pattern as that of the average agglomerate size as determined by the Fisher sub-seive size. The graph is not given (it should be noted that many graphs are not shown in this report for the sake of conciseness, but the discussion in the text is based on actual plots of the relationships involved), but the scatter seems to be greater than that of the data determined by Fisher sizing. Since the sedigraphic method involves a liquid dispersion, the increased scatter could be due to the susceptibility of the method to the strength of bonding in the agglomerates. This is reinforced below, where a better correspondence between the two agglomerate size measurements is obtained when powders from the same vendor, and therefore of a similar history, are compared.

The correspondence in the physical features of the "as received" materials to the processed materials is further illustrated in Figures 5 through 7. Here the physical characteristics of surface area and agglomerate size (measured by two techniques) of the unprocessed and of the processed materials are compared.

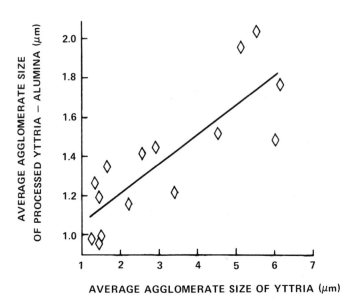

Figure 5. Average agglomerate size of processed Y_2O_3:0.31 m/o Al_2O_3 vs average agglomerate size of "as received" yttria in µm.

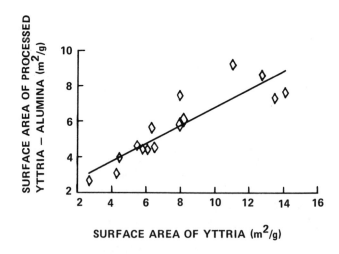

Figure 6. Surface area of processed Y_2O_3:0.31 m/o Al_2O_3 vs surface area of "as received" yttria in m^2/g.

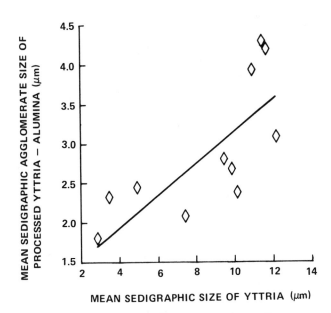

Figure 7. Mean sedigraphic agglomerate size of processed Y_2O_3: 0.31 m/o Al_2O_3 vs mean sedigraphic agglomerate size of "as received" yttria in μm.

Through this study, additional relationships between physical parameters and sinterability have been found. Although some of these may follow directly from the above surface area and agglomerate size results, others do not.

Since the equivalent BET diameter is calculated from the measured surface area, a relationship between this average crystallite diameter and sintered density corresponds inversely to that of Figure 4. (The closer the correspondence, incidentally, the more valid the assumption is that the individual crystallites are in the form of nonporous spheres.)

In view of the agglomerate size and surface area relationships, one could anticipate that a correspondence between agglomeration factor and sintered density should also exist. However, over the narrow range of agglomeration size involved, no clearly discernable relationship is evident. This is due to the fact that the uncertainties in agglomerate size and surface area measurements are compounded over a relatively small range of agglomeration factors. The small range in the agglomeration factor results from an increase in crystallite size and concomitant decrease in agglomerate size, as a result of the processing steps of calcination and air impact milling, respectively. When this interval is expanded, a distinct trend becomes apparent. This will be demonstrated below in the case of the undoped yttria (see Figure 19).

A relationship opposite of that expected from sintering science was found between the green density of the pressed discs and the resultant density of the discs after sintering. The more usual behavior would be a plateau in the 40% to 60% green density range[6] or a slope opposite of that shown in Figure 8. However, this revelation, together with a proportional function found for sintering shrinkage vs sintered density (i.e., greater the shrinkage of a green disc, the more highly sintered is the resultant disc), can be explained on the basis of agglomerate size, surface area (or equivalent crystallite diameter) and the size distribution. The smaller the agglomerate or crystallite size, the more significant are the repulsive coulombic forces compared to the particle weight; whereas, with larger agglomerates or crystallite sizes there is less repelling force between particles compared to their weight. Thus, a lower green density is obtained with smaller sized particles. In addition, the narrow size distribution prevents packing of smaller particles in the interstices, thereby reducing the green density. On the other hand, the higher the surface area, the greater the thermodynamic driving force for reducing the surface area. The combination of small crystallite size, small agglomerate size and green density high enough for good crystallite contact leads to the highest sintered density, which emphasizes that smaller crystallites and agglomerates dominate the sintering behavior over packing density. A powder with constant powder properties, cold pressed at different pressures to achieve green densities over the same range, may have shown different and perhaps the classic plateau behavior.

Finally, no distinct correlation was found for the remaining physical parameters for which correlations were sought with sintered density; i.e., the average sintered grain size and powder bulk density.

Figures 9 through 12 illustrate once again how the initial specifications set on a powdered yttria lot correspond, in general, to those of the processed alumina-doped yttria. These figures demonstrate why contingent specifications are required in addition to those initially set before a specific lot of material is committed; i.e., the order of a given physical parameter is not necessarily the same after processing into Y_2O_3:(Al_2O_3) as it was for the initial Y_2O_3. For example, in Figure 9, the highest in

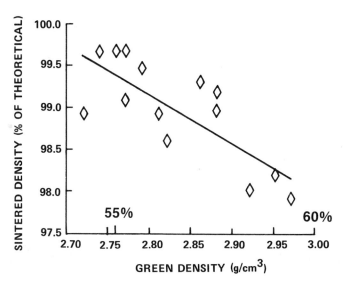

Figure 8. Processed Y_2O_3:0.31 m/o Al_2O_3 – sintered density as percent of theoretical density vs green density of pressed discs in g/cm^3.

agglomerate size of the "as received" powder is not the largest in size after processing. (Note the two points along the horizontals of the plotted percentages of theoretical density.) However, these contingency effects are not as pronounced as the more striking features which include: (1) the narrowing of the overall range of agglomerate size and surface area when the "as received" material is processed with sintering aid and binder, (2) the overall reduction in both parameters due to the processing, and (3) the steeper dependence on these parameters after processing. In addition, much of the initial scatter is lessened by the processing, which increases confidence in these relationships as valid physical laws governing sintering. There is an inverse relationship between sinterability and tap density, probably related to the same factors discussed under green density. However, this parameter changes less with processing than do the surface areas and agglomerate sizes. Although a correlation of sinterability with bulk density is not apparent from our study, this may be due to difficulties in obtaining accurate bulk density measurements on such fine powders.

This apparent lack of relationship between sintered densities and bulk densities explains why the agglomeration factor, if calculated from the tap density, shows a better direct correspondence with sintered density than if the agglomeration factor is calculated from the bulk density.

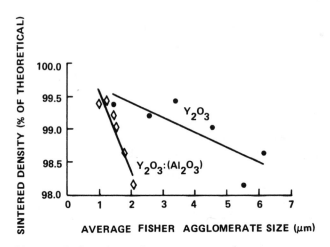

Figure 9. Sintered density of Y_2O_3:0.31 m/o Al_2O_3 as percent of theoretical density vs average Fisher agglomerate size of the "as received" yttria (•) and the processed yttria-alumina (♦) in μm.

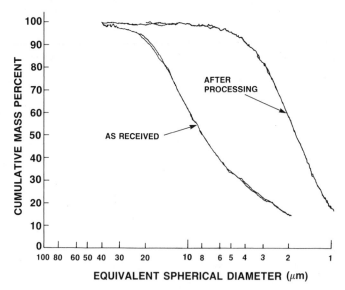

Figure 10(a). Example of change in sedigraphically determined agglomerate size from "as received" Y_2O_3 to processed Y_2O_3:0.31 m/o Al_2O_3.

162

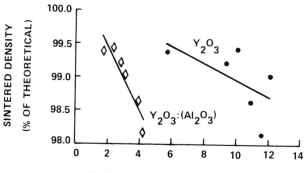

Figure 10(b). Sintered density of Y_2O_3:0.31 m/o Al_2O_3 as percent of theoretical density vs mean sedigraphic agglomerate size of the "as received" yttria (●) and the processed yttria-alumina (♦) in μm.

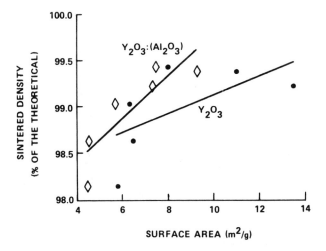

Figure 11. Sintered density of Y_2O_3:0.31 m/o Al_2O_3 as percent of theoretical density vs surface area of the "as received" yttria (●) and the processed yttria-alumina (♦) in m^2/g.

163

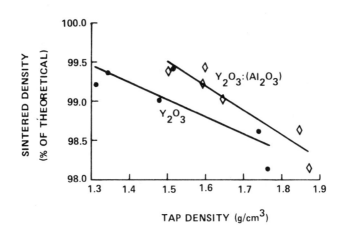

Figure 12. Sintered density of $Y_2O_3:0.31$ m/o Al_2O_3 as percent of theoretical density vs tap density of the "as received" yttria (•) and the tap density of the processed yttria-alumina (◊) in g/cm^3.

Comparison between Processed and "As Received" Powders

Figures 13 through 16 demonstrate even more dramatically that the factors which apply to the "as received" materials also apply to the processed materials. This is initially indicated in Figure 13, which plots the sintered densities of the yttria samples against the sintered densities of the alumina-doped yttria samples derived from the same powder lot. Figures 14 through 16 illustrate the point more strikingly with the more significant physical features. In these figures, the agglomerate sizes, surface areas and mean pore diameters are compared to the sintered densities of the doped and undoped yttria specimens. Again, changes which result from the processing with alumina include an overall shrinkage in the range of agglomerate size and surface area with a concomitant decrease in average values. The effectiveness of Al_2O_3 as a sintering aid is clearly demonstrated.

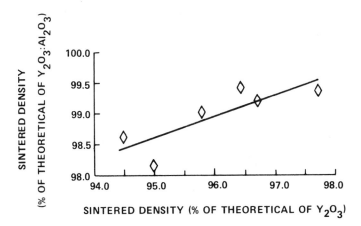

Figure 13. Sintered density of yttria vs sintered density of Y_2O_3:0.31 m/o Al_2O_3 both as percent of theoretical densities.

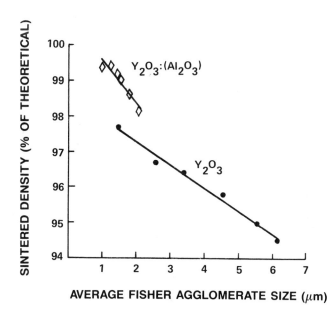

Figure 14. Sintered densities of Y_2O_3 and Y_2O_3:0.31 m/o Al_2O_3 as percent of theoretical density vs average Fisher agglomerate size of the "as received" yttria (•) and the processed yttria–alumina (◊) in µm.

165

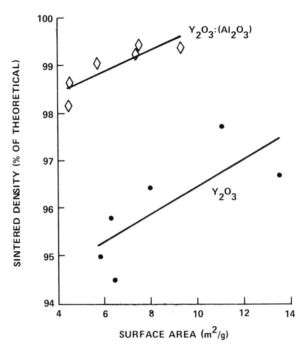

Figure 15　Sintered densities of Y_2O_3 and Y_2O_3: 0.31 m/o Al_2O_3 as percent of theoretical density vs mean pore diameter of the green discs from the "as received" yttria (•) and the processed yttria-alumina (◊) in µm.

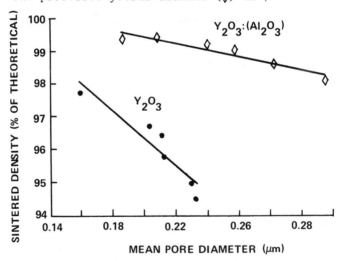

Figure 16　Sintered densities of Y_2O_3 and Y_2O_3: 0.31 m/o Al_2O_3 as percent of theoretical density vs mean pore diameter of the green discs from the "as received" yttria (•)

166

It is of interest to examine whether the pore size is controlled by particle packing of the crystallites or agglomerates,
or by a combination of the two. The observed range of green
densities shown in Figure 8 (54% to 60% of theoretical) suggests
that the powders approximate simple cubic packing, where 48% voids
are calculated for uniform spheres[7]. From the geometry of this
packing arrangement, the minimum pore size was calculated to be
0.4142 times the particle diameter. Calculated mean pore sizes
using BET-determined crystallite sizes range from 0.034 μm to
0.109 μm. If agglomerates are assumed to be the particles controlling mean pore size, values in the range of 0.398 μm to 2.53 μm
are calculated. The range of measured mean pore size is between
the two (0.159 μm to 0.315 μm), which is closer in size to those
calculated from crystallite diameters. Figure 16 shows that mean
pore diameters have increased with powder processing. This is also
true of BET-determined crystallite sizes (inferred from Figure 15),
but is not the case for agglomerate size which decreased with
powder processing, Figure 14. Thus, it appears that the mean pore
size is more closely associated with the packing of the ultimate
crystallites. Figure 17 shows the relationship between the measured
pore sizes and those calculated from the measured crystallite sizes

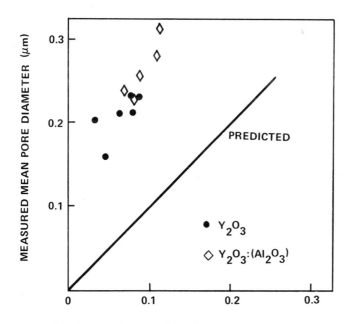

Figure 17   Measured mean pore diameter of pressed discs of yttria
             (•) and pressed discs of Y_2O_3:0.31 m/o Al_2O_3 (◊) vs the
             calculated pore diameter based on crystallite sizes in
             μm.

using the sperical particle model. A one-to-one relationship does
not exist; however, a functional dependence is found. Together,
these facts indicate that mean pore diameters are closely related
to the interstices between crystallites, but that agglomerates
play an important role in preventing pore sizes from being as small
as predicted. Thus, it can be concluded that crystallites are
responsible for the overall sintering activity of the body, but
agglomerates effectively cause bridges which coarsen the packing
structure. This phenomenon is probably responsible for much of the
remaining porosity in the sintered body.

Sintered Yttria (Without Processing)

A discussion follows which relates more directly to a
demonstration that the factors which apply to unprocessed materials
apply to the processed materials.

Figure 18 shows the sintered density of yttria as a function
of the average agglomerate size of the unprocessed yttria and
compares the results using samples from a single supplier and
samples from several suppliers. The steeper dependence of the
latter may be an anomaly, but the lesser scatter from a single
supplier could be expected. In any case, the inverse proportionality between agglomerate size and sintered density of yttria is
unquestionable.

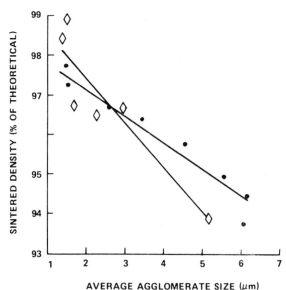

Figure 18. Sintered density of yttria as percent of theoretical
density vs average Fisher agglomerate size of the "as
received" yttria from various sources (◊) and yttria
from a single source (•).

Figures 19 through 23 give some remaining parameters that apply to yttria. In contrast to $Y_2O_3:0.31$ m/o Al_2O_3, the agglomeration factor is an important criterion for determining sintered density in undoped yttria, Figure 19. Discs pressed from these highly agglomerated powders have regions of poor packing as a result of bridging between agglomerates which results in large pores and pore clusters. These pores do not close during sintering because of unfavorable local thermodynamic driving forces and the fact that solid state sintering is the only mechanism available. However, the liquid phase sintering mechanism of the Al_2O_3-doped powder permits particle rearrangement processes to occur which reduces, but does not eliminate, the dependence on agglomerates and their many dependent properties. The role of agglomerates in controlling mean pore size has been discussed, and it is noteworthy that the data accumulated in this study show the agglomeration factor and mean pore size to be directly related. The bulk density, tap density and compaction ratio relationships, Figures 20, 21 and 22 are a consequence of the higher reactivity and better green packed structure associated with smaller crystallites and agglomerates. (Note, however, the relationship already discussed between the green density obtained from small particulates and the sinterability of the processed alumina-doped yttria.) The improved sintering with the higher Hausner ratio powders, Figure 23, is another measure of these same relationships.

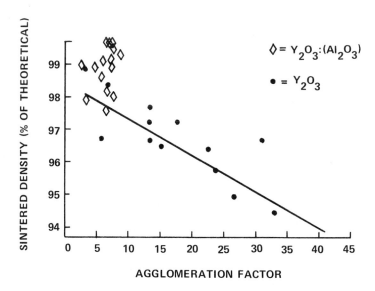

Figure 19. Sintered density of yttria from various sources as percent of theoretical density vs agglomeration factor from average Fisher agglomerate size in μm. Also given are the corresponding data points for the $Y_2O_3:0.31$ m/o Al_2O_3 specimens derived therefrom.

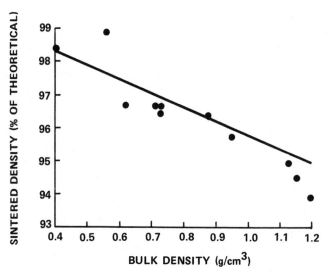

Figure 20 Sintered density of yttria from varipus sources as percent
of theoretical density vs bulk density in g/cm^3.

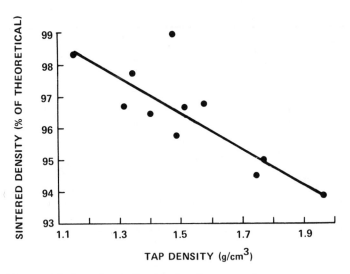

Figure 21 Sintered density of yttria from various sources as percent
of theoretical density vs tap density in g/cm^3.

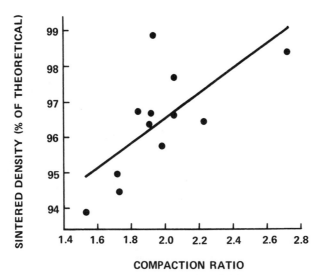

Figure 22   Sintered density of yttria from various sources as percent of theoretical density vs compaction ratio calculated from tap density.

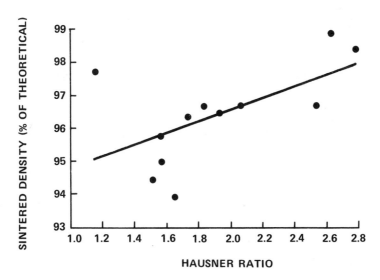

Figure 23   Sintered density of yttria from varipus sources as percent of theoretical density vs Hausner ratio.

171

Final Physical Specifications

On the basis of all the results to this point, a set of physical specifications has been set as follows, assuming a minimum acceptable sintered density of 99.6% of the theoretical for Y_2O_3:0.31 m/o Al_2O_3 when the sintering cycle B is followed:

Table I

	INITIAL SPECIFICATIONS "AS RECEIVED"	CONTINGENT SPECIFICATIONS "AFTER PROCESSING"
Average Agglomerate Size (FSSS)	$\leq$ 1.5 µm	$\leq$ 1.0 µm
Surface Area (BET)	$\geq$ 10 m^2/g	$\geq$ 7 m^2/g

According to these specifications, only 3 out of 15 samples investigated were acceptable. It was later concluded that these were the samples that sintered closest to their theoretical limits and yielded the highest translucencies.

The corresponding minimum acceptable sintered density for the "as received" yttria would be $\geq$97.2% of the theoretical density when sintered directly to Y_2O_3. Such a material would still have to meet the contingent specifications when processed with alumina, however. Two samples did not conform to the results expected from the "initial" specifications. One yielded a higher-than-expected sintered density (98.89%) despite its lower-than-acceptable surface area (2.64 m^2/g). However, it would have been correctly rejected for doping with alumina, yielding only 98.97% of theoretical density as Y_2O_3:0.31 m/o Al_2O_3. (Difficulties in reproducibility had been experienced with this particular sample throughout this study.) A second sample with a surface area of 6.06 m^2/g also gave a higher-than-expected sintered density (98.40%) for undoped yttria. However, in preparing the alumina-doped composition, it would correctly have been rejected on the basis of a higher-than-acceptable contingent agglomerate size of 1.27 µm. (This could be attributable to very strongly bonded agglomerates.) This sample yielded a sintered density of 98.93% of theoretical as Y_2O_3:0.31 m/o Al_2O_3, again when sintering cycle B was followed.

Chemical Purity Specifications

All the samples ordered had been specified at 99.99% purity in rare earths. Emission spectrographic analyses were then performed on all the samples, but the findings were averaged only for the best samples (sintered to >99.6% of the theoretical density) and for the second-best samples (sintered to >99.2% of the theoretical density).

The rationale which was then followed to arrive at a set of chemical specifications included the following:

(1) The limit for each element was set to closely match the levels in the best samples without rejecting all of the second best. While these amounts may not represent the tolerable limits, they do provide the assurances that the sinterability would not be adversely affected by these amounts and that materials of these purities are available from the usual vendors. In addition, and especially for those elements which are likely to serve as useful sintering aids for yttria, the low levels set would allow their study as additives while minimizing any uncertainties due to the background concentrations of these elements.

(2) Other considerations in setting low permissible concentrations included the avoidance of introducing significant amounts of: (a) color centers, e.g., by Cr, Nd, Fe; (b) visible absorption bands, e.g., by Ho, Er, Dy; and (c) nonstoichiometrically induced thermal instabilities by the complex chemistry of multivalent elements, e.g., Ce, Tb, Sm.

(3) By the opposite reasoning, higher levels than normally present were permitted for those elements with electronic configurations that would render them optically inert and/or whose chemistry resembles that of Y so closely that their presence should have no detrimental effect, e.g., Lu, La, Gd.

Analyses using the spark source mass spectrograph were used to supplement the emission spectrographic results. Cations reported by this technique, but not by emission spectroscopy, were Na (3 to 36 ppm), P (0 to 130 ppm), K (0.3 to 17 ppm), V (0 to 4 ppm), Zn (0 to 4 ppm), and Ba (0 to 0.55 ppm). The cations which had been found by emission spectroscopy were also found by this technique and, in general, the agreement was good.

As for anionic impurities, the spark source technique revealed that Cl was the major anion impurity present in levels from 0 ppm to 7400 ppm. Surprisingly, Cl does not completely volatilize during sintering; a value of 300 ppm in the powder and 17 ppm in the sintered product was recorded for one powder. Sulfur was also found in this same powder, which represented a single sample from one supplier. Sulfur was found neither in the sintered product from this sample nor in the powders from the other sources of Y_2O_3.

On the basis of the above experimental results, chemical analyses and rationale, the following chemical specifications are set forth for yttria powders:

Table II

ELEMENTS	TENTATIVE PERMISSIBLE LIMITS (ppm)
Common Elements:	
Si	20
Ca	15
Mg, Ni, Pb	2
Mn, Fe	3
Al	5
Cu, Cr, Co	1
Rare Earth Elements:	
Ce, Pr, Eu, Tm	1
Nd, Sm, Tb, Dy, Ho,	
Er, Yb	5
Gd, Lu, La	10

These limits may not represent the uppermost tolerable limits for these elements, but these levels do provide the assurance that the sinterability and optical properties of the $Y_2O_3:(Al_2O_3)$ would not be adversely affected. In addition, these lower limits will permit any future study of the effect of certain of these elements as additives (e.g., Si, Al, Ca) without introducing uncertainties due to the presence of higher concentrations of these elements at the outset. In any event, the specifications so established represent starting points from which the chemical requirements could be relaxed.

Sintering Schedules Investigated

The sintering parameters included the atmosphere of initial backfill, the concentration of sintering aid, thermal schedules and presintering. It should be noted at the start that all the conclusions which follow have been based on the sintered densities attained, since this information was most conveniently available and because this gravimetric determination is considered the most reliable of all the measurements made. However, justification of this comparison as it applies to the optical properties is demonstrated unequivocally in the composite graphs of Figures 24 and 25. These represent plots of experimental points from 15 different powders and 4 sintering variables; the averaged sintered densities of the fabricated discs are plotted against the optical properties of the corresponding polished discs in terms of total diffuse transmittance and specular transmittance.

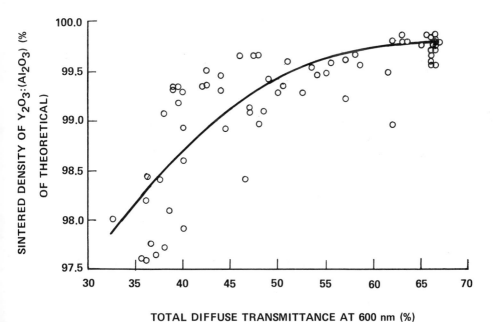

Figure 24.   Processed $Y_2O_3:(Al_2O_3)$ – sintered density as percent of theoretical density vs total diffuse transmittance of sintered discs in percent.

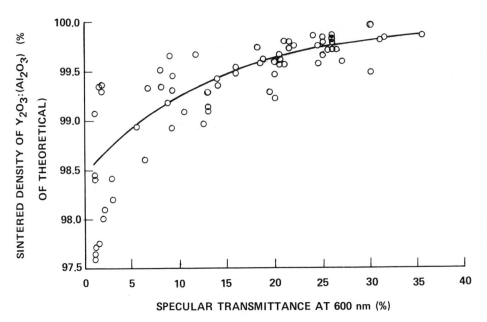

Figure 25. Processed Y_2O_3:(Al_2O_3) – sintered density as percent of theoretical density vs specular transmittance of sintered discs in percent.

The exponential relationships between the transmittances and sintered densities indicated in Figures 24 and 25 are consistent with the Mei theory for light scattering by pores, i.e.:

$$I/I_0 = e^{-KNX}$$

Where

K = Total scattering coefficient

N = Number of scattering centers

X = Thickness of specimen

As will be shown in the section below on ceramographic studies, pores and alumina-rich grain boundaries are the light scattering phases. (It should be noted that yttria has the cubic crystal structure, and therefore there should be no grain boundary birefringent scattering.) The grain boundary phase is present in small concentration (estimated to be >0.8% but <1.3%), and the index of refraction difference between this phase and Y_2O_3 is small compared to the refractive index difference between pores and Y_2O_3. Thus, pores are expected to be the major contributors to light scattering. That the light scattering depends exponentially on total pore volume, pore size and pore size distribution has been confirmed by several investigations[8,9]. These factors apparently apply here as well and are confirmed by the straight line relationships shown in Figure 26.

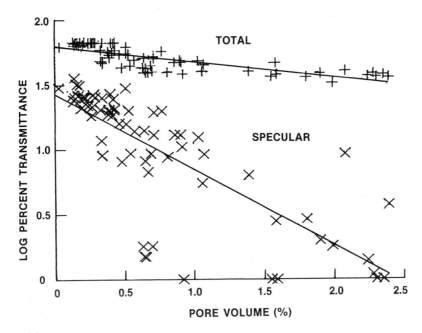

Figure 26. MIE plot of log of percent transmittance vs pore volume in percent.

The scatter in Figures 24 through 26 is largely a result of three major groupings of data corresponding to discs sintered: (1) using a hydrogen backfill, (2) using an argon backfill, and (3) using a higher dopant level of alumina. These groupings, plotted in Figure 27, are probably due to differences in pore sizes, grain sizes and concentrations of grain boundary phases resulting in variations in light scattering. The exponential functionality for each grouping is evident.

There is no obvious effect, either beneficial or detrimental, attributable to the inclusion of a presinter at 1000^0C in the procedure. One could have expected a negative effect without a presinter via decomposition of the PVA to elemental carbon during the sintering operation. Since the optical properties track the sintered densities without any observable darkening, the PVA must volatilize nonpyrolytically during the sintering step.

The results in Figure 27 clearly show the effect of the Ar vs the H_2 backfill. Even though H_2 is purged into the Ar filled furnace chamber prior to initiating the heating cycle, some Ar apparently remains within the pores of green body, and, in fact, is trapped during the final stage of sintering. It is well known that insoluble gases such as Ar can inhibit full densification (H_2 is usually considered to be a soluble gas)[10]. The H_2 has $\cong 1\frac{1}{2}$ hr to displace the Ar prior to reaching closed porosity. Thus, it is surprising to find that the diffusion and permeability rates for H_2 in these small cross section samples are so slow. These results have significant implications for many sintering procedures. For example, green bodies containing air are often placed in H_2 furnaces for sintering. Nitrogen is known to be insoluble in some oxides and to inhibit sintering. It is possible that N_2 is trapped before being flushed completely from the specimen in an analogous manner to the Ar results shown in this study.

A related observation is that larger grain sizes are obtained with the use of an H_2 backfill rather than an Ar backfill (see Figure 28). This is understandable since pores do inhibit grain growth, so that large grain sizes are found in those materials which have lower porosity before the final stage of sintering, as predicted by Zener's relationship. Here also, the optical properties reflect this decrease in porosity. The consequences in the sintered strength over this range of grain sizes have not been determined, but are not anticipated to be significant.

Distinct improvements are realized by increasing alumina concentrations. In one case out of 15, no measurable effect was observed when the Al_2O_3 additive concentration was increased from 0.3 m/o to 0.51 m/o; with no further exception, distinct improvements were obtained in the remaining 14 cases. Although greater improvements can be seen with the samples sintered with an Ar backfill of the furnace, the average sintered densities of these samples are still not as high as they are for those sintered with higher alumina concentrations and H_2 backfill. It is noteworthy, however, that samples which already have high densities resulting from sintering with an H_2 backfill can be improved still further by increasing the alumina concentration as well, although the increase is considerably less.

Once samples are prepared with both an increased alumina concentration and an H_2 backfill, increasing the sintering temperatures has only a small, if any, effect. It is interesting that elevation of only the first step of the sintering cycle yields results similar to those obtained by increasing both the initial and peak temperatures. In fact, the average sintered density for the former is 99.80% and for the latter 99.79% of theoretical. It is, therefore, the solid state sintering mechanism below 1940°C which is primarily affected rather than the liquid phase-assisted sintering at 2150°C.

178

It is also important to note that if the materials are prepared with the 0.31 m/o Al_2O_3 doping level, but are sintered following an initial H_2 backfill, raising the temperature reduces the final densities of both intermediate and "best" samples. It is well known in sintering practice that overfiring could result in increased porosity. This observation may be an indication that such a phenomenon is occurring.

In summary, therefore, the highest density materials are obtained by combining the H_2 backfill, 0.51 m/o Al_2O_3 concentrations and higher sintering temperatures (cycle A). With regard to transmittances of materials analyzed here, a decrease of specular transmittance has been noted when temperatures are elevated for those samples sintered with an Ar backfill. This is true even for samples with higher sintered densities. This is not true of total transmittances, and the effect seems to be eliminated when an H_2 backfill is incorporated in the sintering procedure. In all other cases, specular and total transmittances follow the same trends as the sintered densities.

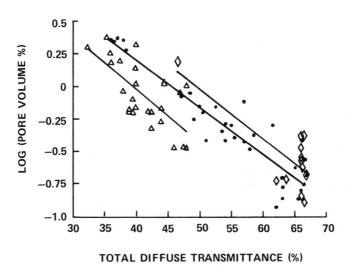

TOTAL DIFFUSE TRANSMITTANCE (%)

Figure 27. Processed Y_2O_3:(Al_2O_3) - sintered density as percent of theoretical density vs total diffuse transmittance of discs prepared with 0.31 m/o Al_2O_3 and H_2 backfill (•); 0.51 m/o Al_2O_3 and H_2 backfill (◊); and 0.31 m/o Al_2O_3 and argon backfill (Δ).

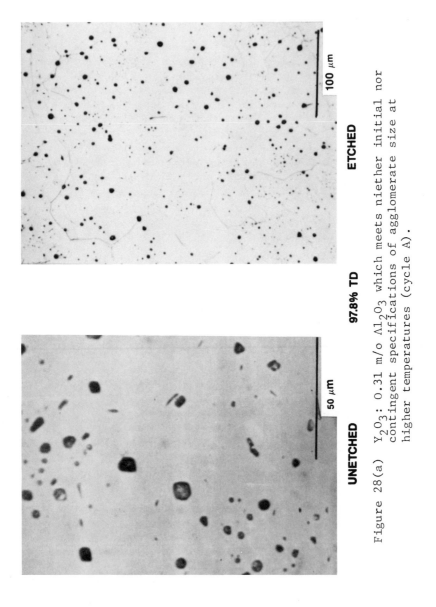

UNETCHED

50 μm

97.8% TD

ETCHED

100 μm

Figure 28(a) Y_2O_3: 0.31 m/o Al_2O_3 which meets niether initial nor contingent specifications of agglomerate size at higher temperatures (cycle A).

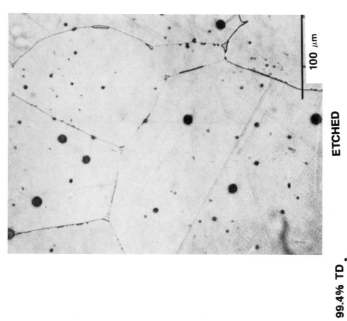

UNETCHED   **99.4% TD.**   **ETCHED**

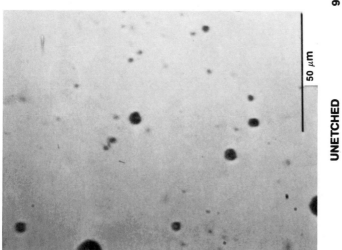

Figure 28(b) Y_2O_3: 0.31 m/o Al_2O_3 from Y_2O_3 which meets only the initial specifications for agglomerate sixe; Ar backfill, then sintered at higher temperatures (cycle A).

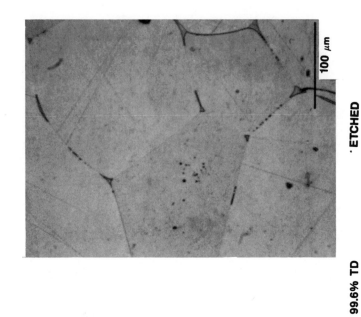

100 μm

ETCHED

99.6% TD

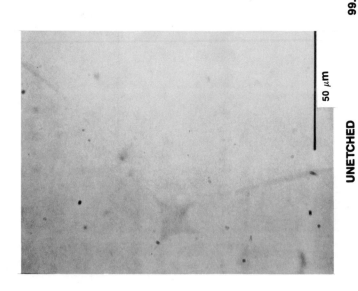

50 μm

UNETCHED

Figure 28(c) Y_2O_3; 0.51 m/o Al_2O_3 from which meets only the initial specifications for agglomerate sixe; in addition to higher Al_2O_3 content and H_2 backfill, specimen was sintered at higher temperatures (cycle A).

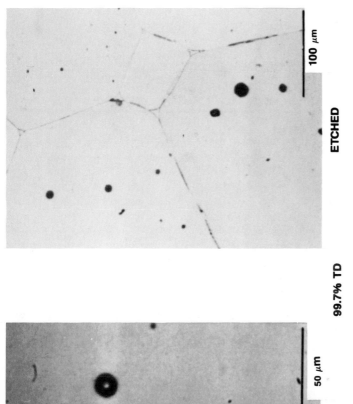

UNETCHED   50 μm   99.7% TD   ETCHED   100 μm

Figure 28(d) Y_2O_3: 0.31 m/o Al_2O_3 from Y_2O_3 which meets all agglomerate size and surface area specifications; sintered to acceptable transparency even with lower Al_2O_3 content, Ar backfill and lower sintering temperatures (cycle B).

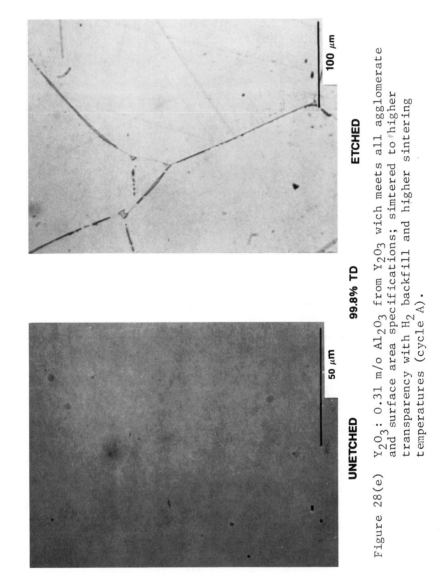

UNETCHED   **99.8% TD**   **ETCHED**

Figure 28(e) Y$_2$O$_3$: 0.31 m/o Al$_2$O$_3$ from Y$_2$O$_3$ wich meets all agglomerate
and surface area specifications; simtered to higher
transparency with H$_2$ backfill and higher sintering
temperatures (cycle A).

Ceramographic Studies

Ceramographic studies have supplemented the measurements of sintered densities and optical properties. These are illustrated in the micrographs of etched and unetched discs fabricated from: (1) materials which miss the physical and chemical specifications set above, (2) materials which miss only some of the specifications set above, and finally, (3) materials which easily meet the specifications established herein.

These micrographs are shown as Figure 28 and include: (a) a sample which misses both surface area and agglomerate size requirements (initial and contingent) and, therefore, results in a porous sintered body even when fabricated using close-to-the-best sintering conditions; (b and c) a sample which meets the size requirements but misses the surface area requirements in both initial and contingent specifications--this sample is not sinterable to acceptable translucency, but comes close when all optimal conditions are combined; (d and e) a sample which meets all the physical requirements--this sample is acceptable even with the less favorable sintering conditions (cycle B).

SUMMARY AND CONCLUSIONS

As a result of the foregoing parametric studies, the following parameters have been identified.

Crucial Parameters

- Initial and contingent specifications of surface area and agglomerate size on "as received" yttria and processed yttria-alumina powders.

- Initial evacuation of pressed $Y_2O_3:(Al_2O_3)$ and backfill with hydrogen.

- Chemical specifications on yttria powders regarding multivalent elements and those which lead to color centers. It must be emphasized, however, that the chemical requirements are assumed to be crucial but that definitive limits on any specific impurity element were not quantitatively established.

Marginal But Positive Parameters

- Use of 0.51 m/o alumina as sintering aid vs 0.31 m/o alumina.

- Elevation of initial and peak sintering temperatures from 1850^0 and 2100^0C to 1900^0 and 2150^0C, respectively.

Parameters With No Obvious Effect

Presintering the processed $Y_2O_3:(Al_2O_3)$ pressed powders at 1000°C.

Final Considerations

All of the above distinctly positive and nondetrimental features were incorporated in the optimal procedures of cycle A, assuming the selection of yttria powder has been made consistent with the size, surface area and chemical considerations given in this text. Therefore, the objective of this study is to establish th important powder parameters and to relate these to the sinterability of $Y_2O_3:(Al_2O_3)$ has been achieved. If translucent $Y_2O_3:(Al_2O_3)$ becomes commercially attractive, the use of powders appropriate for scale-up could be specified and the scale-up accomplished without the complications of powder variabilities. Beyond these practical aspects, relationships have been identified which represent contributions toward a better understanding of the science of sintering. In this regard, the observed relationships between sintered densities and mean pore diameters, green densities and agglomeration factors, and purging insoluble gases from the green body are especially cited.

ACKNOWLEDGMENTS

The authors are grateful to James F. Hopkins for his introduction to many of the physical parameters examined in this investigation and to Leo Fitzpatrick for his contributions toward establishing the sintering procedures used herein. The authors also acknowledge the support of Martha J. B. Thomas and her staff in obtaining many of the surface areas and agglomerate sizes required for this investigation. Gratitude is also expressed to James C. Mathers for discussions on yttria powders and for furnishing the powder specimens, and to James R. McColl for the computer programs used in the generation of the graphic figures of this study.

REFERENCES

1. (a) L. A. Brissette, P. L. Burnett, R. M. Spriggs and T. Vasilos, "Thermomechanically Deformed Y_2O_3," J. Amer. Ceram. Soc. 49, p. 165 (1966).
   (b) R. A. Lefever and J. Matski, "Transparent Yttrium Oxide Ceramics," Mat. Res. Bull. 2, p. 865 (1967).
   (c) S. K. Datta and G. E. Gazza, "Transparent Y_2O_3 by Hot-Pressing," Mat. Res. Bull. 4, p. 791 (1969).

(d) R. C. Anderson, unpublished research; P. J. Jorgenson and R. C. Anderson, "Grain Boundary Segregation and Final-Stage Sintering of Y_2O_3," J. Amer. Ceram. Soc. 50, p. 553 (1967).

(e) A. Muta and Y. Tsukada, "Sintered Polycrystallite Yttrium Oxide," Ger. Offen. 2,056,763 (May 1971).

(f) G. Toda, I. Matsuzama and Y. Tsukuda, "Method for Producing Highly Pure Sintered Polycrystalline Yttrium Oxide Body Having High Transparency," U. S. Patent No. 3,873,657 (March 1975).

2. (a) W. H. Rhodes and F. J. Reid, "Transparent Yttria Ceramics and Method for Producing Same," U. S. Patent No. 4,098,612 (July 1978).

3. (a) W. H. Rhodes and F. J. Reid, "Transparent Yttria Ceramics Containing Magnesia or Magnesium Aluminate," U. S. Patent No. 4,174,973 (November 1979).

(b) W. H. Rhodes, "Transparent Yttria Ceramics and Method for Producing Same," U. S. Patent No. 4,115,134 (September 1978).

4. (a) W. H. Rhodes, "Sintered Yttria for Lamp Envelopes," 2nd International Symp. Incoherent Light Sources, Enschede, Netherlands (April 1979).

(b) W. H. Rhodes, "Controlled Transient Solid Second-Phase Sintering of Yttria," J. Amer. Ceram. Soc. 64, p. 13 (1981).

5. R. L. Fullman, "Measurement of Particle Sizes in Opaque Bodies," Trans. AIME 197, p. 447 (1953).

6. C. A. Bruch, "Sintering Kinetics for the High Density Alumina Process," Bull. Amer. Ceram. Soc. 41, p. 799 (1962).

7. W. D. Kingery, H. K. Bowen and D. R. Uhlmann, Introduction to Ceramics, John Wiley and Sons, NY, p. 47 (1976).

8. D. W. Lee and W. D. Kingery, "Radiation Energy Transfer and Thermal Conductivity of Ceramic Oxides," J. Amer. Ceram. Soc. 43, p. 594 (1960).

9. J. G. J. Peelen and R. Metselaar, "Light Scattering by Pores in Polycrystalline Materials: Transmission Properties of Alumina," J. Appl. Phys. 45, p. 216 (1974).

10. R. L. Coble, "Sintering Alumina: Effect of Atmosphere," J. Amer. Ceram. Soc. 45, p. 123 (1962).

POLYMERS AND POLYMER PRECURSOR CHARACTERIZATION

Gary L. Hagnauer

Polymer Research Division
Army Materials and Mechanics Research Center
Watertown, MA 02172

INTRODUCTION

A polymer may be defined as a chainlike molecule consisting of a large number of relatively simple structural repeating units. There is no fixed lower limit defining the number of repeat units for a molecule to be regarded as polymeric. However polymers usually have molecular weights (MW) greater than about 5000 g/mol or an average of more than 100 repeat units per molecule; and unlike low MW compounds, polymers consist of a distribution of chainlike molecules of differing MW's. Generally the repeat units are held together by covalent chemical bonds in the form of linear or branched open-chain molecules or three-dimensional networks.

Although IUPAC nomenclature has been developed for systematically naming polymers, the common practice is to name polymers after the low molecular weight compounds (monomers) or prepolymers from which they are prepared or the prevalent repeating units. Thus, as shown below, the monomer name is often placed in parenthesis after the prefix "poly".

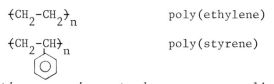

$\{CH_2-CH_2\}_n$ poly(ethylene)

$\{CH_2-CH\}_n$ poly(styrene)

Sometimes generic or trade names are applied to certain types or indeed whole classes of polymers because of their popularity or because the chemical nomenclature is cumbersome; e.g.,

"Teflon"	poly(tetrafluoroethylene)	$+CF_2-CF_2+_n$

"Nylon-6,6"   "a poly(amide)" $+NH(CH_2)_6NHCO(CH_2)_4CO+_n$
          Poly(hexamethylene adipamide)

The subscript $\underline{n}$ signifies that the polymer molecules consist of a large number of repeating units. For polymers with regular and identifiable repeat units, the term "degree of polymerization" denotes the average number of monomer repeat units per polymer molecule.

The unique properties of polymers are chiefly a result of their chainlike structure and high MW and depend primarily upon the following characteristics:

- the chemical structure and conformation of the repeating unit(s).

- the average degree of polymerization or molecular weight (MW) and molecular weight distribution (MWD) of the polymer molecules.

- the type and number of possible branching or crosslinking units.

- the relative ratio and distribution of repeat units for polymer molecules containing more than one type of repeat unit.

- the nature of the polymer chain end group(s).

Ultimate properties of polymer materials are determined by their morphology; i.e., the organization of the polymer molecules on a supramolecular scale; the size, orientation and boundary characteristics of polymer domains (e.g., in the regions of filler particles and inorganic fibers); the degree of crystallinity and the size, shape and orientation of crystalline regions.

This paper reviews state-of-the-art techniques for characterizing the compositions, structures, and properties of synthetic polymers and polymer precursors. Polymer characterization is addressed separately from that of prepolymers. Polymer characteristics and information provided by the characterization techniques are related to approaches for classifying polymers according to their properties and end-use. The importance of representative sampling and applying appropriate procedures for polymer handling, storage, and treatment are discussed. Finally, a scheme for polymer analysis and characterization is recommended; and techniques for the chemical, structural, and bulk characterization of polymers are presented. The final section deals with the development and

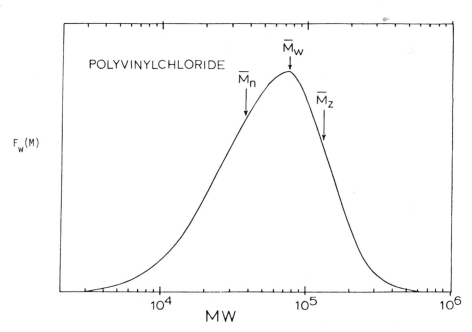

Figure 1. Molecular Weight Distribution of Poly(vinylchloride). Weight Differential Distribution verses log(MW).

$$\bar{M}_n = 3.7 \times 10^4, \ \bar{M}_w = 7.5 \times 10^4, \ \bar{M}_z = 1.2 \times 10^5$$

CLASSIFICATION OF POLYMERS

- ELASTOMERS: WEAK INTERMOLECULAR FORCES

    LOW T_G ($<-30^0$C)

    HIGH EXTENSIBILITY (1000% OR MORE) AND RAPID
    RECOVERY

    INITIAL MODULUS < 1000 PSI (7 MN/m^2)

    THERMOMECHANICAL PROPERTIES

- FIBERS: CRYSTALLINITY - INTERMOLECULAR INTERACTIONS

    LOW EXTENSIBILITY (< 20% TO BREAK) AND POOR RECOVERY

    UNIFORM PROPERTIES OVER BROAD TEMPERATURE RANGE

    T_M-T_G APPROXIMATELY 150^{0}C

    LARGE INITIAL MODULUS 0.4-2.0x10^6 PSI (3-14x10^3 MN/m^2)

- PLASTICS: BROAD RANGE OF PROPERTIES

    INITIAL MODULUS 0.4-4x10^5 PSI (0.3-3x10^{3}MN/m^2)

    EXTENSION TO BREAK 1-400%

- THERMOPLASTICS: OPEN-CHAIN POLYMERS

    T_G OR T_M > 90^{0}C

    MOLDABLE AND SOLUBLE

- THERMOSETS: NETWORK POLYMERS

    T_G > 60^{0}C

    INSOLUBLE AND INTRACTABLE SOLIDS

- PREPOLYMERS: LOW MW, SOLUBLE AND PROCESSIBLE

    POOR MECHANICAL PROPERTIES

Figure 2. Classification of Polymers According to Properties and
End-Use.

application of state-of-the-art techniques for characterizing epoxy resin prepolymers.

Classification of Polymers

Polymers may be grouped according to their end use and then further classified in regard to certain physical or chemical properties; e.g., solubility, glass transition temperature, melting temperature, thermal stability, etc.

Elastomers are characterized by relatively weak intermolecular forces. They recover rapidly and completely when stretched up to 1000% or more, and their initial moduli in tension is generally not greater than 1000 psi (7 MN/m^2). Typically, their glass transition temperatures T_g are below -30^0C and, within a limited temperature range, their strengths increase at high extensions and as temperature is raised.

An elastomer sample may consist of (1) the elastomer polymer or "gum" rubber, (2) the compounded elastomer or (3) the vulcanized elastomer compound. The "gum" may contain low MW synthesis by-products, insoluble "gel", stabilizers, and in the case of natural rubbers, resinous materials. The elastomer compound is formed by mixing and milling the elastomer polymer or a blend of different types of elastomeric polymers with any or all of the following - vulcanizing (crosslinking) agents, accelerators, processing aids (oils, waxes, etc.), stabilizers, carbon black, inorganic fillers, plus a variety of special purpose ingredients such as resins, retardants, pigments, odorants, etc. Degradation processes often occur during compounding and lead to a decrease in the average MW of the polymer(s). During vulcanization, the polymer is crosslinked and thereby becomes intractable to ordinary sample dissolution techniques.

Fibers are formed from polymers with repeat units and structures that enhance crystallization and intermolecular interactions. They have quite large initial moduli in the range 0.4-2X10^6 psi (3-14X10^3 MN/m^2). Upon stretching, fibers seldom fully recover their initial dimensions and usually break at extensions lower than 20%. Characteristically, fibers have a high ratio of length to thickness with a small cross sectional area and their properties do not vary greatly over the temperature range between -50 and 150^0C. Fibers are formed when semi-crystalline polymers are stretched at temperatures between the polymer's T_g and crystalline melting temperature T_m.

Fibers in the form of continuous filaments are often found as multifilament yarns or woven fabrics. Yarns and fabrics are treated with sizing agents and coated with oils, elastomers, powdered plastics, plasticizers, pigments, fillers, flame retardants, etc.

Due to their high T_m's, fibers are often difficult to render soluble in common solvents and may undergo appreciable degradation at the elevated temperatures necessary for dissolution. Many sampling techniques destroy or alter the crystallinity and other morphological features of fibers.

The term plastics covers a broad spectrum of polymeric materials. Their initial tensil moduli are usually in the range $0.4-4 \times 10^5$ psi ($0.3-3 \times 10^3$ MN/m^2) and their recovery with extension is highly variable and dependent upon temperature. The extension at which plastics break varies from about 1% up to 400%. Some polymers can be molded to form a plastic or oriented during processing to form fibers. At elevated temperatures (above T_g), many plastics; e.g., polystyrene, behave like elastomers.

Thermoplastic polymers soften and can be molded when heated and retain their shapes when cooled. If care is taken to prevent degradation, they can be heated and remolded many times. They are generally open-chain polymers with either their T_g or T_m above about 90^{0}C. Polystyrene is a glassy or brittle non-crystalline thermoplastic (T_g = 100^{0}C); while linear polyethylene is a tough, semi-crystalline polymer (T_g =-85^{0}C and T_m=141^{0}C). Most thermoplastics have stabilizers added to minimize polymer degradation during processing at elevated temperatures and high pressures. They may contain some or all of the following - plasticizers, processing aids, fillers, foaming agents, pigments, fire retardants, etc. Since thermoplastics are usually soluble in organic solvents, a variety of techniques are available for sample preparation and characterization.

Thermoset plastics consist of polymer molecules connected to one another by a sufficiently large number of intermolecular bonds to form a network. Thermoset plastics can be defined as network polymers resulting from a chemical reaction which forms permanent chemical bonds, or crosslinks, between certain repeating units. Such network can be formed either by crosslinking open-chain polymers or by introducing crosslinks during polymerization. The type and density of crosslinks have a profound effect on the properties of network polymers. Characteristically, thermoset plastics are hard intractable solids with T_g's above room temperature. They cannot be dissolved or permanently deformed (remolded) even at elevated temperatures without either destroying the crosslinks or degrading the polymer.

Low molecular weight resins or prepolymers used in the manufacture of thermoset plastics often have quite complex compositions. For example, epoxy resin formulations for structural composites may contain several different types of epoxy resins, curing agents, diluents, rubber modifiers, thermoplastic additives, catalysts, fillers, residual solvents and various impurities. Oftentimes, such

194

resins are heterogeneous and may undergo compositional changes
during storage and handling. Since the compositions of precursor
resins affect their processability and the ultimate properties of
thermoset polymers, optimization of sample preparation and
characterization techniques is of paramount importance in the
manufacture of thermoset plastics.

The ASTM (American Society for Testing and Materials) Standard
Guide for Classification System for Plastic Materials (D4000-82)
provides a classification system for tabulating properties and
arranging unfilled, filled, and reinforced plastic materials into
groups, classes and grades based upon fundamental properties[1].
The primary purpose of the Standard is to establish a universal
approach for identifying and designating the properties of plastic
materials. For example, the designation ASTM D4000 PA122 G33
A53380 GA140 indicates the following material requirements for a
reinforced nylon:

ASTM D4000	=	plastic material
PA	=	generic type, poly(amide)
1	=	group, nylon-6,6
2	=	class, heat stabilizer added
2	=	grade, (ref., specification D4066)
G	=	type of reinforcement, glass
33	=	w% reinforcement, 33%
A	=	table (ref. D4066) listing property requirements for poly(amides)
5	=	tensile strength, 175 MPa
3	=	flexural modulus, 7500 MPa
3	=	Izod impact, 75 J/m
8	=	deflection temperature, 235^{0}C
0	=	unspecified
G	=	suffix requirement, specific gravity
A	=	ASTM D792, Method for determining specific gravity (tolerance ±0.02)
140	=	required value for specific gravity, 1.40.

The specification number identifies plastic compounds by a
sequence of letters and numbers from a series of detailed property
tables and thereby serves as a useful aid for polymer characterization. Indeed the Standard may be extended to include other kinds,
as well as future generations, of polymer materials and to provide
pertinent information relating to polymer sampling and characterization. Recently, the ASTM classification system was extended to
include elastomeric materials (ASTM A2000-SAE J200).

Characteristics Relating to Polymer Characterization

When preparing polymer samples for analysis and selecting
techniques for characterization, the following characteristics must

must be considered:

- Inherent stability of the polymer (thermal, mechanical and chemical)

- Polymer chain structure (MW, MWD, open-chain or network polymer)

- Thermal transition behavior (T_g, T_m)

- Solubility

It is advisable to avoid any process which causes the polymer to degrade or otherwise undergo chemical and, for certain techniques, physical changes during sample preparation. Polymer stability is a constitutive property dependent not only on the chemical structure of the polymer, but also upon the presence of other components in the polymer sample (e.g., impurities, catalysts and stabilizers). Polymer degradation involves the breaking of primary chemical bonds and may include the reformation of bonds to produce different chemical compounds. Thermal, chemical (hydrolysis), radiation (photochemical), oxidative (thermal- and photo-oxidation), environmental (ozone), and mechanical processes may all cause or contribute to changes in the polymer structure and lead to polymer degradation. The chemical mechanisms involved in polymer degradation are often quite complex.

Positive steps must be taken to prevent chemical changes and polymer degradation. Knowledge and an understanding of a polymer's chemical structure help by indicating expected problem areas; e.g., thermal, moisture, and UV sensitivity. It also helps to know how the polymer was synthesized and processed since residual catalysts and impurities can adversely affect polymer stability, while additives and processing aids protect the polymer and inhibit certain degradation processes. Polymer texts and handbooks are useful sources of information; and whenever possible, the polymer manufacturer and experts should be consulted to learn more about the polymer's behavior.

The MW and MWD of a polymer depend upon conditions used in its synthesis and may vary over a broad range. An increase in MW generally enhances the strength and toughness of a polymer which in turn may cause problems in sampling and converting the polymer to a useful form for analysis. Higher MW polymers and branched polymers are more likely to undergo chain scission during handling. Also, polymer solubility decreases and the time needed to fully dissolve a polymer increases with increasing MW.

Except for certain microgels, network polymers are insoluble and considered to have an infinitely high MW. However network polymers often can be swollen and extracted by solvents that are good solvents for the respective uncrosslinked, open-chain polymer. A polymer which can be swollen or otherwise deformed is at least tractable for some sample preparation techniques. The extent to which a network polymer is capable of being swollen depends upon its crosslink density or the average polymer chain length between crosslinks. For example, lightly crosslinked elastomers and thermo-set plastics may be swollen, stretched or molded to many times their original dimensions. Highly vulcanized elastomers and cross-linked thermosets are intractable; i.e., they generally fracture when deformed and show little tendency to swell in solvents.

Synthetic polymers are either amorphous or semi-crystalline and usually exhibit a glass transition temperature T_g which is related to the flexibility of the polymer chain backbone. If the temperature of a polymer is below its T_g, molecular motion is restricted and the polymer is stiff and usually brittle. Amorphous polymers are flexible and may display rubber-like properties at temperatures above their T_g. Semi-crystalline polymers are tough and somewhat pliable at temperatures intermediate between their T_g and crystalline melting temperatures T_m. The extensibility, moldability and solubility of such polymers depend upon their degree of crystallinity. Highly crystalline polymers are often difficult to mold and dissolve below their T_m. Above T_m, the polymers exhibit rubbery or liquid-like behavior. Reducing a polymer's MW or adding plasticizers will lower the polymer's T_g but have little effect on its T_m. Polymer chain branching and crosslinking raises T_g and lowers T_m. Polymers with very high T_g and/or T_m values are often intractable and may decompose during sample preparation if excess-ively high temperatures are employed to deform or solubilize them.

Polymer solubility enables a wide variety of techniques to be applied in polymer sample preparation and characterization. A good solvent for a polymer is usually a low MW compound that is a liquid at room temperature and chemically and structurally similar to the polymer chain repeat unit. The ability of one compound to solvate others is described by its solubility parameter $\underline{\delta}$ where

$$\delta = (\Delta E_{coh}/V)^{1/2}$$

and E_{coh} is the cohesive energy density or the heat of vaporization of the low MW compound to a gas at zero pressure and $\underline{V}$ is its molar volume[2]. Solubility parameters of low MW (volatile) compounds or solvents δ_{solv} can be determined directly by measuring their heats of vaporization. Indirect experimental and empirical approaches are used to estimate the solubility parameters of polymers δ_{poly}. Compounds with $\underline{\delta}$ values similar to those of a polymer; i.e., $\delta_{solv} \approx \delta_{poly}$, should be good solvents for the polymer. Conversely, compounds having dissimilar $\underline{\delta}$ values are expected to be nonsolvents.

The solubility parameter concept is strictly valid only for amorphous polymers. However in special cases it can be applied to the crystalline melt and amorphous regions of semi-crystalline polymers. In practice, possible contributions due to hydrogen bonding also need to be considered. Since a compound's potential for hydrogen bonding is directly related to its chemical structure, it is convenient to classify compounds as:

- Poor hydrogen bonding (e.g., hydrocarbons and their halogen-, nitro- and cyano-derivatives)

- Moderate hydrogen bonding (e.g., esters, ethers, ketones, glycol monoethers)

- Strong hydrogen bonding (e.g., alcohols, amines, acids, amides, aldehydes)

and then list the compounds according to their $\underline{\delta}$ values within each class.

Solubility parameters for common polymer solvents and non-solvents are listed in Table 1. All the compounds are nominally liquids at room temperature and available commercially. Compounds with similar $\underline{\delta}$ values or those within the same hydrogen bonding class are miscible and can be used as co-solvents in sample preparation. Sometimes it is useful to take advantage of the reduced viscosity and boiling point provided by mixtures and azeotropes of certain solvents in polymer sample preparation (e.g., filtration and extraction). The identification of nonsolvents is also important, for example, in deciding upon the "best" nonsolvents to use for polymer precipitation and coagulation.

Other parameters listed in Table 1 are also important to consider when selecting polymer solvents and nonsolvents. For high density polyethylene (δ_{poly} = 7.9), hexane has a favorable solubility parameter but would be a poor choice as a solvent since polyethylene is crystalline (T_m = 141°C) and therefore insoluble at hexane's boiling point (b.p., 69°C). Decane is an excellent solvent for polyethylene since solutions can be prepared and handled with little solvent loss at atmospheric pressure and elevated temperatures where the polymer is no longer crystalline. Both the b.p. and vapor pressure are important in handling solutions and applying vacuum techniques for solvent/nonsolvent removal.

Depending upon viscosity $\underline{\eta}$ and density $\underline{d}$, there may be disadvantages in using particular solvent/nonsolvent combinations at room temperature (20°C). For example, filtration is time consuming and errors are introduced in transfer operations if the viscosity of the solvent is too high. If the density of the nonsolvent or solvent/nonsolvent mixture is greater than that of

Table 1. Characteristics of Polymer Solvents/Nonsolvents [ref. 2-4]

Poor Hydrogen-Bonding Solvents	δ $(cal/cc)^{1/2}$	b.p. (°C)	η^{20} (cp)	d^{20} (g/cc)	n_d^{20}	ε^{20}
Aliphatic fluorocarbons	5.5-6.2	--	--	--	--	--
n-hexane	7.3	68.7	0.31	0.66	1.375	1.88
cyclohexane	8.2	80.7	0.98	0.78	1.426	2.02
carbon tetrachloride	8.6	76.5	0.97	1.59	1.460	2.24
decahydronaphthalene (decalin)	8.8	191	2.31	0.89	1.468	2.15^{25}
toluene	8.9	110	0.59	0.87	1.496	2.39^{25}
benzene	9.2	80.1	0.64	0.88	1.501	2.28^{25}
chloroform	9.3	61.7	0.57	1.48	1.446	4.81
tetrahydronaphthalene (tetralin)	9.5	208	2.20^{15}	0.97	1.541	2.77^{25}
methylene chloride	9.7	39.8	0.45^{15}	1.32	1.424	8.93^{25}
o-dichlorobenzene (ODCB)	10.0	181	1.32^{25}	1.31	1.551	9.93^{25}
1,2,4-trichlorobenzene (TCB)	--	213	--	1.57	1.567	--
acetonitrile	11.9	81.6	0.37^{15}	0.78	1.344	37.5
Moderate Hydrogen-Bonding Solvents						
diethyl ether (ethyl ether)	7.4	34.6	0.24	0.71	1.352	4.34^{25}
tetrahydrofuran (THF)	9.1	66.0	0.55^{15}	0.89	1.407	7.58^{25}
methyl ethyl ketone (MEK)	9.3	79.6	0.42^{15}	0.81	1.379	18.5
2-ethoxyethanol (cellusolve)	9.5	136	2.05	0.93	1.408	29.6^{15}
acetone	9.9	56.3	0.32^{15}	0.79	1.359	20.7^{15}
cyclohexanone	9.9	156	2.45^{15}	0.95	1.451	18.3^{25}
1,4-dioxane	10.0	101	1.26	1.03	1.422	2.21^{25}
N,N-dimethylacetamide (DMAC)	10.8	166	2.14^{25}	0.94	1.438^{25}	37.8^{25}
N-methylpyrrolidone (NMP)	11.3	202	1.67^{25}	1.03	1.468^{25}	32.0^{25}
dimethyl sulfoxide (DMSO)	12.0	189	2.00^{25}	1.10^{25}	1.478	46.7^{25}
N,N-dimethylformamide (DMF)	12.1	153	0.92	0.95	1.430	36.7^{23}
propylene-1,2-carbonate	13.3	242	--	1.19	1.421^{40}	69.0^{40}
ethylene carbonate	13.4	238	solid (mp, 36.4°C)	1.32	1.420^{40}	89.6^{40}
Strong Hydrogen-Bonding Solvent						
m-cresol	10.2	202	20.8	1.03	1.541	11.8^{25}
hexamethylphosphoramide (HMPA)	10.5	233	3.47	1.03	1.459	30
trifluoroacetic acid (TFA)	10.6	71.8	0.93	1.49	1.285	8.55
triethylene glycol	10.7	288	49.0^{15}	1.12	1.456	23.7^{25}
i-propanol	11.5	82.3	2.86^{15}	0.79	1.377	19.9^{25}
benzyl alcohol	12.1	206	7.76^{15}	1.04	1.540	13.1
diethylene glycol	12.1	245	35.7^{15}	1.12	1.448	31.7^{16}
formic acid	12.1	101	1.97^{25}	1.22	1.371	58.5
ethanol	12.7	78.3	1.08	0.79	1.361	24.6^{25}
methanol	14.5	64.7	0.55	0.79	1.328	32.7^{25}
formamide	19.2	211	3.76	1.13	1.448	111
water	23.4	100	1.00	1.00	1.33	80.1

the polymer, there will be problems in using precipitation techniques to isolate the polymer and it will be impossible to compact the polymer precipitate by centrifugation. The density and viscosity of a solvent must also be taken into account when preparing polymer samples for dilute solution characterization.

The refractive index n_D (determined at the sodium D line) of both solvent and polymer must be considered if optical techniques are applied during sample preparation to monitor the concentration of polymers in solution and in selecting a solvent for dilute solution polymer characterization techniques such as light scattering, size-exclusion chromatography and sedimentation ultracentrifugation. Similarly, the interactions and electromagnetic absorption characteristics of solvents must be carefully examined when sample solutions are being prepared for spectroscopic analysis (e.g., UV-visible, fluorescence, infrared, raman and nmr).

The dielectric constant ε is related to the refractive index n_D of a non-polar insulator by Maxwell's relationship

$$\varepsilon = n_D^2,$$

where ε is measured at relatively low frequencies. If a material has permanent dipoles, the relationship no longer applies and the experimental value ε_{expt} will be greater than the calculated value, i.e., $\varepsilon_{expt} > \varepsilon_{calc}$, where $\varepsilon_{calc} = n_D^2$. A very large difference $\varepsilon_{expt} \gg \varepsilon_{calc}$ suggests strong hydrogen bonding or semi-conduction. The extent to which a solvent does not behave as an insulator is indicated in Table 1 by comparing the parameters ε and n_D^2. Correlations between ε and both the hydrogen bonding group and δ values of the solvents are evident. The ε value of a solvent has implications regarding its selectivity in the solvation of specific sites on a polymer and interactions with other materials (e.g., metals) with which it may come in contact during sample preparation. Also, ε must be considered when selecting a solvent for use in characterization techniques that depend upon measuring electrical properties (e.g. conductance).

Characteristics of common polymers are listed in Table 2. Although the polymers are classified according to their most general application area, it is recognized that many polymers can be included in more than one category depending upon how they are compounded and/or processed. For example, thermoplastics like polyethylene with relatively high T_m values can be oriented during processing to form fibers; or the chemical composition and processing of polyesters can be varied to generate fibers, thermoplastics, or thermoset plastics. Polyurethanes may have properties characteristic of elastomers, plastics or fibers depending upon how they are synthesized and processed.

Table 2. Characteristics of Common Polymers

	T_g (°C)	T_m (°C)	d (g/cc)	n_D^{25}	ε	δ (cal/cc)$^{1/2}$	solvents	non-solvents
ELASTOMERS								
1. poly(dienes)								
cis-1,4-poly(butadiene) −CH$_2$CH=CHCH$_2$−	−102	−−	0.90	1.516	2.51	8.1−8.6 (8.6)	cyclohexane, THF toluene, benzene chloroform, dioxane	acetone, methanol water
cis-1,4-poly(isoprene) −CH$_2$C=CHCH$_2$− CH$_3$	−73	−−	0.91	1.513	2.41	7.9−10.0 (8.5)	"	"
poly(isobutylene) CH$_3$ −CH$_2$C− CH$_3$	−75	−−	0.86	1.508	−−	7.8−8.1 (8.0)	"	"
poly(chloroprene) [neoprene] −CH$_2$C=CHCH$_2$− 2Cl	−45	−−	1.23	1.558	−−	8.2−9.7 (9.4)	THF, benzene, MEK dioxane, chloroform ethyl acetate,	hexane, acetone methanol, water
2. copolymers								
poly(butadiene-co-styrene) SBR rubber (23.5% styrene monomer)	−60	−−	0.93	1.54	2.5	8.54	cyclohexane, THF toluene, chloroform	"
poly(butadiene-co-acrylonitrile) nitrile rubber (25% acrylonitrile)	−23	−−	−−	1.52	−−	9.5	toluene, benzene THF, chlorform	hexane, water methanol
poly(isobutylene-co-isoprene) butyl rubber (2% isoprene)	−−	−−	−−	−−	−−	7.8	toluene, THF carbon tetrachloride	"
poly(ethylene-co-propylene-co-diene) EPDM rubber (< 2% diene + 26% propylene)	−−	−−	0.86	−−	2.5	−−	cyclohexane	methanol, water

(continued)

Table 2. (Continued)

3. poly(siloxanes)								
Poly(dimethyl siloxane) CH$_3$ -Si-O- CH$_3$	-127	--	0.98	1.404	2.75	7.4	cyclohexane, THF benzene, toluene, ethyl acetate, ethyl ether, chloroform	methanol, ethanol water, acetonitrile, benzyl alcohol
4. poly(sulfides)								
poly(alkylene sulfide) Thiokol™ $(-CH_2CH_2 \rightarrow (-S-)_{2-4}$	-50	--	1.2	1.596	7.5	9.0-9.4 (9.2)	poorly soluble or insoluble, possible solvents – dioxane, THF, TCB	
5. poly(urethanes)								
poly(ester or ether urethanes) ether, HO$\{$R'-O-C-NH-R-NH-C-O$\}_n$-R'-OH O O	--	--	--	--	--	10	THF, DMF, DMSO, m-cresol, formic acid, dioxane(hot)	methanol, ethyl methylene chloride
FIBERS								
1. vinyl polymers								
poly(acrylonitrile) -CHCH$_2$- CN	97	319	1.18	1.518	6.5	12.5-15.4 (12.6)	DMF, DMAC, DMSO, propylene carbonate ethylene carbonate	hexane, methanol ethyl ether acetone methylene chloride
2. poly(esters)								
poly(ethylene- terephthalate) -OCH$_2$CH$_2$OC—⟨⟩—C- O O	(amorph) 67 (cryst) 81 (oriented)125	265	--	1.58	3.25	9.7-10.7 (10.0)	m-cresol, HFIP o-chlorophenol, DMSO (hot)	hexane, acetone methanol, benzene methylene chloride
3. poly(amides)								
Nylon-6 poly(ε-caprolactam) -NHC$\{$CH$_2\}_5$ O	40	210	1.084 (amorph)	1.53	3.8	12.7	m-cresol, HMPA, formic or acetic acid, o-chlorophenol, HMTP ethylene carbonate	hexane, methanol cyclohexane, ethyl ether water

Polymer							Solvents / Nonsolvents
Nylon-6,6 poly(hexamethylene-adipamide) $-NH(CH_2)_6NHC(CH_2)_4C-$ (with two C=O)	50	265	1.07 (amorph)	1.53	4.0	13.7	m-cresol(130°C), DMSO, formic acid, phenol, trifluoroethanol, diethylene glycol (T > 120°C)
Nylon-12 poly(dodecamethylene-adipamide) $-NH(CH_2)_{12}NHC(CH_2)_4C-$ (with two C=O)	41	210	0.99 (amorph)	--	4.2	--	DMF, DMSO, HMTP(85°C), HFIP (25°)
THERMOPLASTICS							
1. poly(alkenes)							
poly(ethylene) $-CH_2CH_2-$	-78	141 hd / 1121d	0.97 hd / 0.91 ld	1.49	2.3	7.7-8.4 (7.82)	ODCB, TCB, decalin, tetralin, xylenes (T > 135°C); acetone, n-butanol, 2-ethoxyethanol, triethylene glycol
poly(propylene) $-CHCH_2-$ CH_3	-20	176 iso / 138 syn	0.94 iso / 0.90 syn	1.49	2.2	8.2-9.2 (8.31)	"
2. vinyl polymers							
poly(styrene) $-CHCH_2-$ (phenyl)	100	--	1.05	1.591	2.55	8.5-9.3 (9.3)	cyclohexane(35°C), THF, MEK, toluene, ethyl acetate, chloroform; hexane, methanol, i-propanol, water, ethyl ether, acetone
poly(vinyl chloride) $-CHCH_2-$ Cl	83	212	1.4	1.539	3.5	9.4-10.8 (9.6)	THF, DMF, MEK, DMSO, cyclohexanone; hexane, methanol, acetone, water
poly(vinylidene chloride) $-CCl_2CH_2-$	-18/15	190	1.7	1.61	4.5	9.9-12.2 (10.1)	THF(hot), dioxane, DMF, ODCB, TCB, cyclohexanone; THF(cold), water, methanol, hexane, chloroform

(continued)

Table 2. (Continued)

Polymer						Solvents	Nonsolvents
poly(vinyl fluoride) -CHCH₂- F	41	200	1.4	--	--	--	DMF, DMSO (hot), cyclohexanone(hot)	hexane, methanol, toluene, water
poly(vinylidene fluoride) -CF₂CH₂-	-40/13	171	1.7	1.42	8.4	--	NMP, cyclohexanone, DMSO	cyclohexane, hexane, methanol
poly(vinyl butyral) -CH HC-CH₂- O O CH C₃H₇	105	--	1.1	1.49	5.0	--	cyclohexanone, i-propanol	hexane, water, cyclohexane
poly(vinyl acetate) -CHCH₂- O O=C-CH₃	32	--	1.19	1.467	3.25	9.1-11.0 (9.6)	THF, acetonitrile, toluene, acetone, chloroform, DMF, ethyl acetate, MEK	hexane, methanol, ethyl ether, water
poly(vinyl alcohol) -CHCH₂- OH	85	232	1.3	1.5	10	12.6-14.2 (12.6)	DMF, DMSO (hot) water, glycols(hot)	hexane, methanol, THF, acetone, methylene chloride, ethyl ether
3. poly(acrylates)								
poly(methyl acrylate) -CHCH₂- C=O O-CH₃	10	--	1.22	1.479	4.5	9.7-10.4 (9.7)	THF, chloroform, ethyl acetate, acetone	hexane, methanol, carbon tetrachloride, ethyl ether
4. poly(methacrylates)								
poly(methyl methacrylate) CH₃ -C-CH₂- C=O O-CH₃	104	--	1.19	1.490	3.5	9.1-12.8 (9.3)	THF, MEK, acetone, ethyl acetate, acetonitrile, benzene, toluene	hexane, water, ethyl ether, ethanol

Polymer							Soluble in	Insoluble in
5. poly(carbonates) poly(4,4'-isopropylidene-diphenylene carbonate) $-O-\!\!\bigcirc\!\!-\overset{CH_3}{\underset{CH_3}{C}}-\!\!\bigcirc\!\!-O-\overset{O}{C}-$	141	--	1.24	1.585	3.0	9.93	THF, chloroform, methylene chloride, DMF, dioxane	hexane, water, acetone, carbon tetrachloride
6. poly(tetrafluoroethylene) Teflon™ $-CF_2CF_2-$	127	330	2.29	1.376	2.1	6.2	insoluble, except in solvents like perfluorokerosene $C_{21}F_{44}$ at 300°C	
7. poly(alkyl oxides) poly(formaldehyde) or poly(methylene oxide) $-O-CH_2-$	-30	60	1.42	1.51	3.1	10-11 (10.2)	DMF, formamide, benzyl alcohol, ODCB, phenol, ethylene carbonate (T > 130°C)	hexane, methanol, ethyl ether
poly(ethylene oxide) $-OCH_2CH_2-$	-67	66	1.2	1.46	4.5	--	THF, acetonitrile, cyclohexanone, DMF, toluene, ODCB, chloroform, water(cold)	hexane, ethyl ether, water (hot)
poly(propylene oxide) $-OCH_2CH_2CH_2-$	-78	75	1.1	1.46	--	7.5-9.9 (9.2)	THF, MEK, acetone, benzene, toluene	ethyl ether, ethyl acetate, water
poly(tetramethylene oxide) $-OCH_2CH_2CH_2CH_2-$	-84	37	1.0	--	--	8.3-8.6 (8.5)	THF, ethyl acetate, MEK, i-propanol(46°C), benzene, ethanol, chloroform, methylene chloride	" + methanol
8. poly(phenylene oxides) poly(2,6-dimethyl-phenylene oxide) $-O-\!\!\bigcirc\!\!\overset{CH_3}{\underset{CH_3}{}}-$	183	--	1.075	--	2.65	--	THF, toluene, DMSO, chloroform, methylene chloride	methanol

(continued)

205

Table 2. (Continued)

9. copolymers poly(styrene-co-acrylonitrile) SAN	--	--	1.07	1.57	3.0	--
poly(acrylonitrile-co-butadiene-co-styrene) ABS	--	--	--	--	--	--
10. poly(sulfones)	>200	--	--	--	3.14	--	solubility depends upon structure, some are soluble in THF
11. poly(imides)	--	--	1.9	--	3.4	--	insoluble, solvents for polyamic acid precursors include DMF, DMAC, NMP
THERMOSET PLASTICS						
1. epoxies precursor -R-CH-CH2 (O)	--	--	--	1.58	3.5-5	10.9	insoluble; however precursors are solubles in a wide variety of solvents such as THF and chloroform
2. phenolics phenol-formaldehyde			1.3	1.6	7	9.5	insoluble
urea-formaldehydes	variable properties					
melamine-formaldehydes	"		"			

206

3. poly(esters)
maleic and phthalic anhydrides
with propylene gylcol
-R-O-C-R-
     ‖
     O

variable properties   insoluble

4. poly(ureas)
-R-NH-C-NH-R'-
       ‖
       O

60-70 -- 1.48 1.54 -- -- "

5. poly(urethanes)
-R-O-C-NH-R'-
    ‖
    O

-- -- 1.49 1.50 -- 9.0 solubility depends upon monomer
type and functionality

HFIP = 1,1,1,3,3,3-hexafluoro-2-propanol
T_m values are not listed for vinyl polymers where the most common form are atactic, nor for elastomers which are generally non-crystalline in the unstretched state
hd = high density
ld = low density
iso = isotactic
syn = syndiotactic
amorph = amorphous
cryst = crystalline
References: 2,3,5,6

Parameters relating specifically to polymer stability and chain structure are not included in Table 2. There are many ways to measure and express polymer stability and such parameters are usually highly dependent upon test conditions. MW and other parameters relating to polymer chain structure depend upon polymerization conditions and therefore may vary over a broad range. Thermal transition temperatures, density, refractive index, dielectric constant and solubility parameters are listed because of their fundamental importance. Both ranges and specific values are noted in cases where parameters are sensitive to processing or synthesis conditions; and unless otherwise indicated, values are quoted at 20^0C and 1 atm. The solvents and nonsolvents are representative of those that have been successfully applied in sample preparation and characterization.

Unlike solvents, common polymers are insulators and do not depart markedly from behavior expected for dielectrics. Polymers with the highest dielectric constants tend to be those with highly polar structural repeat units that can be oriented to form fibers. Such polymers often contain strong, intermolecular hydrogen bonding groups (e.g., hydroxyl, acids, amines, amides, aldehydes) which enhance tensile strength and other physical properties.

Polymer Sampling, Handling and Storage

Procedures for sampling polymers are prescribed in the following ASTM standards:

· D1898   Standard Recommended Practice for Sampling of Plastics[7]

· D3896   Standard Method for Rubber from Synthetic Sources – Sampling[8a]

· D1485   Standard Methods for Rubber from Natural and Synthetic Sources – Sampling and Sample Preparation[8b]

General procedures for random sampling and sampling stratified materials are outlined in ASTM standard D1898[7]. Specific sampling procedures depend upon many factors; e.g., the physical form of the material, the degree of uniformity and the size of the lot. Standard D1898 considers the sampling of molding powders, fabricated stock shapes (blocks, sheets, films, rods, tubes, etc.), molded items and groups of test specimens. Standard D1485 describes procedures for sampling lots of raw rubber and synthetic gum elastomers from bales or packages[8].

Errors in sampling polymers arise from both indeterminate and determinate sources. Sampling molding powders typify a situation where indeterminate or random sources of error can be evaluated statistically[9] and random sampling procedures are applicable[10].

Generally, the weight of sample required increases in proportion
to the density of the powder and decreases with decreasing particle
size. According to ASTM Standard D1898, "randomness is achieved
when every part of the lot has an equal chance of being drawn into
the sample in the first and in successive sampling operations" and
a normal distribution of the sampling error can be assumed in
random sampling.

Determinate sampling errors cannot be estimated statistically
and can be major sources of error if care is not taken in developing specific sampling procedures. Determinate errors occur when
the composition or some property of the polymer material varies
nonrandomly throughout the sample lot (e.g., size segregation may
occur within a lot of molding powder if the particles have a broad
size distribution). It is usually difficult both to obtain a
representative sample for the laboratory and to take replicate
specimens for analysis from such a laboratory sample. If a polymer
material is known or suspected to inhomogeneous, stratified sampling
procedures should be employed. The material or lot should be
divided into sections or strata which are then individually sampled
to evaluate average properties and determine the extent of inhomogeneity. Indeed it is recommended that stratified sampling precede
the application of random sampling.

Care must be taken to avoid contaminating or in any way altering samples during handling and storage. The samples should be
placed in clean, dry, sealable containers and carefully labelled.
The containers must not react with the samples nor be exposed to
any conditions that could affect the composition or change the
properties of the samples. For example, ASTM standard D1485
recommends that raw rubber samples should be placed in air-tight
containers having a volume no greater than twice that of the sample.
Handling and storage in direct sunlight and under conditions of
high temperature and humidity should be avoided. Ideally, the
sample containers should be amber-colored and stored in a dry, dark
and cool location. In any case, storage conditions should not
exceed the temperature of 23^0C and 50% relative humidity recommended
as the Standard Laboratory Atmosphere[11].

For moisture-sensitive polymers, special care is required
during sampling, handling and storage. In some instances, it may
be desirable to maintain the samples in a nitrogen or argon atmosphere and to handle specimens in a glove box. Certain materials
(e.g., reactive epoxy resin precursors for thermoset polymers) must
neither be exposed to moisture nor allowed to stand unrefrigerated
for extended periods of time. Such materials "age"; i.e., undergo
chemical reactions, and entrained moisture affects their polymerization behavior and properties. Sampling of hermetically sealed
materials in cold storage can be conducted after the contained
sample is removed from the freezer and attains laboratory temperature.

Treatment of Polymer Samples in the Laboratory

Polymer samples received in the laboratory usually must be modified either physically or chemically to obtain representative specimens and to convert the specimens to a suitable form for characterization. Obtaining representative test specimens can be an impossible task when materials are heterogeneous and undergo chemical or physical changes as homogenizing processes are applied. In such cases, a large number of specimens may need to be evaluated to provide average values and accurately describe variations within a particular sample. The size and number of specimens depend upon the extent and source of heterogeneity which often can be determined only after preliminary analyses are run. The smaller the sample required for a particular analysis the more difficult it becomes to obtain one that is truly representative.

Ultimately, the polymer is either in solution, in the bulk state, or in an alternate state as demanded by the characterization method. A scheme for polymer sample modification and treatment is shown in Figure 3. Common techniques for improving the compositional uniformity of polymer samples are:

- cutting, grinding and impacting

- molding, extrusion and roll milling

- dissolution

Sample preparation conditions should be those least likely to cause polymer degradation; i.e., low temperature, slow mixing, inert atmosphere, anhydrous conditions, antioxidants, etc. Extreme care must be taken in cutting or shaping and conditioning specimens for mechanical testing and morphological studies. Surface flaws and imperfections within samples may invalidate or introduce major errors in certain mechanical tests. Also, it is noted that processes requiring either physical or chemical modification of a sample are likely to introduce bias and uncertainties in the analyses. This is especially a problem in the preparation of polymer samples because of their complex structures.

Solubility is a prerequisite for the application of many test methods. The dissolution process destroys polymer morphology and is quite slow for most polymers compared to low MW materials. Polymer solution techniques require the removal of insoluble components prior to analysis. Care must be taken in handling and selecting solvents, stabilizers, and temperatures to prevent polymer degradation and assure that the polymer is truly in solution.

Sometimes polymers interact with one another to form aggregates and microgels in solution. Such behavior is not always apparent

210

and may not manifest itself until after the test results are evaluated. The presence of soluble non-polymeric contaminants, low MW polymer components, and other types of polymers may also invalidate the results obtained from the analysis of polymers in solution. Special attention must also be paid during sample preparation to prevent contamination and remove components that may compromise the test method. On the other hand, it is possible that operations applied to isolate polymers (e.g., dialysis, extraction, precipitation, recrystallization and chromatography) may fractionate them to the extent that the sample is no longer representative.

Procedures for Polymer and Prepolymer Characterization

The following questions deserve careful consideration when developing procedures for preparing and characterizing polymer and polymer precursor samples:

- What are the inherent characteristics of the polymer or prepolymer?

- Will certain operations cause irreversible changes in the sample?

- What requirements does the characterization technique impose upon the sample?

- Is it necessary to isolate the polymer or prepolymer from other sample components?

Also, it should be recognized that the properties of polymer compounds and prepolymer formulations are often quite different from those of the pure polymers and polymer precursors. Polymer properties are greatly influenced by the presence of other components, e.g. fillers, additives, processing aids, dyes, residual catalysts, impurities, solvents, and other polymers, low MW oligomers and monomers.

One must decide whether the sample needs to be modified or specially treated for a particular analysis (Figure 3). Characteristics such as chemical structure of the polymer repeat units, thermal transition behavior, and solubility determine what can be done with a sample. Operations, such as heating or extraction, may alter morphology or change the chemical composition of a polymer sample and thereby affect its properties and compromise the validity of certain tests. Many characterization techniques require polymer samples to be modified or have a particular shape or form. If samples do not conform precisely to test criteria, the test may be invalid. On the other hand, in order to apply certain techniques (e.g., light scattering and membrane osmometry for MW analysis), it is essential that the polymer be totally isolated from nonpolymeric components.

POLYMER SAMPLE RECEIVED IN LABORATORY
(RECORD OF SAMPLE HISTORY SHOULD BE AVAILABLE AND INCLUDE INFORMA-
TION ON SAMPLING, HANDLING AND STORAGES CONDITIONS)

SAMPLE MODIFIED FOR LABORATORY OPERATIONS
(AIM IS TO CONVERT SAMPLE INTO MOST APPROPRIATE FORM(S) FOR RE-
QUIRED ANALYSES AND, IF POSSIBLE, TO RENDER THE SAMPLE MORE UNIFORM
SO SMALLER QUANTITIES CAN BE HANDLED)

CASE 1.   POLYMER IN LIQUID FORM
(E.G., EMULSIONS, PAINTS OR COATINGS)

A. USE LIQUID (WARING) BLENDER TO HOMOGENIZE THE
    SAMPLE.

B. POUR LIQUID INTO A MOLD OR ONTO A SMOOTH SURFACE
    (GLASS OR TEFLON) AND REMOVE VOLATILES, IF ANY, TO
    FORM A THIN FILM.

C. REMOVE VOLATILES IN A VACUUM OVEN AND, IF THE RESIDUE
    IS SOLID, APPLY RECOMMENDED PROCEDURES FOR SOLIDS.

D. PRECIPITATE AND COAGULATE POLYMER WITH A NONSOLVENT.

CASE 2.   SOLID POLYMER SAMPLE

A. CUT, MACHINE AND SHAPE.

B. CONVERT TO A POWDER BY GRINDING OR IMPACTION.

C. MOLD TO UNIFORM THICKNESS OR PRESS INTO FILM FORM
    (IF TRACTABLE).

D. DISSOLVE IN SUITABLE SOLVENT AND CAST FILM OF
    POLYMER (IF SOLUBLE).

SAMPLE TREATMENT
(OPTIMIZE SAMPLE FOR PARTICULAR ANALYSIS WHILE RECOGNIZING THAT
CHARACTERISTICS OF THE POLYMER MAY PLACE LIMITATIONS ON SAMPLE
TREATMENT)

CASE 1.   CONDITIONING THE SAMPLE

CASE 2.   ISOLATING THE POLYMER

CASE 3.   POLYMER DISSOLUTION

Figure 3. Scheme for Polymer Sample Modification and Treatment.

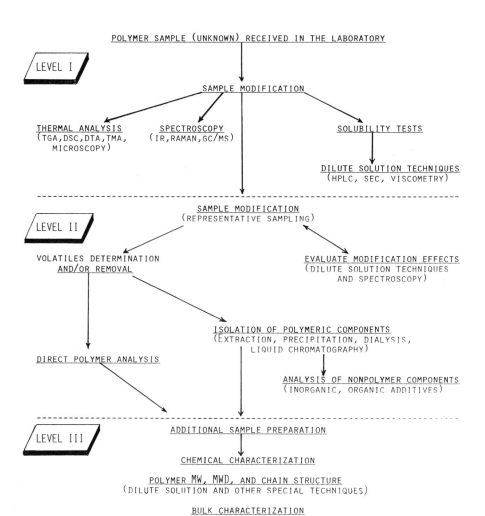

Figure 4.   Polymer Characterization Scheme

Pre-knowledge of the type of polymer or prepolymer is important in developing characterization procedures. If the material is unidentified, a simple series of tests (LEVEL I in Figure 4) may be applied, first to answer the question of whether the sample actually contains polymer and then to determine its characteristics and hopefully identify the polymer or prepolymer.

Sample modification for LEVEL I merely involves breaking or cutting a small section from the sample and, if possible, further reducing sample size by grinding. To facilitate thermal and spectroscopic analysis and solubility testing, the sample should have a large surface area. Liquid and heterogeneous samples should be thoroughly mixed before removing an aliquot for analysis. Each test can be run using as little as 10 mg sample.

Test 1   Thermal Analysis

TGA (thermal gravimetric analysis) provides an indication of the polymer's thermal decomposition temperature T_d and is used to estimate the relative amounts of volatiles, polymers, nonpolymeric additives, and inorganic residues. DSC (differential scanning calorimetry) or DTA (differential thermal analysis) is applied to determine the sample's glass transition temperature T_g and, if the polymer is semi-crystalline, its crystalline melting temperature T_m. Procedures for measuring T_g and T_m are given in ASTM Standards D3417 and D3418[12,13]. TMA (thermal mechanical analysis) can also be used to determine T_g and to obtain further information relating to a polymer's thermomechanical behavior; e.g., its heat distortion temperature and thermal expansion coefficient. For pelletized or molded samples, a razor blade or microtome can be used to cut sample to approximately fit the dimensions (thickness and diameter) of the sample holder. If the sample has been cut or already in film or sheet form with a thickness no greater than 0.015 in (0.04mm), a punch or cork borer may be used to cut disks of an appropriate size.

Alternatively, a hot stage microscope may be used to observe the heat distortion temperature and onset of flow of powdered samples. Initially, the powder particles have sharp, rough edges. However as the sample is heated and the heat distortion temperature is approached, the edges first become blurred and then the particles start to agglomerate. Finally, at T_m, for semi-crystalline polymers, or T_g, in the case of glassy polymers, flow occurs and a clear melt or liquid forms. Microscopes equipped with cross polarizers are useful for defining crystal-crystal transitions and the onset of melting of semi-crystalline polymers.

Test 2 Spectroscopy

Infrared spectroscopy (IR) provides more useful information
for identifying polymers than any other absorption or vibrational
spectroscopy technique and is generally available in most laboratories. Advances like Fourier transform IR (FTIR) spectroscopy
have broadened the areas of application while simplifying the requirements for sample preparation and interpretation of spectra.
IR yields both qualitative and quantitative information concerning
a polymer samples chemical nature; i.e., structural repeat units,
end groups and branch units, additives and impurities[14]. Computerized libraries of spectra for common polymeric materials exist
for direct comparison and identification of unknowns. Computer
software allows the spectrum of a standard polymer to be substracted
from that of the unknown to estimate its concentration and perhaps
to determine whether another type of polymer is also present in
the sample.

The quality of a polymer's IR spectrum is directly related to
care taken in sample preparation. Impurities, solvent residue,
nonuniformity, interference fringes, incorrect concentration or
film thickness may cause poor results and lead to misinterpretation.
For transmission IR the concentration of the polymer solution
should be the highest possible to minimize the contribution of
solvent. Polymer films should be large enough to occupy the entire
cross-section of the light beam and range from about 0.001 to 0.02
mm in thickness. The following procedures are recommended -

1. Soluble polymers - solvent casting to form thin films.

2. Unvulcanized elastomers and thermoplastic polymers -
    molding or melt casting to form thin films.

3. Vulcanized elastomers, brittle and thermosetting polymers
    - pressed-disk or liquid dispersion technique.

4. Fibers and other polymers - direct analysis using IR
    microscope or pressing a grid of fibers into a thin
    film (for fibers with diameters between 0.015 and 0.03
    mm) and microtoming (for thicker fibers >0.03 mm).

Although not as popular as IR, laser Raman spectroscopy compl-
ments IR as an identification technique and is relatively simple
to apply[14]. As long as the sample is stable to the high intensity
incident light and does not contain species that fluoresce,
little or no sample preparation is necessary. Solid samples need
only be cut to fit into the sample holder. Transmission spectra
are obtained directly with transparent samples. For translucent
samples, a hole may be drilled into the sample for passage of the
incident light and a transmission spectra obtained by analyzing

light scattered perpendicular to the incident beam. The spectra of a turbid or highly scattering sample is obtained by analyzing the light reflected from its front surface. Powdered samples are simply tamped into a transparent glass tube and fibers can be oriented in the path of the incident beam for direct analysis.

A powerful, but technically more demanding technique for directly analyzing polymers is pyrolysis GC/MS (gas chromatography/ mass spectroscopy). In this case, the sample only needs to be rendered sufficiently small to fit onto the pyrolysis probe. Not only can the polymer type be identified by comparing its spectrum with standards, but volatiles and additives can be identified rapidly and quantitatively, and polymer branching and crosslink density can cometimes be measured.

Test 3 Solubility Characteristics

Information on the solubility characteristics of a polymer sample is extremely useful in applying or developing more refined sample preparation techniques. Thermal analysis defines T_g and T_m which relate to solubility. Spectroscopy may identify the polymer and thereby aid in selecting a suitable solvent (Table 1), but does not necessarily tell whether the polymer is fully soluble or has solution properties amenable to the application of certain sample preparation techniques or methods of analysis. For example, spectroscopy may show that a sample contains poly(isobutylene), but the sample's solubility behavior (Table 2) is indicative of whether or not the polymer is crosslinked.

If the chemical structure of a polymer sample is unknown, a simple test for solubility is to place 10 to 100 mg aliquots of the sample (preferably powdered or thin film) into small test tubes or screw-capped (Teflon liners) vials each containing 2 to 4 mL of a solvent listed below:

1. diethyl ether
2. toluene or THF or chloroform
3. TCB
4. DMP or DMF
5. m-cresol
6. HMPA
7. acetic or formic acid
8. benzyl alcohol
9. methanol or ethanol
10. water

The tubes or vials are then set in a thermostated shaker bath or ultrasonic cleaner and vigorously mixed for about 5 minutes.

Observations of polymer swelling and solubility are made. If
the polymer is insoluble at ambient temperature, the temperature
of it is slowly raised until the polymer either fully dissolves or
the bath has achieved its maximum temperature. If the polymer
remains insoluble (assuming the solvent does not evaporate or the
polymer decompose), the polymer is either crosslinked or intractable
(e.g., vulcanized rubber or Teflon). A turbid solution indicates
partial solubility and perhaps the presence of "gel" particles.
Soluble non-polymeric components can be ascertained by performing
the solubility test with a nonsolvent for the polymer, filtering
the extract, and examining residue remaining after the liquid is
evaporated from the extract.

The solubility test is also useful for estimating a sample's
density range since the densities of the solvents/nonsolvents are
known. Finally, polymers can sometimes be identified by their
solubility characteristics[15].

Test 4   Dilute Solution Characterization

Once the solubility characteristics of a polymer are known, a
suitable solvent can be selected for dilute solution characterization. THF is most often the solvent of choice for SEC. Toluene,
chloroform, TCB, DMF (or DMP) and m-cresol are also used. If the
polymer's Mark-Houwink constants, $\underline{K}$ and $\underline{a}$, in the solvent are
known, size-exclusion chromatography (SEC) can be applied to det-
ermine the polymer's average MW's and MWD[16]. If the constants
are unknown or the polymer has a complex structure (e.g., branched,
a copolymer or mixture of polymers), SEC still may be used to
estimate the MWD and other parameters relating to the structure and
composition of the polymer. Although SEC indicates the presence of
soluble non-polymeric components, high performance liquid chromatography (HPLC) is the better technique for characterizing residual
monomers, oligomers and other soluble, low MW sample components.

Dilute solution viscometry is a simple technique for determining the limiting viscosity number or intrinsic viscosity[7] of
soluble polymers[17]. The apparatus is inexpensive and simple to
assemble and operate. The[7] of a polymer depends upon its hydro-
dynamic volume in the solvent and is related to the MW of the
polymer.

Structural and compositional information obtained by the tests
in LEVEL I (Figure 4) is used to help develop more sophisticated
sample preparation schemes and provide information necessary for
the application of more detailed or specialized characterization
techniques. The major concern of LEVEL II is representative sampling
and insuring that sample modification procedures (cutting, grinding,
molding, etc.) do not compromise polymer characteristics to be
evaluated. LEVEL II also addresses the "quantitative" aspects of

217

SCHEME FOR POLYMER ANALYSIS

POLYMER SAMPLE (FINE POWDER OR THIN FILM)

VOLATILES REMOVAL AND/OR DETERMINATION

    WEIGHT LOSS ON DRYING
    TGA (THERMAL GRAVIMETRIC ANALYSIS)
    HEAD-SPACE ANALYSIS (GC/MS)
    MOISTURE ANALYZER

ISOLATION OF POLYMERIC COMPONENT(S)

    EXTRACTION, DISSOLUTION, FILTRATION, DIALYSIS
    PRECIPITATION, CENTRIFUGATION, LC

CHEMICAL CHARACTERIZATION TECHNIQUES

    ELEMENTAL ANALYSIS
    FUNCTIONAL GROUP ANALYSIS
    SPECTROSCOPIC ANALYSIS
    CHROMATOGRAPHIC ANALYSIS

POLYMER MW, MWD, AND CHAIN STRUCTURE

    DILUTE SOLUTION TECHNIQUES
    OTHER SPECIAL TECHNIQUES

BULK CHARACTERIZATION TECHNIQUES

    THERMAL ANALYSIS
    MICROSCOPY
    RHEOLOGICAL ANALYSIS
    MECHANICAL TESTING
    MISCELLANEOUS

Figure 5. General Scheme for Polymer Analysis

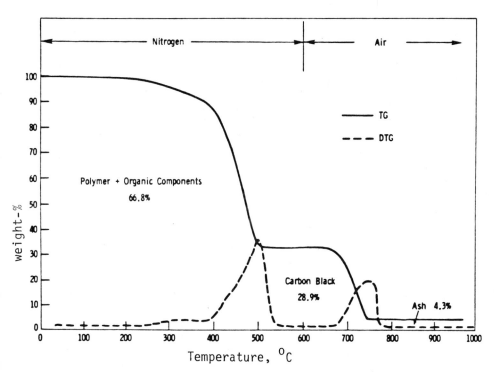

Figure 6. Thermal Gravimetric Analysis (TGA) of a Styrene-
Butadiene Rubber (SBR) Vulcanizate.

sample composition (percent polymer, additives, volatiles, inorganic and other organic residues) and, if necessary, deals with the identification of nonpolymeric components.

A general scheme for polymer analysis is illustrated in Figure 5. The polymer sample should be uniform and have a large surface area. Once volatile components are removed, the polymer can be directly analyzed, or a variety of techniques (e.g., extraction, precipitation, filtration, liquid chromatography) may be applied to isolate the polymer. If required, special procedures are applied to prepare the polymer sample for chemical characterization; MW, MWD, and chain structure evaluation; and bulk characterization (LEVEL III in Figure 4).

Figure 6 illustrates the application of thermal gravimetric analysis (TGA) to determine the percentages of organic components, carbon black, and inorganic components in a vulcanized styrene-butadiene rubber (SBR). State-of-the-art thermal analysis equipment was used to monitor sample weight loss (solid line) with temperature increasing at a constant heating rate (10^0C/min). Typically, the analysis is run with the sample (initial weight 5-10 mg) in a flowing nitrogen atmosphere until the temperature reaches 600^0C. At 600^0C, the nitrogen is displaced by an air atmosphere to facilitate oxidation and thereby volatilization of the carbon black. The first differential of the weight loss curve (dashed line) clearly indicates the temperatures where maximum weight loss occurs and can be used to determine the rate of weight loss. Percent weight loss, rate of weight loss, and weight loss temperatures describe fundamental characteristics relating to the composition and thermal stability of the SBR sample.

Chemical characterization techniques are listed in Table 3. Elemental analysis and functional group analysis provide basic and quantitative information relating to chemical composition. The analysis of reactive functional groups is particularly important in determining equivalent weights of prepolymers. Spectroscopic analysis provides detailed information about the molecular structure, conformation, morphology and physical-chemical characteristics of polymers. Chromatographic techniques separate sample components from one another, and thereby simplify compositional characterization and make a more accurate analysis possible. Employing spectroscopic techniques to monitor components separated by gas or liquid chromatography greatly enhances characterization, providing a means to identify and quantitatively analyze even the most minor components

Figure 7 illustrates the application high performance liquid chromatography (HPLC) using UV and fluorescence detectors to monitor components separated from an SBR compound by acetone extraction. In this case, the extract contains low MW acetone soluble compounds. State-of-the-art HPLC instrumentation and commercially available

220

Table 3.　Techniques for Chemical Characterization

ELEMENTAL ANALYSIS - CONVENTIONAL ANALYTICAL TECHNIQUES
                    X-RAY FLUORESCENCE
                    ATOMIC ABSORPTION (AA)
                    ICAP
                    EDAX
                    NEUTRON ACTIVATION ANALYSIS

FUNCTIONAL GROUP ANALYSIS - CONVENTIONAL WET CHEMICAL TECHNIQUES
                           POTENTIOMETRIC TITRATION
                           COULOMETRY
                           RADIOGRAPHY

SPECTROSCOPIC ANALYSIS - INFRARED (PELLET, FILM, DISPERSION, REFLECTANCE)
                             FOURIER TRANSFORM IR (FTIR), PHOTOACOUSTIC FTIR
                             INTERNAL REFLECTION IR, IR MICROSCOPY, DICHROISM
                         LASER RAMAN
                         NUCLEAR MAGNETIC RESONANCE (NMR) ^{13}C, ^{1}H, ^{15}N
                             CONVENTIONAL (SOLUBLE SAMPLE)
                             SOLID STATE (MACHINED OR MOLDED SAMPLE)
                         FLUORESCENCE, CHEMILUMINESCENCE, PHOSPHORESCENCE
                         ULTRAVIOLET-VISIBLE (UV-VIS)
                         MASS SPECTROSCOPY (MS), ELECTION IMPACT MS, FIELD
                             DESORPTION MS, LASER DESORPTION MS, SECONDARY ION
                             MASS SPECTROSCOPY (SIMS), CHEMICAL IONIZATION MS
                         ELECTRON SPIN RESONANCE (ESR)
                         ESCA (ELECTRON SPECTROSCOPY FOR CHEMICAL ANALYSIS)
                         X-RAY PHOTOELECTRON
                         X-RAY EMISSION
                         X-RAY SCATTERING (SMALL ANGLE-SAXS)
                         SMALL-ANGLE NEUTRON SCATTERING (SANS)
                         DYNAMIC LIGHT SCATTERING

CHROMATOGRAPHIC ANALYSIS - GAS CHROMATOGRAPHY (GC) OR GC/MS (LOW MW COMPOUNDS)
                           PYROLYSIS-GC AND GC/MS (PYROLYSIS PRODUCTS)
                           HEADSPACE GC/MS (VOLATILES)
                           INVERSE GC (THERMODYNAMIC INTERACTION PARAMETERS)
                           SIZE-EXCLUSION CHROMATOGRAPHY (SEC), SEC-IR
                           LIQUID CHROMATOGRAPHY (LC OR HPLC), HPLC-MS,
                               MULTI-DIMENSIONAL/ORTHOGONAL LC, MICROBORE LC,
                           SUPERCRITICAL FLUID CHROMATOGRAPHY (SFC)
                           THIN-LAYER CHROMATOGRAPHY (TLC), 2-D TLC

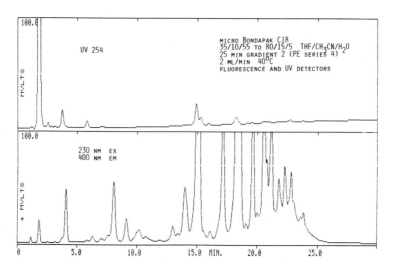

Figure 7. High Performance Liquid Chromotography (HPLC) Analysis
of an SBR Acetone Extract Using UV 254 nm (top) and
Fluorescence (bottom) Detection.

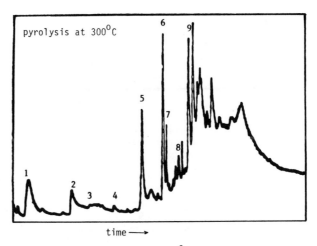

Analysis of Pyrolyzate at 300°C

1. Morpholine (m/e 87)
2. Phenylamine (m/e 93)
3. N-Methylphenylamine (m/e 107)
4. 1,2-Benzisothiazole (m/e 135)
5. 1,2-Dihydro-2,2,4-trimethyl quinoline (m/e 173)
6. N-phenyl-p-phenylamine (m/e 169)
7. N-methyl-N-phenyl-phenylamine (m/e 183)
8. Tetradecanoic Acid (m/e 228)
9. 9,10-Dihydro-9,9-Dimethyl acridine (m/e 209)

Figure 8a Pyrolysis GC/MS Analysis of a SBR Vulcanizate.
Pyrolyzate at 300°C

222

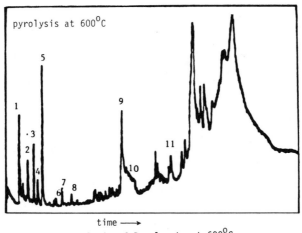

pyrolysis at 600°C

time ⟶

Analysis of Pyrolyzate at 600°C

1. 1,3-Butadiene (m/e 54)
2. Benzene (m/e 78)
3. Vinylcyclohexene (m/e 108)
4. Ethylbenzene (m/e 106)
5. Styrene (m/e 104)
6. 1-Vinyl-4-methylbenzene (m/e 118)
7. 1-Propenyl-Benzene (m/e 118)
8. 1-Vinyl-3-Ethylbenzene (m/e 132)
9. 1,2-Dihydro-6-Methyl naphthalene (m/e 144)
10. 1,2-Dihydro-2,24-trimethyl quinoline (m/e 173)
11. 9,10-Dihydro-9,9-Dimethyl acridine (m/e 209)

Figure 8b   Pyrolysis GC/MS Analysis of a SBR Vulcanizate.
Pyrolyzate at 600°C.

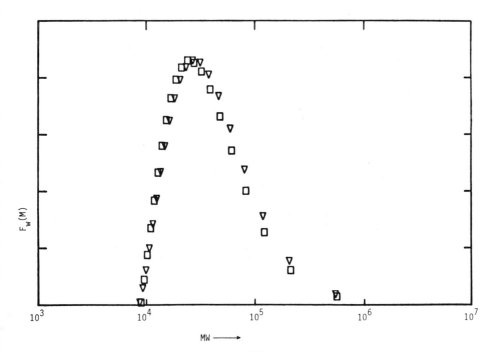

$F_w(M)$

MW ⟶

Figure 9. MWD Analysis of Kevlar[TM] in Concentrated Sulfuric Acid
Using Dynamic Laser Light Scattering.

223

HPLC grade solvents are used. Although taken from a larger
sample in solution, the actual sample size required for an analysis
is 0.5 mg. The size and location of each peak shown in the
chromatograms indicate the concentration and molecular structure of
a particular component in the extract. Since the UV and fluorescence spectra for each component are unique, quite different HPLC
"fingerprints" are obtained not only with the different detectors
but also for different excitation and monitoring wavelengths. Under
specified operating conditions, HPLC can provide a characteristic
fingerprint and quantitative information relating to polymer,
polymer additive, and prepolymer composition.

Pyrolysis-GC/MS (gas chromatography/mass spectroscopy) is an
excellent technique for analyzing the chemical composition of
vulcanized SBR. Again state-of-the-art equipment and small sample
sizes (approx. 0.5 mg) are used. The sample is partially pyrolyzed
at 300^0C (Figure 8a) to analyze the lower MW, more volatile compounds; and then heated to 600^0C (Figure 8b) to pyrolyze the higher
MW and more stable polymeric components. At each temperature,
pyrolyzed components are separated by GC to produce chromatograms
as illustrated in Figure 8. Following GC separation, each component
immediately undergoes analysis by MS; and a computer automatically
compares the mass spectrum for each compcnent against those of
compounds stored in the computer's library to identify the components. Peaks in the GC chromatograms indicate the concentrations
of specific SBR sample components or pyrolyzate products as indicated in Figure 8.

Techniques for evaluating polymer MW's, MWD, and chain structure
are listed in Table 4. Size-exclusion chromatography (SEC) is the
most versatile and widely used method for analyzing polymer MW's
and MWD. Fully automated state-of-the-art instrumentation is
available from a number of manufacturers. The same instrumentation
used for HPLC is also applicable for SEC analysis. The MWD of a
poly(vinylchloride) sample plotted in Figure 1 as the weight
differential distribution vs log(MW) was determined by SEC. The
indicated number-, weight-, and z-average MW's were determined by
a universal calibration procedure[16]. Light scattering, osmometry,
and viscometry are also used to analyze polymer MW's. Although
seldom applied to synthetic polymers, sedimentation is an excellent
technique for characterizing the MW's of polymers having very large
MW's. The "special" techniques tend to be somewhat empirical or
have limited utility and therefore are used less frequently.

Table 4. Polymer Molecular Weights, Molecular Weight Distribution and Chain Structure

Standard Techniques	Parameters Measured	Principle
size-exclusion chromatography (SEC)	mol.wgt. averages and MWD, also provides information relating to polymer chain branching, copolymer composition and polymer shape.	liquid chromatography technique. separates molecules according to their size in solution and employs various detectors to monitor concentrations and identify sample components. requires calibration with standard polymers.
light scattering (Rayleigh)	weight-average mol.wgt. M_w (g/mol) virial coefficient A_2 (mol·cc/g^2) radius of gyration $\langle R_g^2 \rangle_z$ (Å) polymer sturcture, anisotropy, polydispersity.	measurement of scattered light intensities from dilute polymer solutions dependent upon solute concentration and scattering angle. requires solubility, isolation and in some cases fractionation of polymer molecules.
membrane osmometry	number-average mol.wgt. M_n (g/mol) virial coefficient A_2 (mol·cc/g). good for polymers with MW's in the range $5000 < MW < 10^6$, lower MW species must be removed.	measurement of pressure differential between dilute polymer solution and solvent separated by a semi-permeable membrane. colligative property method based upon thermodynamic chemical potential for polymer mixing.
vapor phase osmomety	same as membrane osmometry except that the technique is best suited for polymers with MW<20,000 g/mol.	involves isothermal transfer of solvent from a saturated vapor phase to a polymer solution and measurement of energy required to maintain thermal equilibrium. a colligative property.
viscometry (dilute solution)	viscosity-average mol.wgt. M_v (g/mol) as determined by intrinsic viscosity $[\eta]$ (ml/g) relationship $[\eta] = KM_v^a$ where K and a are constants.	employs capillary or rotational viscometer to to measure increase in viscosity of solvent cause by the presence of polymer molecules. not an absolute method, requires standards.
ultracentrifugation or sedimentation	sedimentation-diffusion average mol.wgt. M_{sd} as defined by the relationship $M_{sd} = S/D$. number-and z-average mol.wgt., M_n and M_z. MWD determined by the relation $S = kM^a$ where k and a are constants. also provides information on the size and shape of polymer molecules.	strong centrifugal field is employed with optical detection to measure sedimentation velocity and diffusion equilibrium coefficients S_w and D_w. Sedimentation transport measurements of dilute polymer solutions corrected for pressure and diffusion provides the sedimentation coefficient S. permits analysis of gel containing solutions.

(continued)

Table 4. (Continued)

Special Techniques	Parameters Measured	Principle
ebulliometry	number-average mol.wgt. M_n (g/mol) for $M_n < 20,000$ g/mol.	measures boiling point elevation by polymer in dilute solution. a colligative property.
cryoscopy	number-average mol.wgt. M_n (g/mol) for $M_n < 20,000$ g/mol.	measures freezing point depression by polymer in dilute solution. a colligative property.
end group analysis	number-average mol.wgt. M_n (g/mol) generally for $M_n < 10,000$. upper limit depends on the sensitivity of the analytical method employed.	the number or concentration of polymer chain end groups per weight or concentration of polymer are determined by specific chemical or instrumental techniques.
turbidimetry	weight-average mol.wgt. M_w (g/mol) and MWD based upon solubility considerations and fractional precipitation of polymers in very dilute solutions	optical techniques are applied to measure the extent of precipitation as polymer solution is titrated with a non-solvent under isothermal conditions or as the solution prepared with a poor solvent is slowly cooled.
chromatographic fractionation	molecular weight distribution An absolute MW technique is needed to analyze fractions.	polymer is coated onto silica particles packed in thermostated column and separated according using solvent gradient elution. polymer solubility decreases with increasing MW.
melt rheometry	weight-average mol.wgt. M_w (g/mol) and weight-fraction differential molecular weight distribution. semi-empirical method.	dynamic melt rheological method involving measurement of spectrum of diffusional relaxation times for polymer during oscillatroy deformation.
gel-sol analysis of crosslinked polymers	gel fraction cross-link density	extraction, filtration, and centrifugation are employed to isolate soluble polymer from gel. MW of soluble polymer is determined separately
swelling equilibrium	network structure, crosslink density, number-average mol.wgt. of chains between crosslinks M_c.	molar volume of crosslinked polymer immersed in swelling liquid and density of the swollen polymer are determined. theory of partial molar free energy of mixing is applied.

226

Promising Techniques	Parameters Measured	Principle
laser light scattering (quasi-elastic, line-broadening or dyanmic)	same as Rayleigh light scattering plus translational diffusion coefficient, molecular weight distribution and information relating to gel structure.	same as above but also involves measurement of the low-frequency line broadening of the central Rayleigh line of the scattered light. the stucture of polymers in both dilute and concentrated solutions can be analyzed.
field flow fraction-ation (FFF)	mol.wgt. averages and MWD. requires calibration.	separates polymers according to their size and shape in solution. an elution technique, like chromatography, except that a field/gradient (thermal, gravitational, flow, electrical, etc.) is applied perpendicular to the axis of solution flow through a capillary or ribbon-shaped channel and a single phase is employed.
non-aqueous reverse-phase high performance liquid chromatography HPLC and thin-layer chromatography TLC	mol.wgt. averages and MWD. requires calibration	liquid chromatography technique based upon equilibrium distribution of polymer molecules between a non-aqueous binary solvent mobile phase and a nonpolar stationary (packing) phase.
supercritical fluid chromatography (SFC)	mol.wgt. averages and MWD. requires calibration.	liquid chromatography technique involving the use of a mobile phase under supercritical conditions (100 bars, 250°C).
neutron scattering small angle (SANS)	weight-average mol. wgt. M_w (g/mgl) virial coefficient A_2 (mol cc/g^2) radius of gyration $\langle R_g \rangle_z$ (Å)	measurement of amplitude of neutron scattering momentum vector for polymer in dilute solution or blend with another polymer. Scattering angle and polymer concentration are varied. Deuterated solvents are used. Dilute solid solutions and polymer blends have been studied

227

Table 5. Techniques for Bulf Characterization

THERMAL ANALYSIS - TGA (THERMAL STABILITY AND COMPONENT ANALYSIS)
DSC (DIFFERENTIAL SCANNING CALORIMETRY- C_P, T_G, T_M)
DTA (DIFFERENTIAL THERMAL ANALYSIS- T_G, T_M)
DMA (DIFFERENTIAL MECHANICAL ANALYSIS- T_G, T_M)
TMA (THERMAL MECHANICAL ANALYSIS- T_G, T_M, EXPANSION)
TBA (TORSIONAL BRAID ANALYSIS- T_G, T_M)
DILATOMETRY (THERMAL EXPANSION COEFFICIENT, T_G, T_M)

MICROSCOPY - OPTICAL (MORPHOLOGICAL STUDIES)
ELECTRON (TRANSMISSION AND SCANNING - MORPHOLOGY)

RHEOLOGICAL - ROTATING DISC RHEOMETER
CAPILLIARY RHEOMETER
MELT VISCOSITY
RELAXATION (VISCOELASTIC BEHAVIOR)

MECHANICAL TESTING - DIELECTRIC SPECTROSCOPY
RELAXATION TIMES
MODULI
COMPLIANCES
IMPACT TESTING
HARDNESS

MISCELLANEOUS - DENSITY
BIREFRINGENCE RELAXATION
REFRACTIVE INDEX
TRANSPARENCY
PARTICLE SIZE ANALYSIS
GAS/LIQUID DIFFUSION/PERMEATION BEHAVIOR
DIPOLE MOMENT
SOLUBILITY STUDIES
CHEMICAL REACTIVITY

New techniques which show great promise for characterizing polymer chain structure also are listed in Table 4. One of the most promising new techniques is dynamic laser light scattering. Unlike SEC, dynamic light scattering can be applied to any soluble polymer, regardless of temperature or solvent, and does not require polymer standards for calibration. Figure 9 illustrates the MWD of poly(1,4-phenyleneterephthalamide) (i.e., Kevlar™) measured by the laser light scattering[18]. As indicated, the polymer's MWD can be fully characterized using very little sample and a single solution with concentrated sulfuric acid as the solvent. Pertinent references for the other characterization techniques are given at the end of this paper.

A number of techniques and approaches used to characterize the bulk properties of polymers are listed in Table 5. A variety of physical properties can be measured by thermal and microscopic analysis. One of the most promising new techniques for characterizing polymers is thermomicrophotometry (TMP) - a combination of thermomicroscopy and thermophotometry[19]. Like DSC and DTA, TMP determines glass transition temperature (T_g), crystalline melting temperature (T_m), thermal decomposition temperature, and polymorphic transformations. However TMP is unique in that it can also detect changes in residual stress and polymer orientation. TMP measurements employ unpolarized, linearly polarized, or circularly polarized light. Each state of polarization measures the change in a specific property of the polymer as a function of temperature. Very fast, as well as very slow, changes in state can be monitored; and the high sensitivity of TMP permits subtle changes to be detected. Although TMP shows great promise for characterizing polymer materials, no instrument has been developed commercially.

Characterization of Epoxy Resin Prepolymers

The most popular and versatile chemical characterization techniques for epoxy resin prepolymers used in adhesive, coating, and structural composite applications are high performance liquid chromatography (HPLC)[20-23] and Fourier transform infrared (FTIR) spectroscopy[22,24]. These techniques provide the capability for characterization and quality control fingerprinting of individual resin constituents as well as resin formulations. In the case of HPLC, dilute solutions of resin samples are prepared and injected into a liquid mobile phase which is pumped through column(s) packed with a stationary phase to facilitate separation and then into a detector. The detector monitors concentrations of the separated components, and its signal response recorded as a function of time after injection provides a "fingerprint" of the sample's chemical composition. Quantitative information may be obtained if sample components are known and sufficiently well-resolved and if standards for the components are available. Recent advances have resulted in improved and automated HPLC instrumentation that is relatively low cost and simple to operate and maintain.

FTIR is not limited by the solubility of the resin. IR spectroscopy is sensitive to changes in the dipole moments of vibrating groups in molecules and as such yields useful information for the identification of polar groups. FTIR spectroscopy is a computer-supported technique for rapidly scanning and storing the infrared spectra of liquids, solids, and gases. Multiple scans and Fourier transformation of the infrared spectra enhance the ratio of signal-to-noise and provide improved spectra for interpretation. FTIR is used to identify and in certain cases quantitatively analyze resin components. The FTIR attenuated total reflection (ATR) technique is particularly useful for characterizing the state of resin cure (i.e., residual epoxy concentration).

Gas chromatography (GC) and GC-mass spectroscopy are useful for analyzing residual solvents and some of the more volatile resin components. Other chromatographic and spectroscopic techniques have also been considered[25-28]. Elemental analysis, titration and wet chemical analysis for specific functional groups are useful techniques for characterizing individual epoxy components but have limited application and may provide misleading results when complex resin formulations are analyzed. When necessary, ion chromatography, atomic absorption (AA), x-ray fluorescence or emission spectroscopy are applied to analyze specific elements, such as boron or fluorine.

Thermal analytical techniques, such as differential scanning calorimetry (DSC), dynamic mechanical analysis (DMA), torsional braid analysis (TBA) and thermal gravimetrical analysis (TGA), are not chemical analysis techniques; however they provide useful information relating to the composition and processability of epoxy resins[22,25,29,30]. Similarly, rheological and dielectric techniques are used frequently to evaluate the chemoviscosity properties of epoxy resins during cure, and there is an increasing interest in applying such techniques for process monitoring and process control[31-33].

Perhaps the most widely used thermal technique is DSC. DSC measures the glass transition temperature T_g and heat of reaction H of the prepreg resin but does not directly provide information about chemical composition. Since the average sample size used in DSC is only 5-10 mg, special care must be taken to obtain representative samples. Sometimes there are problems both in directly sampling the prepreg and in sampling resin from the extraction or press extrusion of the prepreg. Multiple sample runs are advisable.

Whenever possible, complementary techniques should be used to characterize the chemical composition of epoxy resins. Techniques, such as LC and IR, are fundamentally different from one another and provide different types of information about a resin's composition. If appropriate test methods are applied, LC and IR are usually powerful enough to detect differences or changes in the chemical compositions of epoxy resins. DSC complements HPLC and

230

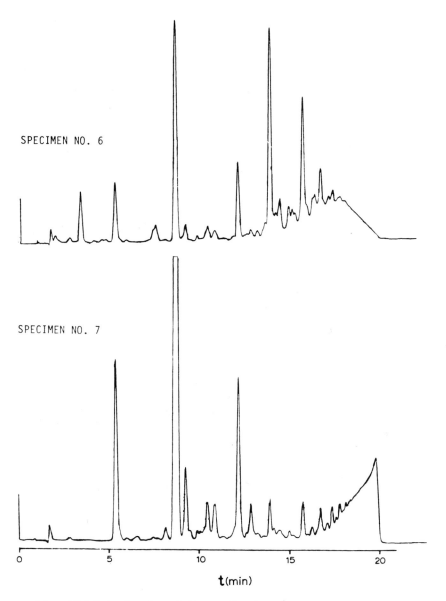

Figure 10. HPLC Analysis of Epoxy Resin-Glass Fiber Prepreg Resin.
UV 280 nm Detection.

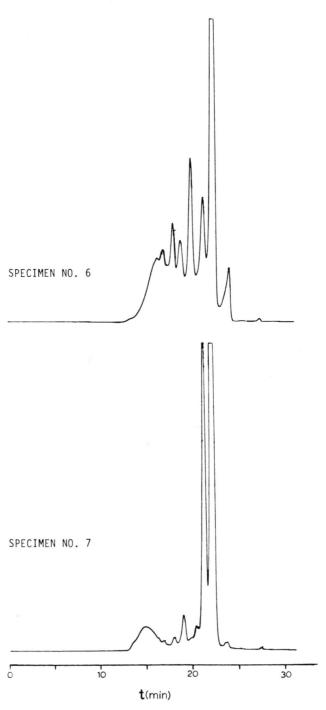

Figure 11. SEC Analysis of Epoxy Resin–Glass Fiber Prepreg Resin.
UV 254 nm Detection.

FTIR analyses by providing information on the extent of reaction of the resin and handleability of the prepreg. TGA and GC headspace analysis techniques for volatile components are secondary, but important, techniques. Special techniques for analyzing specific components or elements should be used if knowledge of their concentration is critical for processing the resin or if their presence could adversely affect the performance and durability of the cured material.

Chromatograms in Figures 10 and 11 illustrate the application of HPLC and SEC to characterize the composition of the epoxy resin from a glass fiber-epoxy resin prepreg used in the manufacture of helicopter rotor blades. The chemical compositions of the first six (6) specimens were essentially identical with a variance in composition of only 2% for major components. HPLC and SEC however showed that the composition of specimen No. 7 was different. Both specimens No. 6 and No. 7 have over 30 HPLC resolvable components. Although both specimens apparently contain many of the same components, differences in peak heights indicate substantial differences in the relative amounts of each component. Such differences could cause problems during manufacture or may effect properties of the cured resin. The epoxy resin monomer p,p'- diglycidyl ether of bisphenol-A (DGEBA) was identified by its peak retention time and detector response and was verified by running a standard. SEC analysis indicates the presence of a high MW polymeric material (2000<MW<32,000 g/mol). A separate HPLC method was used to identify and quantitatively analyze the epoxy resin's curing agent – dicyandiamide (dicy)[34].

Figures 12 and 13 show how HPLC and FTIR techniques complement one another in characterizing the composition of an epoxy-aramid fiber prepreg resin. Under the particular conditions used for HPLC analysis, the major epoxy resin component was determined to be the epoxy novalak DEN 438 (Dow Chemical). FTIR idenfified the curing agent as nadic methyl anhydride and confirmed the HPLC analysis of DEN 438.

Selected References for Polymer Characterization Techniques

Recent reviews of state-of-the-art characterization techniques and texts dealing with polymer characterization are listed in Table 6.

ACKNOWLEDGMENTS

The author thanks coworkers Messrs. D. P. Macaione, D. A. Dunn, and Mr. A. Deome for providing Figures 6, 7 and 13, respectively, and gratefully acknowledges Prof. B. Chu, State University of New York, Stony Brook for providing Figure 9.

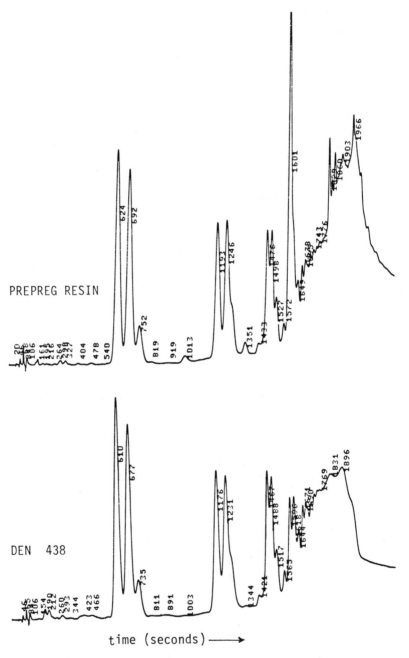

Figure 12. HPLC Analysis of Aramid Fiber Prepreg Resin and Major
Epoxy Component (DEN 438). UV 280 nm Detection.

234

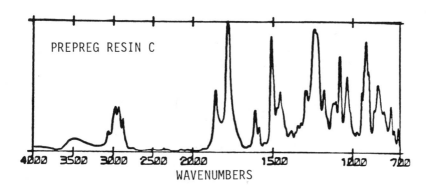

PREPREG RESIN C

WAVENUMBERS

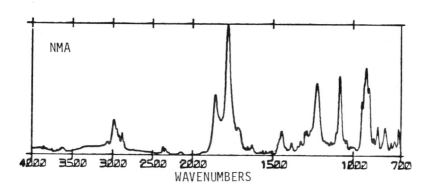

NMA

WAVENUMBERS

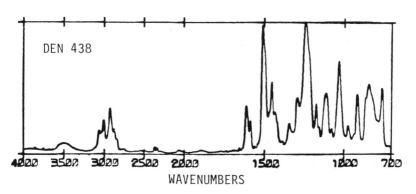

DEN 438

WAVENUMBERS

Figure 13.   FTIR Spectrum of Aramid Fiber Prepreg Resin Compared
with Spectra of Curing Agent Nadic Methyl Anhydride
(NMA) and Major Epoxy Component (DEN 438).

235

BIBLIOGRAPHY

Thermal Analysis

J.F. Johnson and P.S. Gill, eds., "Analytical Calorimetry", Vol. 5, Plenum: New York, 1984.

J.A. Reffner, "Characterization of Polymers by Thermomicro-photometry", Am. Lab., 29-33 (April 1984).

P.S. Gill, "Thermal Analysis Developments in Instrumentation and Applications", Am. Lab., 39-49 (Jan. 1984).

E.A. Turi, ed., "Thermal Characterization of Polymeric Materials", Academic Press: New York, 1982.

J. Chiu, "Polymer Characterization by Thermal Analysis", Marcel Dekker: New York, 1974.

P.E. Slade, Jr. and L.T. Jenkins, eds., "Techniques and Methods of Polymer Evaluation: Thermal Characterization Techniques", Vol. 2, Marcel Dekker: New York, 1970.

P.E. Slade, Jr. and L.T. Jenkins, eds., "Techniques and Methods of Polymer Evaluation: Thermal Analysis", Vol. 1, Marcel Dekker: New York, 1966.

Spectroscopy

J.L. Koenig, "Fourier Infrared Spectroscopy of Polymers", Adv. Polym. Sci., 54, 87-145 (1984).

R.S. McDonald, "Infrared Spectrometry", Anal. Chem., 56, 349R-372R, (1984).

P.C. Painter, M.M. Coleman, and J.L. Koenig, "The Theory of Vibrational Spectroscopy and Its Application to Polymeric Materials", John Wiley & Sons: New York, 1982.

"An Infrared Spectroscopy Atlas for the Coating Industry", Federation of Societies for Coating Technology: Philadelphia, PA, 1980.

"The Infrared Spectra Atlas of Monomers and Polymers", Sadtler Research Laboratories: Philadelphia, PA, 1980.

H.W. Siesler and K. Holland-Moritz, "Infrared and Raman Spectroscopy of Polymers", Marcel Dekker: New York, 1980.

F. Adar and H. Noether, "Raman Microprobe Characterization of Polymeric Fibers", Microbeam Analysis, 269-273 (1983).

F. Adar, "Developments in Raman Microanalysis", Microbeam Analysis, 67-73 (1981).

S.C. Israel, "Mass Spectroscopy of Polymers" in Flame-Retard. Polym. Mater.", M. Lewin, S.M. Atlas and E.M. Pearce, eds., Plenum: New York, 1982, 201-232.

F.A. Bovey, "NMR of Polymers", Pure Appl. Chem., 54, 559-568 (1982).

H. Sillescu, "NMR - Polymers", Pure Appl. Chem., 54, 619-626 (1982).

V.J. McBrierty and D.C. Douglass, "NMR - Polymers", J. Polym. Sci., Macromol. Rev., 16, 295-366 (1981).

F. Bovey, "High Resolution NMR of Macromolecules", Academic Press: New York, 1972.

D. Hummel and F. Sholl, "Infrared Analysis of Polymers", Halsted: New York, 1969.

Chromatography

M.G. Rawdon and T.A. Norris, "Supercritical Fluid Chromatogr. as a Routine Analytical Technique", Am. Lab., 17-23 (May 1984).

W. Jennings, "The Renaissance in Analytical Chromatography", Am. Lab., 14-32 (Jan. 1984).

J.C. Touchstone and M.F. Dobbins, "Practice of Thin Layer Chromatography", 2nd Ed., John Wiley & Sons, Interscience: New York, 1983.

F.P. Schmitz and E. Klesper, "Supercritical Fluid Chromoatography", Polym. Commun., 24, 142 (1983).

G.L. Hagnauer, "Size Exclusion Chromatography", Anal. Chem., 264R-276R (1982).

D.W. Armstrong and K.H. Bul, " Nonaqueous Reverse Phase HPLC", Anal. Chem., 54, 706-708 (1982).

C.G. Smith, N.E. Skelly, R.A. Solomon, and C.D. Chow, "CRC Handbook of Chromatography: Polymers", CRC Press: Boca Raton, FL, 1982.

T. Provder, ed., "Size Exclusion Chromatography: Methodology and Characterization of Polymers and Related Materials", <u>ACS Symp. Ser.</u>, <u>245</u>, Am. Chem. Soc.: Washington, D.C., 1984.

T. Provder, ed., "Size Exclusion Chromatography (GPC)", <u>ACS Symp. Ser.</u>, <u>138</u>, Am. Chem. Soc.: Washington, D.C., 1980.

J. Cazes, ed., "Liquid Chromatography of Polymers and Related Materials", <u>Chromatogr. Sci. Ser.</u>, Vol. 8, Marcel Dekker: New York, 1977.

J. Cazes and X. Delamare, eds., "Liquid Chromatography of Polymers and Related Materials II", <u>Chromatogr. Sci. Ser.</u>, Vol. 13, Marcel Dekker: New York, 1980.

J. Cazes, ed., "Liquid Chromatography of Polymers and Related Materials", <u>Chromatogr. Sci. Ser.</u>, Vol. 19, Marcel Dekker: New York, 1981.

G.L. Hagnauer and D.A. Dunn, "High Performance Liquid Chromatography: A Reliable Technique for Epoxy Resin Prepreg Analysis", <u>Ind. Eng. Chem. Prod. Res. Dev.</u>,21, 68-73 (1982).

G.L. Hagnauer, "Analysis of Commercial Epoxies by HPLC and GPC", <u>Ind. Res. Dev.</u>, <u>23</u>(4), 128-133 (1981).

W.W. Yau, J.J. Kirkland, and D.D. Bly, "Modern Size Exclusion Chromatography", John Wiley & Sons, Interscience: New York, 1979.

L.R. Snyder and J.J. Kirkland, "Introduction to Modern Liquid Chromatography", 2nd Ed., John Wiley & Sons, Interscience: New York, 1979.

M.P. Stevens, "Characterization and Analysis of Polymers by Gas Chromatography", Marcel Dekker: New York, 1969.

Dilute Solution Techniques

J.J. Hermans, ed., "Polymer Solution Properties, Part II: Hydrodynamics and Light Scattering", Dowden, Hutchinson & Ross: Stroudsburg, PA, 1978.

P.E. Slade, Jr., ed., "Techniques and Methods of Polymer Evaluation: Polymer Molecular Weights, Part I", Vol. 3, Marcel Dekker: New York, 1975.

P.E. Slade, Jr., ed., "Techniques and Methods of Polymer Evaluation: Polymer Molecular Weights, Part II", Vol. 4, Marcel Dekker: New York, 1975.

H. Cantow, "Polymer Fractionation", Academic Press: New York, 1967.

H. Morawetz, "Macromolecules in Solution", John Wiley & Sons, Interscience: New York, 1966.

Light Scattering

V.B. Elings and D.F. Nicoli, Submicron Partical Sizing by Dynamic Light Scattering", Am. Lab., 34-40 (June 1984).

B. Chu, "Laser Light Scattering", Academic Press: New York, 1974.

M.B. Huglin, ed., "Light Scattering from Polymer Solutions", Academic Press: New York, 1972.

L. Alexander, "X-Ray Diffraction Methods in Polymer Science", John Wiley & Sons: New York, 1969.

Rheology

S. Wu, "Characterization of Polymer Molecular Weight Distribution by Dynamic Melt Rheometry", ACS Polym. Mat. Sci. Eng., 50, 43-47 (1984).

J.D. Ferry, "Viscoelastic Properties of Polymers", 3rd Ed., John Wiley & Sons: New York, 1980.

N.G. Kumar, "Viscosity-Molecular Weight-Temperature-Shear Rate Relationships of Polymer Melts: A Literature Review", J. Polym. Sci., Macromol. Rev., 15, 255-325 (1980).

F.R. Eirich, ed., "Rheology: Theory and Applications", Vol. 5, Academic Press: New York, 1969.

Mechanical

C. Gramelt, "A New Approach for Determining Mechanical Properties of Thermoset Resins During the Cure Cycle", Am. Lab., 102-109 (Jan. 1984).

S. Wu, "Characterization of Polymer Molecular Weight Distribution by Dynamic Melt Rheology", Am. Chem. Soc., Polym. Mat. Sci. Eng. Proc., 50, 43 (1984).

B.E. Read and G.D. Dean, "The Determination of Dynamic Properties of Polymers and Composites", John Wiley & Sons: New York, 1978.

L.E. Nielsen, "Mechanical Properties of Polymers and Composites", Vol. 1, Dekker: New York, 1974.

L.E. Nielsen, "Mechanical Properties of Polymers and Composites", Vol. 2, Dekker: New York, 1974.

I.M. Ward, "Mechanical Properties of Solid Polymers", John Wiley & Sons: New York, 1971.

General References

R.A. Dickie and S.S. Labana, eds., "Characterization of Highly Crosslinked Polymers", ACS Symp. Ser., in press.

C.G. Smith, N.H. Mahle, and C.D. Chow, "Analysis of High Polymers", Anal. Chem., 55, 156R-164R (1983).

C.D. Carver, ed., "Polymer Characterization", Adv. Chem. Ser., 203, Am. Chem. Soc.: Washington, D.C., 1983.

J.V. Dawkins, ed., "Developments in Polymer Characterization", Vol. 3, Applied Science Publishers: London, 1982.

J.V. Dawkins, ed., "Developments in Polymer Characterization", Vol. 4, Applied Science Publishers: London, 1983.

A.H. Landrock, "Handbook-Plastics Flammability and Combustion Toxicology", Noyes Publishing: Park Ridge, NJ, 1983.

L.S. Bark and N.S. Allen, eds., "Analysis of Polymer Systems", Applied Science Publishers: London, 1982.

"Proceedings of the Critical Review: Techniques for the Characterization of Composite Materials"AMMRC MS82-3 (1982).

F.A. Bovey, "Chain Structure and Conformation of Macromolecules", Academic Press: New York, 1982.

D.O. Hummel and F. Scholl, Atlas of Polymer and Plastics Analysis. Vol. 3, Additives and Processing Aids", Verlag Chemie International: New York, 1981.

C.F. Cullis and M.M. Hirschler, "The Combustion of Organic Polymers", Clarendon Press: Oxford, UK, 1981.

J.F. Rabek, "Experimental Methods in Polymer Chemistry", John Wiley & Sons, Interscience: New York, 1980.

J.L. Koenig, "Chemical Microstructure of Polymer Chains", John Wiley & Sons, Interscience: New York, 1980.

Z. Tadmor and C.G. Gogos, "Principles of Polymer Processing", John Wiley & Sons, Interscience: New York, 1979.

A.R. Blythe, Electrical Properties of Polymers", Cambridge University Press: Cambridge, 1979.

"Proceedings of the TTCP-3 Critical Review: Techniques for the Characterization of Polymeric Materials", AMMRC MS 77-2, ADA036082 (Jan. 1977).

R.J. Samuels, "Structured Polymer Properties", John Wiley & Sons: New York, 1974.

J.M. Schultz, "Polymer Materials Science", Prentice-Hill: Englewood Cliffs, NJ, 1974.

M. Ezrin, ed., "Polymer Molecular Weight Methods", Adv. Chem. Ser., 125, Am. Chem. Soc.: Washington, D.C., 1973.

J. Green and R. Dietz, eds., "Industrial Polymers: Characterization by Molecular Weight", Transcrypta: London, 1973.

"Characterization of Macromolecular Structure", Pub. 1573, Natl. Acad. Sci.: Washington, D.C., 1968.

W. Wake, "The Analysis of Rubber and Rubber Like Polymers", John Wiley & Sons: New York, 1968.

M.L. Miller, "The Structure of Polymers", Reinhold: New York, 1966.

H.E. Haslem and H.A. Willis, "Identification and Analysis of Plastics", Van Nostrand: Princeton, NJ, 1965.

REFERENCES

1. ASTM D4000 "Guide for Classification System for Plastic Materials," 1984 Annual Book of ASTM Standards, Vol. 08.03, ASTM: Philadelphia, PA (1984).
2. D. W. Van Krevelen, "Properties of Polymers - Their Estimation and Correlation with Chemical Structure," Elsevier, New York, (1976).
3. J. Brandrup and E. H. Immergut, Editors, "Polymer Handbook," 2nd Edition, John Wiley & Sons, New York, (1975).
4. J. A. Riddick and W. B. Bunger, "Organic Solvents," 3rd Edition, John Wiley & Sons: Interscience, New York, (1970).
5. "Encyclopedia of Polymer Science and Technology," H. F. Mark, N. G. Gaylord and M. M. Bikales, Editors, Interscience Publishers: John Wiley & Sons, Inc., New York, (1967).
6. "Handbook of Plastics and Elastomers," C. A. Harper, Editor, McGraw-Hill, New York, (1975).
7. ASTM D1898 "Recommended Practice for Sampling of Plastics," 1984 Annual Book of ASTM Standards, Vol. 08.02; ASTM: Philadelphia, PA (1984).
8a ASTM D3896 "Standard Method for Rubber from Synthetic Sources - Sampling," 1984 Annual Book of ASTM Standards, ASTM: Philadelphia, PA (1984).
8b ASTM D1485 "Standard Methods for Rubber from Natural and Synthetic Sources - Sampling and Sample Preparation," 1984 Annual Book of ASTM Standards, ASTM: Philadelphia, PA (1984)
9. W. E. Harris and B. Kratochvil, Anal. Chem., 46, 313 (1974).
10. ASTM E105 "Recommended Practice for Probability Sampling of Materials," 1984 Annual Book of ASTM Standards, ASTM: Philadelphia, PA (1984).
11. ASTM E171 "Specification for Standard Atmospheres for Conditioning and Testing Materials," 1984 Annual Book of ASTM Standards, Vol. 08.03; ASTM: Philadelphia, PA (1984).
12. ASTM D3417 "Test Method for Heats of Fusion and Crystallization of Polymers by Thermal Analysis," 1984 Annual Book of ASTM Standards, Vol. 08.03; ASTM: Philadelphia, PA (1984).
13. ASTM D3418 "Test Method for Transition Temperatures of Polymers by Thermal Analysis," 1984 Annual Book of ASTM Standards, Vol. 08.03; ASTM: Philadelphia, PA (1984).
14. H. W. Siesler and K. Holland-Moritz, "Infrared and Raman Spectroscopy of Polymers," Marcel Dekker, New York (1980).
15. E. L. McCaffery, "Laboratory Preparation for Macromolecular Chemistry," p. 19, McGraw-Hill, New York (1970).
16. ASTM D3593 "Test Method for Molecular Weight Averages and Molecular Weight Distribution of Certain Polymers by Liquid Size-Exclusion Chromatography (Gel Permeation Chromatography GPC) Using Universal Calibration," 1984 Annual Book of ASTM Standards, ASTM: Philadelphia, PA (1984).

17. ASTM D2857 "Test Method for Dilute Solution Viscosity of Poly-
    mers," 1984 Annual Book of ASTM Standards, Vol. 08.02;
    ASTM: Philadelphia, PA (1984).
18. B. Chu, State University of New York at Stony Brook.
19. J. A. Reffner, "Characterization of Polymers by Thermomicro-
    photometry," Am. Lab., 29-33 April (1984).
20. G. L. Hagnauer and D. A. Dunn, "High Performance Liquid
    Chromatography: A Reliable Technique for Epoxy Resin
    Prepreg Analysis," Ind. Eng. Chem. Prod. Res. Dev., 21,
    68-73 (1982).
21. G. L. Hagnauer, "Analysis of Commercial Epoxies by HPLC and
    GPC," Ind. Res. Dev., 23(4), 128-133 (1981).
22. G. L. Hagnauer, J. F. Sprouse, R. E. Sacher, I. Setton and M.
    Wood, "Evaluation of New Techniques for the Quality Control
    of Epoxy Resin Formulations," AMMRC TR 78-8, (1978).
23. G. L. Hagnauer, "Quality Assurance of Epoxy Resin Prepregs
    Using Liquid Chromatography," Polymer Composites, 1, 81-
    87, (1980).
24. J. L. Koenig, "Quality Control and Nondestructive Evaluation
    Techniques for Composites Part II: Physiochemical Charac-
    terization Techniques; A State-of-the-Art Review,"
    AVRADCOM TR 83-F-6, Contract No. DAAG29-81-D-0100, May
    (1983).
25. D. W. Hadad, "Chemical Quality Assurance of Epoxy Resin Formu-
    lations by Gel Permeation, Liquid, and Thin Layer Chroma-
    tography," SAMPE J., 14(4), 4-10 (1978).
26. C. S. Lu and J. L. Koenig, "Raman Spectra of Epoxies," ACS
    Div. Org. Coat, Plast. Chem., 32(1), 112 (1972).
27. C. F. Poranski, Jr., W. B. Moniz, D. L. Birkle, J. T. Kopfle
    and S. A. Sojka, "Carbon-13 and Proton NMR Spectra for
    Characterizing Thermosetting Polymer Systems I. Epoxy
    Resins and Curing Agents,: NRL Report 8092, June (1977).
28. J. S. Chen and A. B. Hunter, "Development of Quality Assurance
    Methods for Epoxy Graphite Prepreg," NASA CR-3531, NAS
    1.26: 3531, (1982).
29. J. K. Gillham, "Characterization of Thermosetting Materials
    by TBA,: Polym. Eng. Sci., 16, 353 (1976).
30. N. S. Schnieder, J. F. Sprouse, G. L. Hagnauer and J. K. Gillham,
    "DSC and TBA Studies of the Curing Behavior of Two Dicy-
    Containing Epoxy Resins," Polym. Eng. Sci., 19, 304 (1979).
31. D. Crozier and F. Tervet, "Rheological Characterization of
    Epoxy Prepreg Resins," SAMPE J, 12-16, Nov/Dec (1982).
32. W. E. Baumgartner and Tom Ricker, "Computer Assisted Dielectric
    Cure Monitoring in Material Quality and Cure Process
    Control," SAMPE J., 6-16, Jul/Aug (1983).
33. S. D. Senturia, N. F. Sheppard Jr., H. L. Lee and S. B. Marshall,
    "Cure Monitoring and Control with Combined Dielectric/
    Temperature Probes," SAMPE J., 22-26, Jul/Aug (1983).
34. G. L. Hagnauer and D. A. Dunn, "Dicyandiamide Analysis and
    Solubility in Epoxy Resins,: J. Appl. Polym. Sci., 26,
    1837-1846 (1981).

APPLICATION OF CHEMICAL CHARACTERIZATION FOR

PRODUCT CONTROL AND IMPROVED RELIABILITY

W. Evan Strobelt

Quality Assurance Technology

Boeing Aerospace Company, Seattle, WA   98124

ABSTRACT

Traditional testing of incoming materials relies extensively on physical/mechanical evaluations which require long
flow time and are labor intensive.

Boeing Aerospace Quality Assurance has been developing and
implementing Chemical Characterization techniques for the last
nine years. The sophisticated analytical methods, which fingerprint materials using advanced instrumental techniques, provide rapid material identification and formulation verification.

The advanced instruments utilized and the types of materials characterized are discussed with emphasis on applications.
The several benefits from implementing Chemical Characterization, including costs savings, are also presented.

The principal methods for qualifying materials during
receiving inspection rely extensively on physical/mechanical
evaluations, such as tensile testing. These traditional tests,
which require long flow time for sample preparation, conditioning, machining, coupon preparation and data analysis are very
labor intensive. In addition, inspection results provide only
subjective pass/fail information which does not identify
changes in material formulations that can detrimentally impact
product quality and reliability.

Boeing Aerospace Quality Assurance is vitally concerned
about changes in the formulation of materials. Several in-
cidents in the past 10 years have emphasized this concern.
Cured sealants and electrical potting compounds have reverted

CONVENTIONAL TEST METHOD:

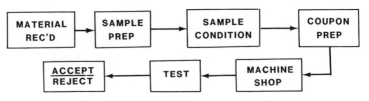

- 21 DAY FLOW TIME
- 16 HOUR TEST TIME
- SUBJECTIVE ACCEPT/REJECT CRITERIA
- NO FORMULATION CONTROL

Figure 1.   Problem

to their precured monomeric components while in service.
Additionally, problems have been experienced with composite
materials adhesion, cure and fire retardant properties.   In
some cases, conventional test methods passed marginal or
unacceptable materials because they lacked the sensitivity to
detect formulation changes; changes which were ultimately found
to be responsible for hardware failure.   Consequently corrective action often required extensive retrofit of in-service
hardware at considerable expense.

The following illustration Figure 1 depicts the
conventional physical test method and associated problems.
As can be seen in this figure, conventional testing requires
excessive flow time for sample preparation and conditioning and
lacks the capability to control material formulation which is
essential for insuring uniform product quality.

The challenge at Boeing was to develop and implement more
efficient test methods that would (1) quantitatively reflect
material performance characteristics and (2) detect critical
formulation changes prior to costly manufacturing processing or
incorporation of a marginal material into production hardware,
Figure 2.

DEVELOPMENT OF MORE EFFICIENT TEST METHODS

- PROVIDE TEST METHODS WHICH QUANTITATIVELY
  REFLECT MATERIAL PERFORMANCE CHARACTERISTICS

- PREVENT INCORPORATION OF MARGINAL
  MATERIALS INTO PRODUCTS

Figure 2.   Challenge

APPROACH

For the past nine years, the Boeing Aerospace Company has funded an improved process development program designed to document and demonstrate the effectiveness of controlling purchased non-metallic materials by Chemical Characterization (CC) techniques. Chemical Characterization is defined as the methodology of controlling non-metallic materials by finger-printing each material utilizing advanced instrumental analytical methods. In this approach the material undergoes extensive analysis by a variety of instrumental techniques simultaneous with qualification by conventional testing. At the conclusion of successful qualification by the conventional methodology, certain instrumental fingerprints are selected to become the standard reference for future materials receival tests. Subsequent receivals can then be rapidly analyzed for assurance that no formulation changes have been made by the supplier either accidentally or intentionally.

The chemical characterization control operation, which utilizes selected analysis methods, is shown in Figure 3. This new concept provides rapid testing with associated shortened flow time, and most important, significantly improved formulation control of the material.

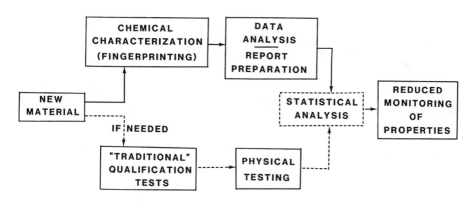

ADVANTAGES:

- ONE DAY FLOW TIME
- 2 HOUR TEST TIME
- 5:1 TEST COST REDUCTION
- FORMULATION CHANGES RAPIDLY IDENTIFIED

Figure 3. Control Operation

Current activities on this program are directed toward the correlation of mechanical/physical testing with chemical characterization results to assure that physical test specification requirements are being met. Characterization of new materials for advanced technology products is continuing as are discussions on implementing Chemical Characterization criteria as Mil-Spec requirements. This approach is summarized in Figure 4.

- CORRELATION OF MECHANICAL/PHYSICAL TESTING WITH CHEMICAL CHARACTERIZATION (CC) RESULTS

- CHARACTERIZATION OF NEW MATERIALS

- IMPLEMENTATION OF CC CRITERIA AS MIL-SPEC REQUIREMENTS

Figure 4. Approach

APPLICATIONS

The implementation of CC does require the availability of some rather sophisticated analytical instrumentation, as well as knowledge and expertise to apply this technology for maximum benefit. The Boeing Aerospace Quality Assurance Engineering Laboratories have been equipped for this purpose. The basic advanced instrumental equipment inventory required to conduct CC includes the following instrumental capabilities, Figure 5:
1. Thermal analysis including:
    a. Thermal Mechanical Analysis (TMA)
    b. Thermal Gravimetric Analysis (TGA)
    c. Differential Scanning Calorimetry (DSC)
    d. Dynamic Mechanical Analysis (DMA)
    e. Kinetics - TGA and Reaction
2. Gas Chromatography/Mass Spectrometry (GC/MS)
3. Liquid Chromatography (LC)
4. Fourier Transform Infrared Spectroscopy (FTS)
5. Ion Chromatography
6. Elemental Analysis

ADVANCED INSTRUMENTAL CAPABILITIES:

- THERMAL ANALYSIS
    - THERMAL MECHANICAL ANALYSIS (TMA)
    - THERMAL GRAVIMETRIC ANALYSIS (TGA)
    - DIFFERENTIAL SCANNING CALORIMETRY (DSC)
    - DYNAMIC MECHANICAL ANALYSIS (DMA)
    - KINETICS (TGA AND REACTION)

- GAS CHROMATOGRAPHY/MASS SPECTROMETRY (GC/MS)

- LIQUID CHROMATOGRAPHY (LC)

- FOURIER TRANSFORM INFRARED SPECTROSCOPY (FTS-IR)

- ION CHROMATOGRAPHY

- ELEMENTAL ANALYSIS

Figure 5. Applications

Thermal analysis, Figure 6 is used to measure cure cycle weight and dimensional changes in a material consequent to changes in temperature and pressure. A typical thermal analysis fingerprint is shown in Figure 7. This differential scanning calorimeter (DSC) plot quantitatively displays the amount of heat gained or lost by a material during temperature cycling, and provides useful fingerprinting information for manufacturing processing. Gas chromatography/mass spectrometry Figure 8 is used primarily to identify the presence of various components within a material. Liquid chromatography, Figure 9, is used to separate a material into its component parts and to provide molecular weight distribution information. Fourier transform infrared spectroscopy, Figure 10, is used to identify components as well as monitor chemical reactions during the curing of a material. A special temperature-programmable sample cell, Figure 11, has been developed for this purpose. Ion chromatography, Figure 12, is used for quantitative analysis of anions, such as chlorine, fluorine, bromine and sulfates in a material. (It can also provide cation analysis, if needed.) An elemental analyzer, Figure 13, recently implemented, provides rapid analysis of hydrogen, carbon, sulfur, oxygen and nitrogen in composites. This analyzer has significantly improved the accuracy of analytical results and has reduced test time for these hard-to-analyze elements.

Figure 6.   Thermal Analyzer

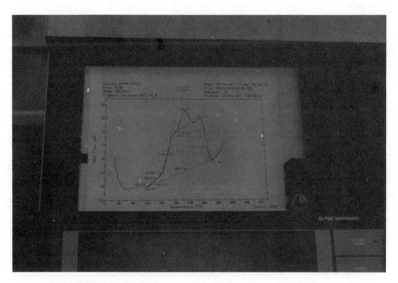

Figure 7.   Typical Thermal Analysis Plot

Figure 8.   Gas Chromatograph – Mass Spectrograph

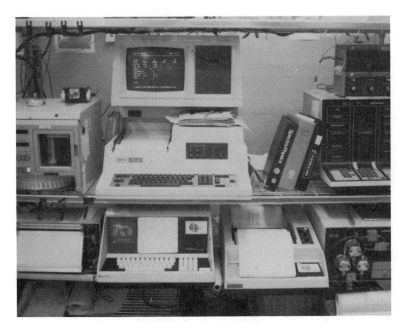

Figure 9.   Liquid Chromatograph

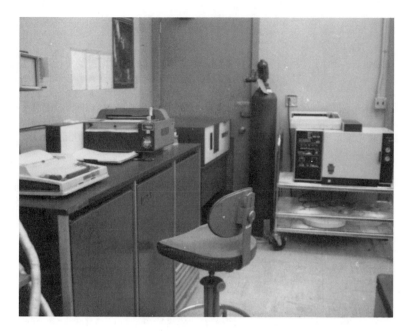

Figure 10.   Fourier Transform Infrared Spectrometer

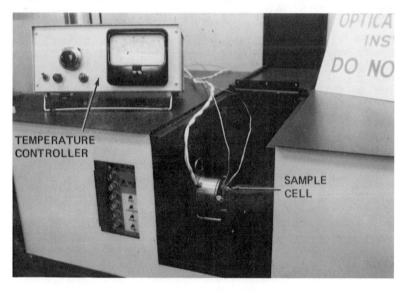

Figure 11.   Temperature – Programmable Sample Cell

Figure 12.   Ion Chromatograph

Figure 13.   Elemental Analyzer

Composites and resinous materials often contain fillers and inorganic components; these must be quantitatively and qualitatively analyzed. This analysis can be nondestructively evaluated by energy dispersive X-ray techniques. Once elements have been identified, the quantitative evaluation can be made very accurately with atomic absorption spectrophotometry.

Currently over 300 materials have been characterized by the "CC" program, Figure 14, with many of the resulting specifications incorporated into company material requirements. For example, extensive chemical characterization data was defined for Advanced Epoxy Graphite Prepreg for composite application. The results were recently published in NASA Report #3531 titled "Development of Quality Assurance Methods for Epoxy Graphite Prepreg."[3] The study was designed to ensure that new material concerns were met, and provide the impetus required to achieve an improved level of confidence in advanced composite primary structures. The chemical characterization methods studied in that specific project included:

1. Liquid Chromatography
2. Differential Scanning Calorimetry
3. Gel Permeation Chromatography
4. Dynamic Mechanical Analysis

Also, over 50 of the aerospace material analytical specifications have been transferred to commercial aircraft materials, providing improved manufacturing control and reliability for Boeing aircraft. Fingerprinting of advanced materials, such as polysulfones, polyimides and high temperature resins is continuing as materials are identified for use by Engineering.

BENEFITS

The overall benefits of controlling materials in this way can be placed into three categories, Figure 15. However, the following listing does not reflect any implicit priority. One, there are economic benefits; reduction of testing costs, improved test flow time which minimizes delay caused by long tests, better trouble-shooting techniques, and expedited solution of material problems. Additionally, there is improved quality; formulation changes can be readily detected, information gained can prevent performance and service problems, and verification that sample material matches parts is simplified. Further, chemical characterization data supplements and supports manufacturing in optimizing fabrication processes, such as in defining proper time/temperature cure cycle for adhesives, paints and polymers. These improvements have resulted in cumulative cost savings over and above implementation/equipment costs of over $886K since the program was begun in 1974. These benefits have been ample justification for the Boeing Aerospace Company to invest in CC and to use these developed techniques to assure the quality of purchased non-metallic material to the benefit of its customers. We expect that all future Mil-Spec's for non-metallic materials will contain CC test techniques, and that many existing specifications will be revised to be consistent with this new test methodology. Thus, chemical characterization has significantly changed the way we control material at Boeing and we believe it will continue to play an even more important role for material control in our future hardware, resulting in increased reliability and performance of our products.

- OVER 300 ANALYTICAL "SPECIFICATIONS DEVELOPED FOR:

• PLASTICS	• FOAMS
• TEXTILES	• COMPOSITES
• RUBBERS	• LUBRICANTS
• SEALANT	• SOLVENTS
• POLYBLENDS	• GASES

- AEROSPACE MATERIAL ANALYTICAL METHODS APPLIED TO OVER 50 AIRCRAFT MATERIALS

- "FINGERPRINTING" OF ADVANCED MATERIALS IN PROGRESS
    - GRAPHITE-EPOXIES
    - THERMOPLASTICS (POLYSUFONE)
    - HIGH TEMPERATURE RESINS
    - POLYIMIDE

Figure 14.   Applications

- INSPECTION FLOW TIME REDUCED FROM 5.5 TO 1.2 DAYS

- IMPROVED INSPECTION CAPABILITY THROUGH MATERIALS FORMULATION CONTROL

- IMPROVED MANUFACTURING PROCESS CRITERIA

- CUMULATIVE COST SAVINGS OF OVER $886K SINCE PROGRAM START IN 1974

Figure 15.   Benefits

REFERENCES

1. A. B. Hunter, "Chemical Characterization," Society of Plastics Engineers, Seattle Section, November (1974).
2. A. B. Hunter, "Automated Materials Analysis via Chemical Characterization," International Conference on Automated Inspection and Product Control, November (1978).
3. J. Chen, A. B. Hunter, and M. Katsumoto, "Development of Quality Assurance Methods for Epoxy-Graphite Prepreg," NASA Composites Conference, Los Angeles, CA, August (1980).
4. A. B. Hunter, "Chemical Characterization for Receiving Inspection," National SAMPE Conference, Seattle, WA, September (1980).
5. A. B. Hunter, "Reaction Mechanism of PMR-15," Gordon Research Center Conference, New Hampshire, June (1981).
6. A. B. Hunter, "Chemical Characterization of Composites," American Chemical Society, New York, NY, August (1981).
7. W. E. Strobelt, "New Quality Technology and Productivity," Forum 81, American Defense Preparedness Assn., Washington, DC, October, (1981).
8. W. E. Strobelt, "Inspection Methods Development for Improved Quality Productivity," Quality Assurance Productivity Improvement Workshop, New Orleans, LA, May (1982).
9. W. E. Strobelt and J. P. Hartman, "Automated Inspection Methods for Improved Quality and Productivity," Quality Management Symposium, American Society for Quality Control, Seattle Section, March (1983).
10. A. B. Hunter, "Characterization Methodology for PMR-15," High Temperature Polymer Matrix Composites Conference, NASA Lewis, MD, March (1983).
11. A. B. Hunter, "Analysis of Reaction Products of Polyimide by High Pressure DSC with GC-MS," National Analysis and Calorimetric Conference, Williamsburg, VA, September (1983).
12. C. H. Sheppard, "Program Comparison of PMR-15 Graphite Composite Cure Cycles," SAMPE Conference, Las Vegas, Nevada, April (1984).

CERAMIC MATERIALS CHARACTERIZATION USING SMALL ANGLE

NEUTRON SCATTERING TECHNIQUES

K. A. Hardman-Rhyne and N. F. Berk

Center for Materials Science
National Bureau of Standards
Gaithersburg, MD 20899

INTRODUCTION

Small angle neutron scattering (SANS) techniques are used to study microstructural phenomena in the range of 1 to 10^4 nm in size. Since they cover a wide range of sizes, these techniques are particularly useful in studies of ceramic processing and distributed damage in ceramics. While many metal and alloy systems have been studied using SANS techniques, few experiments have been published on ceramic materials. This is not surprising considering the difficulties inherent in analyzing SANS data on these materials. Often ceramics have several microstructural components such as residual voids from the sintering process, inclusions or impurities from starting materials, second phases, and microcracks or cavities from temperature and/or pressure treatments, as well as dislocations present in the material. All these effects will contribute to the observed small angle scattering of neutrons. It is important to either eliminate all effects except the one of interest or to identify the effects through complimentary studies that use other techniques such as electronic or optical microscopy. While these complementary techniques can identify defects, voids and second phases, SANS can quantify these effects throughout the bulk of the material in a nondestructive way.

Neutrons are an excellent nondestructive probe of microstructure because the neutrons are uncharged, the energies of thermal neutron beams are very low, and the neutrons are not absorbed in most materials. The neutrons primarily interact with the nucleus of the atom, and the neutron beam is highly penetrating without disturbing or damaging the sample. These attributed place neutrons in the category of a bulk probe of the material, whereas x-ray and many other conventional techniques are more sensitive to surface phenomena. One strength of neutron scattering is its dependence on the discrete chemical elements in the material through a quantity called the coherent scattering length, b[1]. Since b values vary in an unsystematic way from one element to another, differences between elements with similar atomic numbers can be detected (e.g. aluminum and magnesium or iron and manganese). Voids and microcracks can be detected as well, so that studies of powder consolidations are possible.

EXPERIMENT

The SANS instrument at the National Bureau of Standards reactor is described in detail in reference 2; and a schematic diagram of the major features of this instrument is shown in Figure 1. The neutron wavelength λ, can be varied from 0.4 to 1.5 nm by selecting the appropriate speed of a rotating helical-channel velocity selector. This is particularly important in beam broadening (multiple scattering) experiments, for example on ceramic materials, because the functional form of the wavelength dependence of the scattering is a critical element in the analysis. The longer wavelengths (> 10 nm) are also useful in diffraction measurements where larger size (> 0.5 nm) particles or voids are being examined and where multiple Bragg scattering from the crystal structure of the material is to be avoided. A cold source is important for a SANS facility because measurements can be obtained at the longer wavelengths in a reasonable time interval. The function of a cold source is to lower the thermal neutron effective temperature which shifts the peak flux to higher wavelengths.

There are two types of collimating aperatures which define the beam direction and divergence. One type consists of a pair of

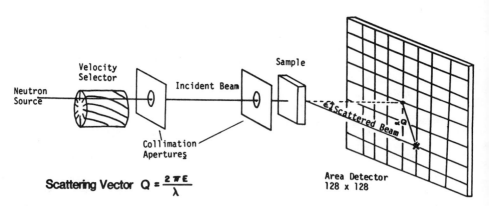

Scattering Vector $Q = \dfrac{2\pi\varepsilon}{\lambda}$

Figure 1.   Schematic Diagram of Key Components Used for the Small Angle Scattering Measurements.

cadmium pin holes irises, one after the velocity selector and
another before the sample. The other collimation system is designed for higher resolution measurements and consists of a set of
channels in cadmium masks which effectively converge the neutron
beam to a point at the center of the detector. The multiple sample
chamber is computer controlled, and can be used under vacuum.
Single samples can be studied as a function of temperature from below 4K to 1600K. Sample sizes are usually 1.0 to 2.5 cm diameter
disks 2 to 30 mm thick. The scattered neutrons are detected on a
64 cm x 64 cm position-sensitive proportional counter divided into
128 columns and 128 rows with a spatial resolution of 8 mm in each
direction. The angle between the incident beam and the scattered
beam is the scattering angle ε. The magnitude of the scattering
vector Q in the small angle limit is $Q \approx k\varepsilon$, where $k = 2\pi/\lambda$.

Single particle diffraction and beam broadening measurements
are distinct and usually require different configurations of the
SANS instrument. Diffraction experiments probe for microstructural
phenomena in the range of 1 to 100 nm. For this case the SANS
detector is usually centered directly in the incident beam. To
minimize multiple scattering events the sample thickness is usually
kept in the 2 to 6 mm range. It is also desirable to use materials

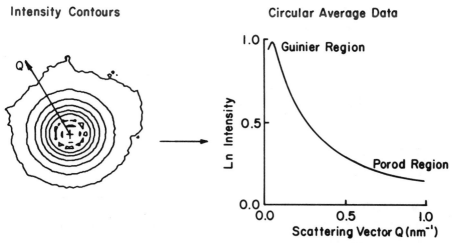

Figure 2.   Schematic of Single Particle Diffraction Data from the
            Two Dimensional Detector (left), and the Circularly
            Averaged Intensity as a Function of the Scattering
            Vector Q (right).

with a minimum of incoherent scattering and absorption. An example of single particle diffraction is shown in Figure 2. Since the constant scattering intensity contours are generally circular, the intensity data can be circularly averaged and plotted as a function of the scattering vector, Q (see the right side of Figure 2). The small-Q range of the data is generally called the Guinier region, and the large-Q data are called the Porod region. The shape of the scattering curve is wavelength independent when single particle scattering is dominant.

The beam broadening effect is wavelength dependent, and the resulting widths in Q of the scattering pattern show the most sensitivity to wavelength for the longer wavelengths. For the analysis it is imperative to obtain as much data at very small Q values as possible, and to determine the exact center of the scattering distribution on the detector. Normally the detector is operated with a centered beam-stop to prevent damage from the intense unscattered incident beam; however, in the beam broadening studies the detector is operated off-center with no beam stop.

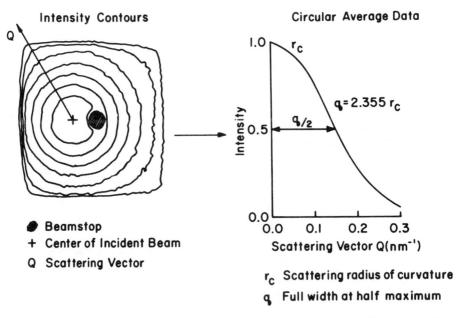

Figure 3. Schematic of Multiple Scattering Data from the Two Dimensional Detector (left). Note the Center of the Incident Beam is not at the Detector Center in this Case. (Right) Circularly Averaged Intensity Versus Q.

This can be done without damage to the detector because the incident beam is virtually entirely scattered in passing through the sample. In cases where the overall scattering intensity is too high for reliable operation of the detector, cadmium foil attenuators are placed in the incident beam. The experimental configuration for the beam broadening experiments is shown in Figure 3 which is a schematic of the constant intensity contours on the detector face with each contour representing a power of two in scattering intensity. The resulting data array from the detector is circularly averaged for evenly spaced values of Q. The shape of the scattering distribution can be then described in terms of a scattering radius of curvature, r_c, for $Q \approx 0$ which is related to the full width at half maximum of the distribution, q, as $q = 2.355 \, r_c$. Qualitatively, the shape dependence of the multiple scattering curve in Q can be called beam broadening.

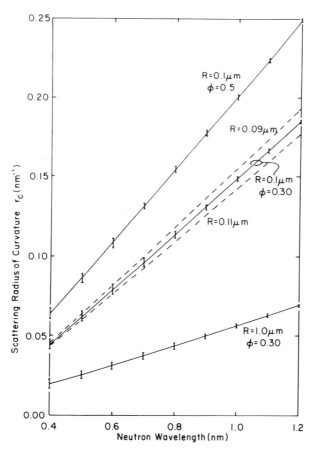

Figure 4. Theoretically Determined $r_c(Q)$ as a Function of λ for Various Radius and Volume Fraction ϕ Conditions with all Other Parameters Held Constant.

The radius of curvature obtained from the small Q data is wavelength dependent as mentioned above. In addition, as shown in Figure 4, significant variations in the r_c values occur with particle size and volume fractions ϕ. The lines in Figure 4 are calculated from the theory and the error bars represent the statistical accuracy and resolution of the data. Wavelength resolution has little effect on the scattering data at small Q, but the instrumental collimation (Q resolution), and most particularly the limited number of data points in the low Q regime at short wavelength, do have a significant effect on the determination of r_c. The sensitivity of the values of r_c to particle radius is shown by the dotted lines in Figure 4, which represent the results which would be produced by a 10% variation in particle size. Clearly the technique is more accurate at the longer wavelengths ($\lambda > 0.7$ nm) where deviations in particle size can be determined to better than 2% of the actual value. At the shortest wavelengths the calculated errors increase to above 5%. Examples of results obtained from both single particle diffraction and multiple scattering follow.

Fe INCLUSIONS IN Si_3N_4 (SINGLE PARTICLE DIFFRACTION)

Although optical and electron microscopy can identify small defects (< 10 μm) in ceramic materials, SANS can quantify the size, shape, and distribution of these defects in the bulk of the material. A complementary study using SANS and transmission electron microscopy, TEM, has been made on hot-pressed MgO doped Si_3N_4. TEM studies of this material clearly showed small, approximately spherical, inclusions in Si_3N_4 that were identified as Fe. There was no evidence of pores, microcracks, or microvoids from the TEM observations on this sample which had not been heat treated at temperatures where voids are most likely to appear.

Using SANS pores larger than about 90 nm can be detected by analyzing the scattered data collected at two or more wavelengths (the longer wavelength data have a higher sensitivity to porosity effects). The wavelength dependence of the neutron scattering under high resolution conditions (focussing collimation) is shown in Figure 5. No wavelength-dependent beam broadening was observed. This is consistent with observations from TEM and other characterization methods which suggest that this sample of Si_3N_4 is fully dense with little or no porosity. Therefore we assume the SANS data reflect information relevant to the Fe inclusions that are present.

The SANS data from the inclusions is isotropic in Q so that the data have been circularly averaged. Background and transmission corrections have been made to the raw data. The logarithm of the circularly averaged and corrected data are plotted versus the scattering vector, Q, in Figure 6. The scattering vector has units of inverse nanometers (1 nm^{-1} = 0.1 $\mathring{A}^{-1}$). Region 1 in Figure 6 is

the Guinier region where the logarithm of the intensity has a Q^2
dependence. A Guinier fit to these data at λ = 0.76 nm yields a
radius of gyration[3,4] of 18.6 nm which is related to the average
radius of the Fe inclusions. If we assume the inclusions are mono-

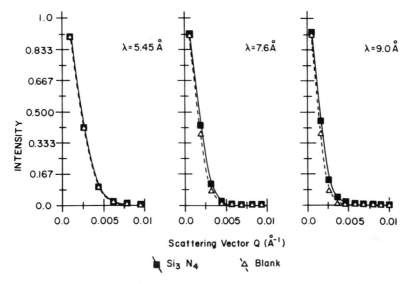

Figure 5.   Comparison of the Incident Beam Scattering (blank sam-
            ple) with that from the Si₃N₄ Sample at Several Wave-
            lengths. Note, the Scattering Vector Q is in Units of
            Reciprocal Angstroms.

dispersed and spherical in shape, $R_G = (3/5)^{1/2}R_S$ so that the
average particle radius is 24.0 nm. In this region the neutron
intensity is dominated by the scattering particles of larger dimension. The Guinier approximation is valid over a range of $Q_{max}R_G <$
1.2. In our case the maximum Q is 0.17 nm and $Q_{max}R_G \sim$ 3 which
extends outside the Guinier approximation range, although the
logarithm of the intensity still retains a Q^2 dependence.

Region 2 in Figure 6 is called the Porod region and has a Q^{-4}
dependence. The Porod region is more sensitive to smaller dimensions of the scattering centers and yields a characteristic Porod
length, ℓ_p , which is related to the total surface area of the
scattering center if absolute scattering intensities can be determined. For a population of spheroidal particles $\ell_p \approx (4/3)R_S$ and
in this case ℓ_p is 21.2 nm. Therefore, the average radius of the
Fe inclusions, as determined from the Porod region, is approxi-

mately 16 nm. The difference in the two average Fe particle radius
values indicates a distribution of sizes in the sample.

YCrO$_3$ (MULTIPLE SCATTERING)

Small angle neutron scattering (SANS) techniques can be used

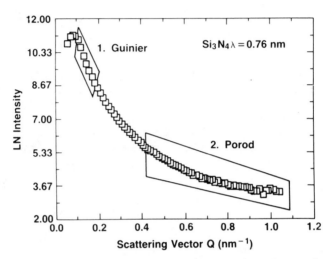

Figure 6. Logarithm of the Neutron Scattering Intensity from Fe
and W Inclusions in Si$_3$N$_4$ Versus Scattering Vector Q.

to study microstructural phenomena in the range of 1 to 10^4 nm in
size. This is particularly useful for powder characterization and
quantitative defect analysis. The defects include the initial
porosity, particle agglomeration, and impurity effects in the compacted powder ("green" state) before firing has occurred and during
the early firing stages. With the use of multiple scattering
methods recently developed at NBS[5,6], particle and/or void sizes
(0.08 to 10 μm) and volume fraction can be studied. This technique
is possible because the NBS SANS facility has variable neutron
wavelength capabilities and will soon have a cold source which will
allow data to be collected with more intensity at wavelengths in
the range (0.5 < λ < 1.5 nm).

Three stages of YCrO$_3$ consolidation were measured with multiple

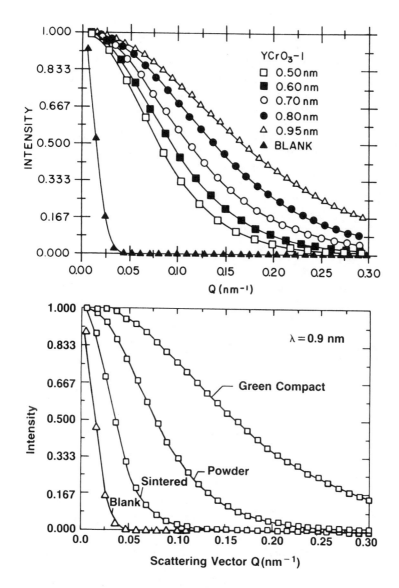

Figure 7. (a) Wavelength Dependence of the Circularly Averaged Neutron Scattering Intensity Versus Q for the "green" State (unfired) Ceramic of $YCrO_3$. (b) Circularly Averaged Intensity Versus Q for the Three Forms of $YCrO_3$. These Data were Collected at λ = 0.9 nm.

scattering SANS techniques. Two samples of $YCrO_3$ were fabricated from pure powders by isostatic pressing at 5,000 psi; one sample was then sintered. The starting ceramic powder was approximately 30% of the theoretical density. The porosity of the green state ceramic was 49% and of the sintered material 8%. The neutron scattering resulted from the particles in the case of the powder and the voids in the sintered material. The voids were assumed to be the source of neutron scattering in the green state material in that the powder particle size could not be obtained from the green state measurements assuming 51% of theoretical density. The neutron scattering curves differed greatly for these three forms as shown in Figure 7b. This figure shows the volume fraction dependence of r_c. The sintered material exhibits slight wavelength dependence but the green state material reveals dramatic beam broadening which is strongly wavelength dependent as seen in Figure 7a. Although the qualitative aspects of the data clearly demonstrate a strong pro-

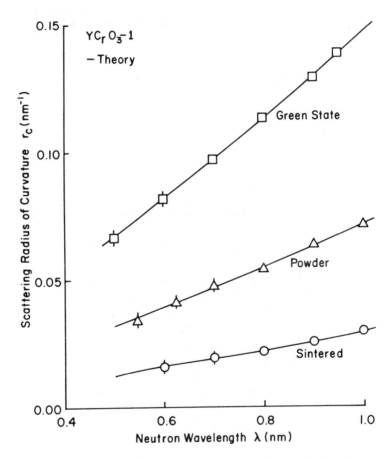

Figure 8.   Theoretical and Experimental Values of the Scattering Radius of Curvature $r_c(Q)$ as a Function of λ for the Three $YCrO_3$ Materials.   The Experimental Error Bars are Given or are Less Than the Size of the Symbol.

cessing effect on the population of scattering centers in these materials, quantitative measures of the particle or void size, shape, and size distribution are less straightforward. In the past, SANS theory has been limited to single particle diffraction with minor corrections for multiple scattering and multiple refraction. Thus, the theory had to be expanded to include all phase shifts (i.e. diffraction scattering effects perturbed by refractive components) and neutron scattering dominated by multiple scattering effects. This neutron scattering theory is used to quantitatively analyze the SANS data for densified ceramics and metals. The straight lines in Figure 8 show the derived scattering radius of curvature values r_c with the parameters for the three forms of $YCrO_3$ given in Table 1. Excellent fits to the data are observed

TABLE 1

Material	Thickness	Size (diameter)	Volume Fraction
$YCrO_3$ Powder (particles)	5.0 mm	0.886 μm	0.29
$YCrO_3$ Green State (voids)	12.2 mm	1.086 μm	0.49
$YCrO_3$ Sintered (voids)	6.9 mm	2.064 μm	0.08

with the above parameters which agreed well with values obtained from other techniques such as x-ray sedimentation, TEM, and density measurements.

A simulated consolidation experiment with spinel powder, $MgAl_2O_4$, has been undertaken to study the porosity changes in the material as a function of temperature. Measurements were taken at 975C, 1300C, 1500C, and 1550C. Although the analyses are not complete, striking changes in the neutron scattering curves occur between 1300 and 1500C, where sintering begins, Figure 9. Nevertheless the material is not fully sintered in that its full width at half maximum is significantly larger than the intrinsic instrumental beam width (blank sample).

CONCLUSIONS

Often ceramic materials have several microstructural components such as residual voids from the sintering process, inclusions or impurities from starting materials, second phases, microcracks, or cavities from temperature and/or pressure treatments as well as dislocations present in the material. Intensive efforts in synthesizing new reproducible ceramics have resulted in fewer microstructural defects. Nevertheless, the effects of temperature and pressure on these defects are not understood. Single particle diffraction SANS experiments are concerned with inhomogenieties such as voids, cavities, microcracks, precipitates, sintered poros-

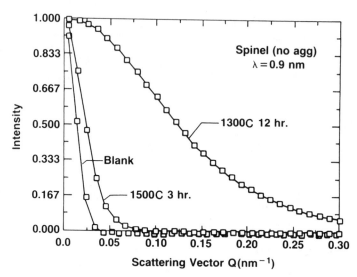

Figure 9. Dramatic Change in Multiple Scattering as a Function of Temperature Indicating a Rapid Change in Porosity While Sintering as seen in the Spinel Material at λ = 0.9 nm.

ity, inclusions, and nucleation and growth of second phases. In particular, microstructural values can be determined throughout the bulk of the material quantitatively such as particle size, shape, size distribution and total surface area.

The future of new high-technology ceramic materials is dependent on understanding the densification process and developing new alternative procedures to eliminate defects at earlier stages of the processing. These defects include the initial porosity, agglomeration, and impurity effects in the compacted powder (green state ceramics) before firing has occurred. Few techniques exist which can monitor a fragile green state ceramic through the densification process in a nondestructive manner. With the use of theoretical and experimental multiple SANS methods developed at NBS, particle and/or void sizes (0.001 to 10 μm) and volume fractions can be studied. These methods require the largest possible neutron wavelength range available with a neutron cold source.

268

This SANS analysis technique is presently being expanded to include determinations of shape, surface area, and possible size distribution effects.

REFERENCES

1. G. Kostorz, Treatise on Materials Science and Technology, Vol. 15, Editor, G. Kostorz, Academic Press, New York, S-8, (1979).
2. C. J. Glinka, AIP Conference Proceedings Series No. 89, 395, (1981).
3. J. R. Weertman, Nondestructive Evaluations: Microstructural Characterization and Reliability Strategies, Editors, O. Buck and S. M. Wolf, Amer. Inst. of Mining, Metallurgical and Petroleum Engineers, New York, pp. 147-168, (1981).
4. G. Kostorz, Treatise on Materials Science and Technology, Vol. 15, Editor, G. Kostorz, Academic Press, New York, pp. 227-289, (1979).
5. N. F. Berk and K. A. Hardman-Rhyne, (to be published).
6. K. A. Hardman-Rhyne and N. F. Berk, (to be published).

THE ROLE OF MATERIALS AND SURFACE SCIENCE IN QUALITY ASSURANCE

John D. Venables

Martin Marietta Corporation
Martin Marietta Laboratories
1450 South Rolling Road
Baltimore, MD 21227

ABSTRACT

As a manufacturer of large high-technology aerospace structures, Martin Marietta Corporation is dedicated to applying advanced analytical techniques such as scanning transmission electron
microscopy (STEM), X-ray photoelectron microscopy (XPS) and specular
reflectance fourier transform infra-red spectroscopy (SR-FTIR) to
aid in quality assurance. The Corporate Research Center, Martin
Marietta Laboratories, has taken an active role in this effort and
in this paper we discuss examples of how these techniques have been
applied to achieve improved quality control at the manufacturing
level.

The use of adhesive bonding to join metal parts as well as
polymeric composites has increased dramatically in the past several
years. Recognizing that this joining technique can be used successfully only if careful attention is given to surface preparation
and avoidance of surface contamination, we have demonstrated the
value of sophisticated surface science equipment to assist in manufacturing control of these parameters. For example, it has been
shown that the high resolution capabilities of the STEM, when used
in the SEM mode, is invaluable for monitoring the effectiveness of
surface preparation procedures such as the Forest Products Laboratories (FPL) etch and the phosphoric acid anodize (PAA) processes
for treating Al prior to bonding. An extensive review of this work
has appeared recently and should be referred to for more details[1].
Briefly, it has been observed that the major function of these
etching and anodization processes, besides removing rolling mill
oils, etc., is to develop a microporous oxide on the Al surface
which interlocks with the adhesive, forming a much stronger bond

271

than if the surface were smooth. The importance of this microporous oxide cannot be overemphasized; we have shown, for example,
that, all other things being equal, the strength of a bond made
to a microporous oxide surface may be up to five times stronger
than if the oxide is devoid of microporosity. Since important
features of the microporous oxide are less than 200Å, we have found
it necessary to use the high resolution capabilities of a STEM
(30Å resolution in a SEM mode) in order to characterize the oxide
structures. Currently, the STEM is used for in-process control of
these surface preparation treatments and, in fact, has been of
considerable value for quickly detecting variations in processing
procedures that could have resulted in bonding problems had they
not been caught at the surface preparation level.

Advanced Surface analytical techniques have also been very
effective in avoiding problems with bonding composites-to-composites
Specifically, the use of FTIR and XPS has enabled us to identify
certain types of contaminants that can severely degrade bond
strength. The most important contaminants identified in these
studies are mold release agents that are used to allow separation
of a composite part from the tool upon which it is formed and cured.
Using both specular reflectance FTIR and XPS, Matienzo et al[2,3]
have shown that these release agents transfer to the composite
surface and can cause serious problems if attempts are made to bond
one cured composite part to another. In particular, it was shown
that as little as 12 atomic percent Si (as silicone) contamination
from mold release agents can cause significant reductions in lap
shear strength; at the 22% Si level the strength is an order-of-
magnitude less than uncontaminated controls. This work has led to
the use of certain types of peel plies which, when stripped off the
cured composite, leave behind an uncontaminated surface which yields
excellent bond strengths[2].

The above are only a few examples of the role that surface
science and materials science can play in assuring the integrity of
high technology aerospace structures. It has been gratifying to
witness, over the past ten years, the use of these sophisticated
techniques and talents of the surface and materials scientists to
assist the production and quality engineers perform a better job.

REFERENCES

1. Venables, J. D., "Review – Adhesion and Durability of Metal
       Polymer Bonds," J. Mat. Sci., 19, 2431 (1984).
2. Matienzo, L. J., Shah, T. K. and Venables, J. D., "Detection and
       Transfer of Release Agents in Bonding Processes," Proc. 15th
       Nat'l SAMPE Tech. Conf., p. 604, October (1983).

3. Matienzo, L. J., Venables, J. D., Fudge, J. D. and Velton, J. J., "Surface Preparation for Bonding Advanced Composites, Part I: Effect of Peel Ply Materials and Mold Release Agents on Bond Strengths," Proc. 30th Nat'l SAMPE Symposium, p. 302, March (1985).

CURE MONITORING FOR POLYMER MATRIX COMPOSITES

B. Fanconi, F. Wang, D. Hunston and F. Mopsik

Center for Materials Science
National Bureau of Standards
Gaithersburg, Maryland 20899

INTRODUCTION

Polymer matrix composites are desirable engineering materials owing to their high strength and stiffness as well as corrosion resistance. The strength and stiffness achieved in these materials on a per unit mass basis (specific properties) surpasses that for metals by factors of up to 5[1]. This has led to use of polymer composites in weight sensitive applications ranging from commercial and military aircraft to automobiles. Currently, the wider application of these materials is limited by product variability and the labor intensive, time consuming manufacturing processes. These factors are particularly troublesome in large volume production where the trend is towards more rapid processing that tends to exacerbate the problem of high rejection ratios. Recent advances in automated, computer controlled manufacture have been encouraging, although limited by the lack of suitable process monitoring.

The manufacture of polymer matrix composites involves complex chemical and physical processes that must be adequately controlled to produce desirable products[2]. Monitoring techniques and models to correlate monitoring data to improve processing are key aspects to increasing production rates and product quality. The lack of process monitoring methods also affects the rapid introduction of new and improved matrix resins. Processing variables are often selected by empirical trial and error methods which are time consuming and limit the introduction of better resins.

The major chemical change that occurs in processing is the transformation from liquid monomer, or prepolymer, to the solid, cross-linked polymer matrix. Chemical reactions that bring about this transformation are influenced by temperature, pressure, types and concentrations of accelerators, inhibitors, and monomers or prepolymers. The chemical nature of coupling agents on the reinforcement fibers is also involved. In addition, molecular mobility, viscosity, and the glass transition temperature will affect the extent of the cross-linking chemical reactions (degree of cure) that form the three-dimensional structure of the matrix.

The chemical and physical structure at the polymer matrix reinforcement interface is thought to have a significant effect on properties and performance. As the interfacial region encompasses an exceedingly small volume of material that is difficult to probe, there is little known about this important region. De-bonding at the interface can adversely affect mechanical properties of composites, and as this failure mode is influenced by the chemical and physical structure of the matrix at the interface, methods of probing the interfacial region during processing may have a significant impact on durability and performance.

The initial physical change that occurs during processing composites prepregs (fibers impregnated with uncured resin) is a decrease in resin viscosity associated with the elevated temperature of processing and the exothermic nature of the chemical reactions. The temperature increase accelerates the polymerization and cross-linking reactions, and once these processes become dominant the viscosity increases dramatically as shown in Figure 1. During the low viscosity interval it is crucial that cure conditions be carefully controlled to obtain proper flow for proper wetting and consolidation of the reinforcement plies. If the viscosity is too high insufficient flow is achieved for consolidation, if it is too low the resin may flow out of the part.

The narrow range of time over which the viscosity is in the right range describes the processing window at the temperature in question. Processing windows are described in terms of viscosity-time-temperature diagrams. Information of this sort is available from laboratory studies on neat resins and torsion braid analysis of the curing phenomena, but there are few, if any, on-line techniques currently available. Proper cure cycle design is a problem presently addressed empirically by developing cure procedures for each new resin and sometimes each new product. This approach is time consuming and not entirely satisfactory as evidenced by high rejection ratios. Failure to achieve proper flow during the early stages of cure is a major reason that many resin systems, otherwise superior to existing materials, are not used. In the latter stages of processing the viscosity and elasticity increase as shown in Figure 1. This behavior can also lead to problems due to high internal stresses at the fiber-matrix interface as a result of densification of the resin and differences in the thermal expansion coefficients of resins and fibers. If cure is too rapid these stresses may lead to microvoid formation, a cause of product variability.

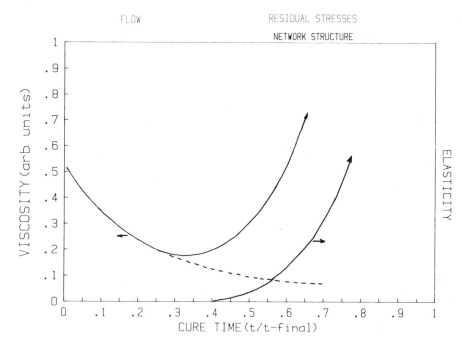

Figure 1. Viscosity and Elasticity Behavior of Curing Resin.

In this paper, we describe the potential of several techniques that are either directly applicable to process monitoring, or provide ancillary data to gain a better understanding of the chemical and physical processes that occur. The methods which are under investigation include Fourier transform infrared (FT-IR) to monitor chemical reactions that cross-link the monomer or prepolymer, dielectric spectroscopy and excimer fluorescence spectroscopy to examine molecular mobility and microviscosity, and an ultrasonics attenuation probe of the mechanical properties of thin films of curing resins. A brief description will be given of each method along with an analysis of its potential.

FT-IR STUDIES OF CURE CHEMISTRY

Both infrared and nuclear magnetic resonance (NMR) have potential as spectroscopic methods to follow the cure chemistry, as well as quantify the chemical compositions of starting materials, monomers of prepolymers, initiators, and accelerators. The importance of quality control of starting materials is addressed by other contributions to this symposium.   Of the two methods, FT-IR appears to be more sensitive, requires smaller amounts of material as a specimen, and is more rapid.   For these reasons, FT-IR was selected as the technique to monitor the chemistry of curing.   As an example of the chemistry involved, Figure 2   shows the basic cross-linking chemistry of a vinyl type polymerization.

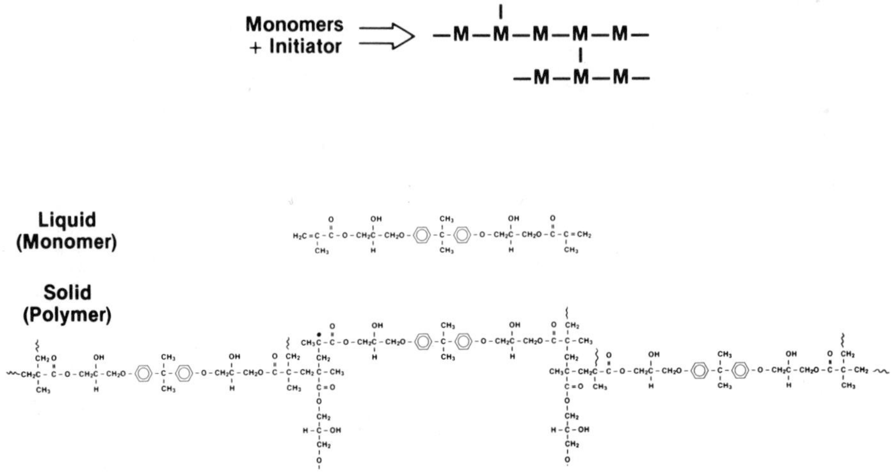

Figure 2.   Chemical Structure of a Curing Bis Phenol-A Glycidyl Methacrylate Resin.

In this case, the monomer, a bis phenol-A glycidyl methacrylate copolymer, cures near room temperature, and has application as a dental restorative material[3].   The cross-linking occurs through the carbon-carbon double bond, so that the concentration of these bonds decreases during polymerization.   The decrease in carbon double bond concentration is monitored by the IR intensity changes of the band associated with the double bond stretching vibrational mode.   Figure 3 illustrates that portion of the infrared spectrum in which the band in question appears at 1638 cm-1.   The two other bands in this portion of the IR spectrum are associated with the phenyl ring and would not be expected to change to any extent during polymerization.   The top spectrum of Figure 3 was recorded subsequent to processing, the middle spectrum was recorded after the specimen was exposed to a simulated use environment for 20 minutes, and the bottom trace is the result of subtracting the top spectrum from the middle spectrum.   The dip in the difference spectrum at 1638 cm-1 indicates that additional carbon double bonds were depleted when the specimen was exposed to the use environment.   The processing conditions were insufficient to complete cure.   Once the viscosity increases significantly, molecular mobility is greatly reduced and the chemical reactions proceed at much slower rates.

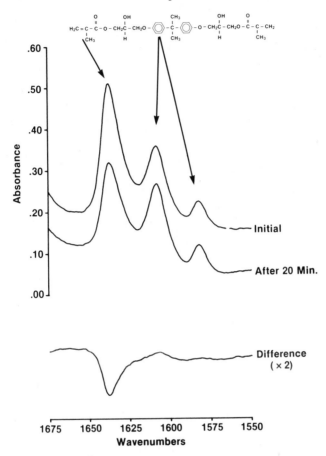

Figure 3.   Portion of the IR Spectrum of a Bis Phenol-A Methacrylate
Copolymer System.   Top Spectrum, Subsequent to Processing;
Middle Spectrum, after 20 Minutes of Post-Cure; Bottom
Spectrum Obtained by Subtracting the Top Spectrum from
the Middle Spectrum.

Through studies of this kind, insight is gained into the
chemical mechanisms associated with polymerization, the effect of
the molecular architecture of the monomer, and the dependence of
the rate of cure on accelerator and initiator types and concentrations.   For example, a higher degree of cure under similar processing
conditions can be achieved by using a monomer that is more flexible
than the rigid bis-phenol-A group.   The effect of substituting the
flexible triethylene glycol group for bis-phenol-A on the post cure
behavior is shown in Figure 4 in which the upper curve corresponds
to a monomer system that is 70% bis-phenol-A monomer and 30%

triethylene glycol derivative and the bottom line is the behavior of the resin system with the reverse composition. The various sets of points are for formulations in which the accelerator type and concentrations were different. It is seen that the flexibility of the monomer molecules has the dominant effect on the cure characteristics.

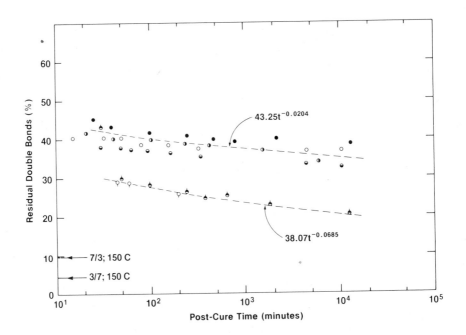

Figure 4. Post Cure Behavior of Bis Phenol-A Glycidyl Methacrylate Copolymer. Top Curve, 70% Bis Phenol-A Glycidyl Methacrylate and 30% Triethylene Glycol Dimethacrylate; Bottom Curve, 70% Triethylene Glycol Dimethacrylate and 30% Bis Phenol-A Glycidyl Methacrylate.

Thus, FT-IR provides quantitative data on the kinetics of the cross-linking reactions and the influence of the various parameters on cure. However, this method is not readily adaptable to on-line process monitoring, and its main contribution is to provide, in concert with other techniques, the type of information needed to establish the relationships between macroscopic effects (e.g. viscosity) and purely chemical changes. This FT-IR example did point out the importance of molecular mobility in achieving high degrees of cure. For this reason, methods that more directly measure molecular mobility were explored.

EXCIMER FLUORESCENCE DETERMINATIONS OF MICROVISCOSITY

In this method, which has considerable promise as an on-line procedure, fluorescence spectroscopy is combined with optical fibers as a means of interrogating the interior of a composite part as it is being manufactured. The basis of the technique is the sensitivity of the fluorescence spectra of some organic dyes to the local viscosity (microviscosity)[4]. Generally, the probe dye molecules would be added to the curing resins in low dilutions, and therefore would not be expected to affect performance properties. In some cases, the fluorescing chromophores may be part of the resin itself, or chemically linked to the resin monomer. Research to date has focused on soluble dyes; however, the other approaches have interesting possibilities, in particular for characterization of the fiber-matrix interface region.

It is not uncommon for organic dyes to have viscosity sensitive fluorescence yields. The decrease in rotational and translational mobility affects the nonradiative transition probabilities. Non-radiative decay of electronic excitations is a competing mechanism for de-excitation of excited electronic states. A more interesting possibility is the so-called excimer formation process[5]. In this process, two identical chromophores are chemically linked through a structure that permits sufficient molecular mobility so that the two chromophores can align themselves in a preferential arrangement to permit formation of a new electronic state in the time interval between electronic excitation of one of the chromophores and emission of fluorescent light. De-excitation from the excimer state results in fluorescence at a different wavelength than that from one of the two chromophores (monomers). Hence, the ratio of monomer to excimer fluorescence intensities is directly related to molecular mobility and local viscosity. An example of an excimer forming dye molecule is 1,3-bis-(1-pyrene) propane, whose structure is shown in Figure 5.

The absorption of light by one of the two pyrene groups produces an electronic excitation that has a lifetime of about 300-500ns. If during this time period the two pyrenes rearrange so that the ring moieties are in a sandwich configuration, an excimer state is formed and fluorescence occurs at a longer wavelength as shown in Figure 5.

The series of emission spectra of Figure 5, recorded as a function of increasing solvent viscosity going from A to E, shows how the excimer emission (longer wavelength band) decreases as the viscosity increases. The spectra have been scaled to constant monomer fluorescence intensity.

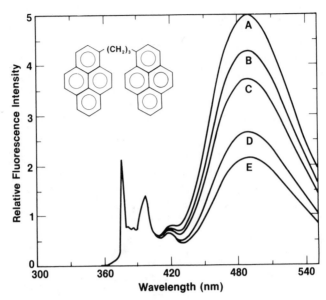

Figure 5. Dependence of the Fluorescence Spectrum of 1,3 Bis-
(1-Pyrene) Propane on the Solvent Viscosity. All Spectra
are Normalized to Constant Monomer Intensity. The
Viscosity Progressively Increases Going from A to E:
A, 0.44cp; B, 0.91cp; C, 1.36cp; D, 2.81cp; E, 4.0cp.

The manner in which such a probe might be utilized in a nondestructive measurement system is illustrated in Figure 6. An
optical fiber is used as a waveguide to propagate laser light of
the frequency needed to excite the probe molecule that is present
at low concentration in the prepolymer. The optical fiber is
envisioned to become part of the reinforcement system once processing
has been completed. Under suitable conditions, one of the glass
fibers of a fiberglass composite could be used as the optical fiber,
and no special fibers would be added to the curing part. The
evanescent wave of the propagating light probes the material surrounding the fiber to a depth that depends on the refractive indices
of the fiber and matrix. A probe molecule in the vicinity of the
fiber would be excited by this evanescent wave and its emission
could be viewed at right angles to the fiber direction, at the end
of a neighboring fiber, or at the propagating fiber end. Monitoring
the fluorescence provides a means of determining the changes in the
microviscosity near the fiber-matrix interface as curing proceeds.
At this point, the procedure as outlined in Figure 6 has not been
fully developed, although results of preliminary studies on component
parts have been encouraging.

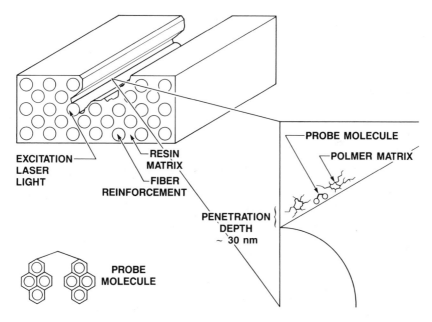

Figure 6. Outline of the Experimental Approach to On-Line Process Monitoring Using Optical Waveguides and a Fluorescing Probe.

The utility of the fluorescent probe technique for monitoring polymerization was demonstrated with poly(methyl methacrylate), PMMA, as shown in Figure 7[6]. These data were obtained using a commercial fluorimeter. The monomer-excimer fluorescence ratio is plotted as a function of polymerization time for three types of probe molecules and for polymerization in the presence of glass rovings. It is seen that the emission intensity ratios rise abruptly as the polymerization time increases.

Polymerization of an epoxy network has also been investigated by a similar method and the results are shown in Figure 8 where the monomer/excimer ratio is plotted for a stoichiometric mixture of EPON 828 and the diamine, 4,4 methylene bis(cyclohexylamine). The initial decrease in the intensity ratio is attributed to the decrease in viscosity that results from the initial rise in temperature as discussed in reference to Figure 1. As cross-linking proceeds, the viscosity increases owing to the growth in molecular weight and this is reflected by the rise in the emission intensity ratio. The dip in the ratio at longer times is attributed to changes in the monomer quantum yield that is also associated with increased viscosity. This effect is detectible only when the excimer emission approaches its minimum value.

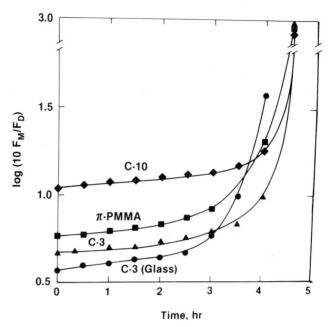

Figure 7. Ratio of Monomer (F_M) to Excimer (F_D) Fluorescence
Intensities in the Polymerization of Poly(Methyl Metha-
crylate). Curves are Labeled by the Dye Type: 1,10-
Bis-(1-Pyrene)Decane, (C-10); Pyrene-Labeled Poly(Methyl
Methacrylate), (π-PMMA); 1,3-Bis-(1-Pyrene)Propane, (C-3).

The use of optical fibers has been explored for an organic
dye, rhodamine-B, dissolved in an epoxy matrix in which a glass
fiber has been embedded. In Figure 9 is shown the emission spectrum
of this dye when viewed normal to the fiber axis. Attempts to
observe the emitted light at the end of the embedded fiber have not
been successful, and this appears to be related to the match in the
index of refraction of the glass fiber and the resin. To explore
the relationships necessary to trap some of the fluorescence by

the fiber, solvent mixtures having different refractive indices were used and the fluorescent light intensity at the end of the fiber was monitored. The other spectrum of Figure 9 was obtained with the optimal refractive index difference between the glass fiber and the solvent mixture. The emission from the dye is clearly seen as is the contribution from a glass filter used to block the laser light at the entrance slit of the monochromator.

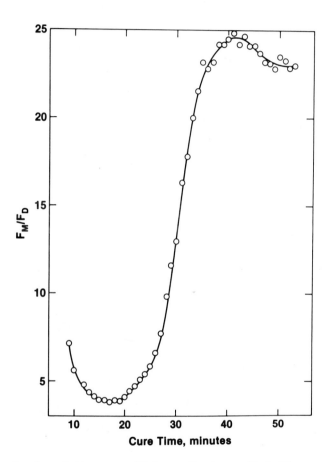

Figure 8. Ratio of Monomer (F_M) to Excimer (F_D) Fluorescence Intensities of 1,3-Bis-(1-Pyrene)Propane in Epoxy as a Function of Cure Time. Spectra were Recorded Using a Commercial Fluorimeter.

In summary, results from preliminary experiments have been encouraging. It has been shown that the emission from an excimer-forming dye is sensitive to viscosity in a polymerizing resin system, and under suitable conditions, glass fibers may be used as

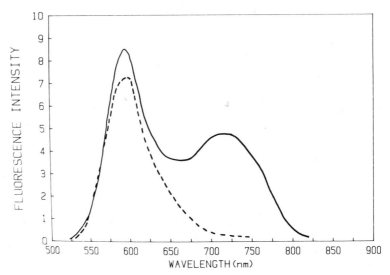

Figure 9. Fluorescence Spectra of Rhodamine B Dye Obtained Using an Optical Fiber. Dashed Curve, Viewed at Right Angles to the Fiber, Solid Curve, Viewed Out the End of the Fiber Emersed in a Solvent Mixture.

waveguides to interrogate in situ the microviscosity near the fiber-matrix interface. The viscosity is of critical importance in obtaining good consolidation of reinforcement plies during processing. The optical fiber-fluorescence technique also appears to be ideally suitable for on-line process monitoring.

DIELECTRIC SPECTROSCOPY

Through measurements of dielectric loss and conductivity, information on molecular mobilities can be obtained during the cure of polymer matrix composites. The information is particularly useful if one can measure dielectric loss over a wide range of frequencies so that a good picture of the molecular relaxations can be obtained. With the conventional approach to dielectric measurements, however, this is not possible because the sample properties change during the long time period required to conduct the tests at many different frequencies. The newly developed NBS Time Domain Dielectric Spectrometer provides a unique capability to overcome this problem[7]. The NBS apparatus measures time response of the charge flowing through a sample that is subjected to a step application of voltage, and then numerically transforms this data into dielectric loss as a function of frequency. In the time required for just one-third of a cycle for the lowest frequency of interest, the NBS device can determine the entire dielectric loss spectrum up to 10^4 Hz. Such a short measuring time means that the dielectric loss spectrum can be measured while the polymer is curing.

The capabilities of the spectrometer is especially valuable in the final phases of polymer cure. Near the end of the cure process, polymer relaxations have moved to lower frequencies and become smaller in magnitude. This occurs because the formation of a molecular network inhibits molecular motion. Although these relaxations are normally difficult to measure, they are very important because incomplete cure is a major problem in composite processing. The Spectrometer's ability to measure small losses accurately at low frequencies means that these relaxation processes can be followed to the completion of cure.

The usefulness of dielectric loss measurements in the initial stages of cure is complicated by the conduction of ionic species that are always present. This conduction will easily dominate any dielectric loss spectra. As a result, in the initial stages of cure a conductivity measurement is more useful. At the appropriate point in the cure, however, it becomes beneficial to shift to the more elaborate and informative dielectric loss measurement. Consequently, it is highly desirable to be able to make both measurements. To address this problem, additional modifications have been made to the Dielectric Spectrometer. These allow the measurement of conductance at a fixed frequency during the initial stages of cure and guide the use of dielectric measurements with this spectrometer. Examples of the dielectric response of a thermosetting resin subsequent to cure and following an extensive post cure treatment are shown in Figure 10. The resin system used in this case is identical to that used to obtain the data shown by the upper cure in Figure 4. It is seen from Figure 10 that the two responses deviate at low frequencies (long times) with the resin specimen prior to post cure showing a higher response. This has been attributed to greater ionic conduction in the incompletely cured resin. Through intercomparisons of the IR determinations of the extent of cure and the dielectric measurements sensitive to molecular mobility, the relationships between the chemistry and mobility can be obtained. In the next section it will be shown how mechanical properties measurements can also be made to compare with the above two determinations.

The mechanical properties of curing resins affect the degree to which consolidation of reinforcement plies and build up of

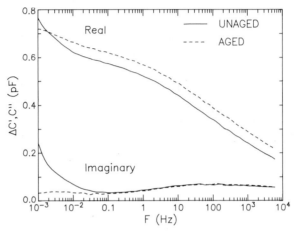

Figure 10. Dielectric Response of Bis Phenol-A Glycidyl Methacrylate (70% by Weight), Triethylene Glycol Dimethacrylate (30%) Copolymer: Solid Curves After Processing, Dashed Curves After Extended Post Cure Treatment.

residual stresses occur during processing. For this reason, it is desirable to measure mechanical properties of the curing resin. An ultrasonics wave propagation method has been developed for this purpose. This technique involves a plate or rod that is coated with a thin film of the sample (one or both major flat surfaces of the plate or one end of the rod)[8]. A shear wave is generated in the plate or rod, and the displacement in this wave causes a similar wave to be propagated in the sample. For anything but a very rigid material the attenuation in the sample is very high, and thus the wave propagates only an extremely short distance in the film. This means that the sample thickness does not matter, and therefore even thin films can be measured. Since the wave propagates such a small distance in the sample, this wave cannot be measured directly.

Instead, the wave in the plate or rod is measured. This wave is characterized first without the sample, and then after the sample is applied and as it cures. From measurements of the changes in the wave produced by the sample, it is possible to calculate the properties of the wave in the sample at any given time during cure. From that information the shear mechanical properties of the sample can be deduced. The shear properties are of most interest since the shear viscosity controls the flow behavior and the shear elasticity is very sensitive to curing. For a polymerizing material the elasticity can change from 0 dynes/cm^2 for a liquid to 10^{10} dynes/cm^2 for a glassy solid although the range is generally less at ultrasonics frequencies. The ultrasonic shear wave method provides a direct measure of the dynamic shear storage (elasticity) and loss (viscous) moduli. If desired, the absolute values of these parameters can be measured and followed during cure. Often, however, it is the relative changes that are of prime interest, and thus a simpler measurement can be made. In this case, only the amplitude attenuation coefficient (loss in wave amplitude per unit distance of wave propagation, α) for the wave in the plate or rod need be measured, and not the propagation velocity. This measurement can be done in a simple and automated manner. As the sample cures, the attenuation increases markedly. Figure 11, for example, shows curves for two different formulations of an epoxy-based film. The results show clear evidence for an induction time, a curing phase, and a fully cured film. The effects of increasing the accelerator concentration on both the rate and induction time can be seen. These parameters are even more obvious in Figure 12 where a first order type plot is made. Figure 13 shows a similar set of results except that, in this case, one sample shows clear evidence of a loss of adhesion between the film and the plate (decreased real contact area). Although the external appearance of this film gave no indication of this loss of adhesion, efforts to peel-off the film confirmed the poor bonding.

Although the ultrasonic shear wave technique shows considerable promise as a method to characterize the mechanical properties and curing behavior of a film, a number of problems remain to be solved. The most important is the current cost of the probe (the plate or rod). At the present time, they are too expensive to be discarded after a test and must be reused. This limits the type of materials that can be examined since many films are difficult to remove after curing. The second important problem area is the material for the probes. Presently, it is quartz, and thus studies are limited to the topic of interactions between the sample and a quartz surface. It would be very useful if other probe materials could be developed. A third major problem area is the current limitations on the range of temperatures and environmental conditions that can be investigated. This limit is set by the type of materials used in the probe and transducer. Expanding these limits would greatly increase the number and the type of samples that could be studied. The final

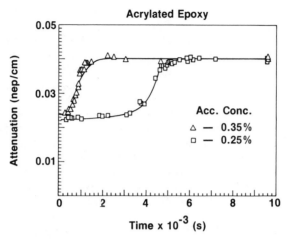

Figure 11.   The Average Attenuation Per cm as a Function of Curing
Time for Epoxy Based Systems with Different Accelerator
Concentrations.

area for future work concerns the variety of samples that have been
examined with the technique. At present the range of resins that
have been studied is very small relative to the potential candidate
materials. It would be beneficial to substantially broaden the
data base by examining a much wider variety of resin systems. All
of these problem areas appear to have solutions, and current work
is pursuing these topics.

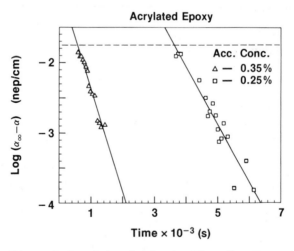

Figure 12.   First Order Data Analysis Plot for Data in Transition
Regions in Figure 11. Dashed Curve is Estimate for
Value of Line with Coating Before Initiation of Cure.
The Amplitude Attenuation Coefficient is α While α_∞ is
the Limiting Value of α at Long Times.

290

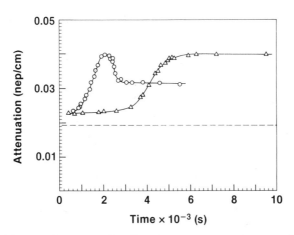

Figure 13. Effect of Loss of Adhesion Between Quartz Plate and
Thermosetting Resin Film on the Ultrasonics Attenuation
(Open Circles) Compared to Good Contact (Open Triangles).

REFERENCES

1. B. D. Agarwal and L. J. Broutman, "Analysis and Performance of
Fiber Composites," Wiley, New York, (1980).
2. R. G. Weatherhead, "FRP Technology, Fibre Reinforced Resin
System," Applied Science, London, (1980).
3. W. Wu and B. M. Fanconi, Polym. Eng. & Sci. 23, 704 (1983).
4. G. Oster and Y. Nishijima, J. Am. Chem. Soc. 78, 1581 (1956).
5. K. A. Zachariasse and W. Kuhnle, Z. Phys. Chem. NF 101, 267
(1976).
6. F. W. Wang, R. E. Lowry, and W. H. Grant, Polymer 25, 690
(1984).
7. F. I. Mopsik, Rev. Sci. Instrum. 55, 79 (1984).
8. D. L. Hunston, in "Review in Quantitative Nondestructive Evaluation," Vol. 2B, p. 1711, edited by D. O. Thompson and D. E.
Chiminti, Plenum, New York, (1983).

PROCESS CONTROL AND MATERIALS CHARACTERIZATION

WITHIN THE STEEL INDUSTRY

James R. Cook and Brian G. Frock

ARMCO Research
703 Curtis Street
Middletown, Ohio 45043

ABSTRACT

Within the steel industry, chemical, thermal and mechanical
processing of materials are the means by which the final properties are developed within the material. Historically, these procedures often evolved as an art perfected by practice over an
extended period of time. The processes can be complex, interactive, and extremely rapid and though the processing steps may not
always be completely understood, the outcome is generally known
and controllable within limits. The control problem is complicated by the unhappy circumstance that important processing
parameters may not be measured and in some cases are
unobservable.

The approach to many of these control problems may lie in the
development of sensor systems which are able to comment more directly on the controlled parameters. Some can be approached by
using multiple sensory information, mathematical algorithms, and
inference. Other problems can benefit from the use of a
reference process model, particularly if imbedded within the control system. Some control problems are best solved by modifying
the process to something which is inherently more controllable.

INTRODUCTION

When we were asked to discuss in-process characterization
within the steel industry we realized that we faced a significant
challenge: our individual expertise is not in non-destructive
evaluation techniques. Fortunately, we did have some in-house
experts with whom we were able to consult. However, after some
discussion we realized that we still had a problem: according to

our experts we do not do any non-destructive evaluation for in-process characterization!

We thought about that presumably facetious comment, and decided that the reasons for not doing significant in-process characterization were both economic and technical in nature.

The attitude which pervaded the steel industry seems to have been that in a competitive environment one could simply not afford the cost associated with that characterization. This is especially true of the portion of the industry producing commodity steel products. Of course that attitude is myopic, as Demming and the Japanese examples are proving.

The technological gaps also have been recognized, and are being systematically and aggressively attacked by various Sensor Development Task Groups under the auspices of the American Iron and Steel Institute (AISI).

We will describe some of the processes, problems, and approaches which are used in the steel industry and mention briefly the efforts of five separate AISI task groups. One of these will receive special note as we describe some of the work underway at the National Bureau of Standards (NBS) and at Battelle Pacific Northwest Laboratory (PNL). The work at PNL is being funded jointly by AISI and the Department of Energy (DOE).

PROCESSES AND PROBLEMS

Within the steel industry, chemical, thermal and mechanical processing are the means by which the final properties are developed within the material. Historically, these procedures often evolved as an art perfected by practice over an extended period of time. The processes can be complex, interactive, and extremely rapid and although the processing steps may not always be completely understood, the outcome is generally known and controllable within limits. The control problem is complicated by the unhappy circumstance that important processing parameters may not be measured and in some cases are unobservable.

Material characterization within the industry involves a number of parameters which must be closely controlled to develop the attributes critical to the material's end use. These final properties are the net result of a series of processing steps aimed at the development of the appropriate microstructure, geometry, and surface characteristics.

There are many processes within the steel industry which benefit from characterization. This audience is most familiar with methods and procedures for the interrogation and qualification of materials well along in the production process. We would like to discuss the techniques, and limitations as applied to the early stages of steel production listed in Table 1.

Table 1.

- Basic oxygen steelmaking
- Secondary steelmaking
- Ingot & strand casting
- Liquid & hot centre rolling
- Direct rolling
- Hot charge & incremental heating
- Thermal/mechanical processing

Table 2.

Direct Sensing
- Speed & reliability
- Accessibility
- Calibration

Driving Forces
- Basic understanding
- Improving controllability
- Computing power

Indirect Sensing
- Inference
- Sensor augmented models
- Model augmented sensors
- Qualify sensory input

Table 3.

Controlling the Process:
- Enhanced quality
- Productivity
- Efficiency

Feasible Strategies:
- Operate more efficiently
- More complex products & processes
- Replication defines properties

The first area is steelmaking, and although there are many different processes, we will concentrate on the Basic Oxygen Process. Irreducible errors in estimating charge weights, and limited accuracies in sampling and analyzing the charge components result in an imperfectly defined initial charge. Methods of measuring, and modifying conditions during the crucial refining stages are important.

The steel proceeds through a number of secondary refining steps in which it is essential to characterize the thermal and chemical states in order to make minute adjustments.

The steel is then cast into ingot, slab or billet form. The motion of the solidifying front and the bulk temperature distribution have an enormous influence on microstructure, surface qualities, inclusion distribution, and mechanical properties. After casting and cooling, much of this material will undergo a series of thermal and mechanical processing steps intended to develop the requisite metallurgical and mechanical qualities. An intimate understanding of the influence of these treatments on microstructure is important, and allows the use of optimized procedures such as liquid centre rolling, direct rolling, and incremental heating[1].

CONTROLLING THE PROCESSES

The future direction of the steel industry is being dictated both by customer demand and foreign competition, Table 2. The route to improved competitiveness and profitability is through enhanced quality, productivity, and efficiency. The only feasible strategy is to operate more efficiently and to move towards more sophisticated products. These products would require more complex processing and the processing steps and the properties we are trying to develop may be close to the limits of controllability.

The material properties developed during the processing of steel are largely determined by the ability to replicate processing parameters which have been established empirically. For highly sophisticated materials with more complex processing, relatively minor variations can have an inordinate influence on ultimate product properties.

Direct measurement of the material parameters which control the properties of interest is not always possible. In many cases a reliable and sufficiently rapid sensing method is simply not available or is made difficult due to problems with accessibility, maintenance, or calibration in a hostile environment, Table 3.

Modernization of steel processing is underway, and in large part is being aided by an improved understanding of the basic mechanisms and an enhanced ability to control the processes. A major impetus is the accessibility to sufficient computing power at the process levels.

Table 4.

- New & improved sensors
- Model reference (inference)
- Faster computers
- Smarter algorithms
- Change the process

Table 5.

- Develop a method of determining thermal state
- Demonstrate a practical device
- Operate in real-time
- Non-contacting/non-disruptive
- Signal processing/information extraction
- Computer aided/algorithms
- Prior information/ancillary measurements
- Process models

Table 6.

- Quantify economic incentives
- Coordinate development
- Generate funding (AISI/DOE/NBS)
- Consultation
- Qualify samples
- Support research associates
- Technology transfer to industry
- Provide test sites

This computing power and the ability to infer information by means of indirect sensing allows the use of sensor augmented process modelling approaches in which neither the model nor the sensory inputs needs to be perfect. The control system uses a reference model to project present and future processing conditions, and rejects or accepts sensory information on the basis of reasonability.

Most of the thermal and mechanical processes used in the industry are sufficiently well understood to be described by process models of one sort or another. These can take many guises from the extremely complex, and costly off-line technical models to the more specialized and restrictive real-time reference models which control engineers use in computer control systems. The control system designer is constrained by several factors:

- his ability to reliably sense by direct means or by inference the parameters to be controlled;
- his ability to comprehend the process well enough to provide an adequate model description;
- and his ability to install that representation in a practical (affordable) computer system.

There are several possible solutions, Table 4:

- one could develop improved sensors which comment more directly on the parameters to be controlled;
- or infer the controlled parameters using model reference by developing off-line models fully embodying all that is known and understood in the process, and use more restrictive real-time models to emulate the process over narrower ranges of interest;
- or possibly use a priori information, multiple sensor input, knowledge-based adaptive algorithms, or other aproaches to enhance computing efficiency;
- or one could alter the process to one that is inherently more observable and controllable and mathematically more tractable (ingots poured into cylindrical molds of semi-infinite length are convenient mathematically).

The recent proliferation of inexpensive, powerful, and intrinsically more reliable computer systems is of enormous consequence: the effort once devoted to describing model fundamentals can now be devoted to completeness. Alternatively, more exhaustive optimization procedures can be attempted using the model reference approach.

Observability problems are being addressed by use of inferential methods (generally model-based) and by the development of improved sensors which comment more directly on the controlled parameter. A new generation of smart sensors which combine multiple sensor inputs with on-board intelligence and powerful self-tuning and adaptive control algorithms are becoming available.

298

Table 7.

Tomographic Method

- Handles general geometry,
  arbitrary temperature distribution

- Sensitive also to material properties,
  porosity, grain size, microstructure,
  thermal-mechanical treatment

- Fails for slab geometry with
  gradient normal to surface

Table 8.

Slab Reheating Process:

- Accommodate hot connection
- Use residual energy
- Reference model approach
- Imbedded & real-time
- In-furnace sensing
- Control firing profile & discharge rate
- Measure slab rollability

Table 9.

Hot & Cold Working

Objectives:

- Geometric..
  - Width   Flatness (shape)
  - Thickness (gauge)   Length
- Metallurgical..
  - Finishing temperature   Grain orientation
  - Coiling temperature   Magnetic quality
  - Inclusion distributions   Core loss
  - Grain size distribution
- Mechanical...
  - Surface condition (roughness)   Coating thickness
  - Surface flaws   Coating uniformity
  - Subsurface flaws   Coating adherence
  - Porosity & laminations   Residual stresses

The salient feature in all of these is the use of computing power to both control the device, and to interpret the indications.

SENSING THE PROCESSES

The General Research Committee of the American Iron & Steel Institute (AISI) recognized that many of the processes used in the steel industry were being compromised by inadequate sensors and developed a list of sensor needs[2]. Some of these were expected to have a substantial impact on productivity or quality and were given special priority; industry supported groups have been established to accomplish sets of tasks such as those listed in Table 5.

One of these groups has been working on a method to determine the temperature distribution within solid and solidifying bodies. The task group was established to quantify the economic incentives, and to coordinate the development. The group was involved in the search for funding, and continues involvement through routine consultation and by means of the Research Associate program at NBS. That program, and the members of the task group provide the linkage by which the important technological advances which this efort is sure to produce can be transferred back to industry. The group will also arrange for a demonstration of the sensing methods in a real-world environment once the device is developed.

The stated aim, Table 6 is the development of a method for determining the thermal state, but the ultimate aim is the demonstration of a practical device. Ideally, this device would operate in real-time without contacting the workpiece, and would not disrupt normal operations. It was expected that the measurement task might be more tractible if combined with signal processing and information extraction methods. Some of these might well involve the use of computers and complex algorithms. Prior information on geometry, and auxiliary measurements of surface temperatures and physical dimensions might be important. Process models of cooling, solidification, and reheating could provide crucial information, and insight, which might make a prospective technique more practical.

In our task group activities we uncovered four basic techniques which held some promise of providing the type of information which we felt essential to the task: eddy current methods, Gamma radiography and X-ray diffraction, neutron thermalization, and ultrasonic methodologies. After considerable deliberation we determined that only ultrasonic methodologies, perhaps coupled with computer aided reconstruction techniques offered any realistic prospect for success.

300

Table 10.

Conclusions:

- Steel industry is evolving
- Adapting new technologies
- Innovative processing methods
- Improved process control
- New materials with better
  or
- New and unusual properties

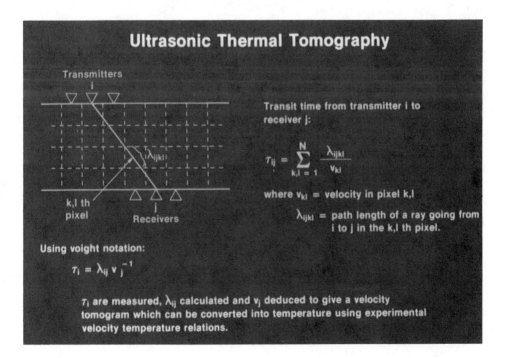

Figure 1.

Ultrasonic pulses propagate through a solid body with a velocity which depends locally on the material properties, microstructure, temperature and stress distributions, Figure 1*. By mathematically decomposing the solid body into elements one can extract information on the average velocity within each element. The temperature distribution in the solid body can be reconstructed if the dependence of ultrasound velocity on temperature can be determined, and provided that dependence is not overwhelmed by variations in the other important material parameters. The temperature dependence should be monotonic and single-valued; however, that requirement could be waived if the expected temperature range is limited.

The reconstruction algorithms are relatively well-understood, but the sensitivity to microstructure, grain size and porosity is an important complication. One AISI task group is taking advantage of that sensitivity in devising methods for the characterization of gross porosity in hot material[3].

A successful method for sensing internal temperature distribution would find application in strand casting to monitor the temperature distributions in the solidifying strands and to control the cooling sprays.

This sensor might also be used in the slab reheating process to control incremental heating and to permit the use of optimum heating and scheduling practices. In slabs, however, temperature gradients are nearly normal to the broad faces, and the temperature isotherms are parallel. In this instance the tomographic method fails because the different ultrasonic paths sample the isotherms in the same proportion and provide no additional information, Table 7.

NBS has been studying a second approach to this problem by a method referred to as Dimensional Resonance Profiling. Variations in the elastic modulus or density of a body can be reconstructed from measurements of resonance frequencies. A simple relationship exists between the Fourier expansion coefficients of these parameters[4] and the measured values of the fundamental and overtone frequencies. For slab geometries this approach should allow the reconstruction of the spacial variations in modulus and density from which the velocity of the ultrasound can be computed. That profile would then be converted to a temperature reconstruction in the manner described earlier.

This method has the advantage that low frequencies with the attendant low attenuation can be used. The theory shows that only even order Fourier coefficients can be extracted, implying that the reconstructed temperature distribution will appear symmetric about the centre whether or not that is the actual case. Another limitation is that the absolute value of the temperature is not recoverable and measurements of surface or average temperature are needed. This is another instance in which a priori information and auxiliary measurements may be critical to the success of the technique.

302

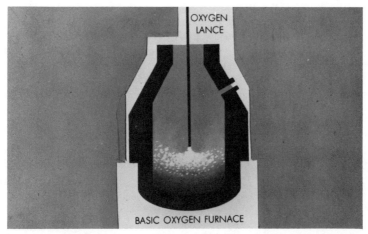

Figure 2.

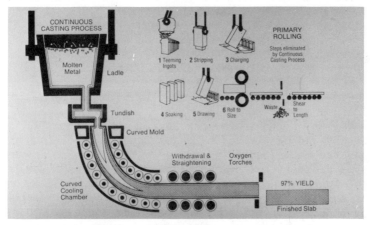

Figure 3.

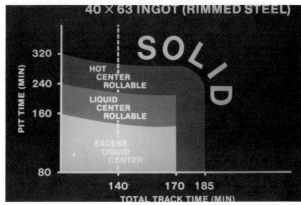

Figure 4.

303

CLOSING THE LOOP

The perception of steel processing as a series of discontinuous operations which incrementally develop the desired properties in the product may be changing as more of those processing steps are combined into single facilities for the continuous processing of material. One of the important factors in this development is an improving ability to sense and control the processes.

STEELMAKING

The steel industry relies heavily on oxygen steelmaking processes; these operations are rapid, comparatively violent, and at the edge of controllability much of the time, Figure 2. Slapping, surges, lance fouling, and ejaculation of material attest to the limits on control, yet the refining trajectory is inherently stable and self-correcting. Even though that trajectory may make large excursions, the melt will reach an endpoint with most of the scrap melted and achieve a reasonably high temperature and low carbon content with the impurities dissolved in the slag system. The control objective is to reach these desirable conditions simultaneously. The type and amount of corrective action critically determine the quality of the product and the efficiency and economics of the process.

The charge is thermally and chemically balanced by means of a charge model with adaptive coefficients to accommodate unknown and unmeasured variances in charged materials. The refining trajectory can be altered by adjusting the partition of oxygen between the metal and slag systems. In the Basic Oxygen Process this is accomplished by controlling the momentum and penetration of the oxygen jet through the slag layer.

A process model fully describing the thermochemical evolution of the steel during the melting and refining stages would be complex and unprovable. Happily, for control purposes, the rate of decarburization and the temperature rise during the latter stages can be represented by simple mathematical expressions. A single simultaneous measurement of bath temperature and carbon is sufficient to estimate the refining state and to project the type and amount of corrective action needed.

One AISI sensor task group is working with DOE support at Los Alamos to perfect a device for the in-process analysis of molten metal. This device will prove invaluable in applications such as secondary refining for which rapid, accurate and repetitive analyses may be required[5].

STRAND CASTING

There has been a great deal of recent effort in process development and refinement for the strand casting of steel;

innovative methods of casting are under study or in limited production use. These advanced processes often lack the means of
sensing critical information rapidly enough to provide even a
modicum of control. Diagnostic information generated after the
material has been produced is the only means of modifying the
process.

In conventional strand casting, intimate control of the flow
of liquid steel between ladle, tundish and mold, Figure 3, is
important in producing inclusion free "clean" steel. One AISI
task group is working towards the development of a method of
detecting inclusions in pouring streams[6]. There has also been
a recent demonstration of the ability of boron nitride ceramic
sheaths to provide protection for directly immersed sensing
devices[7].

Primary heat removal is accomplished in the mold system to
provide a solidified shell of material capable of retaining the
liquid metal. The mold is usually oscillated to improve the heat
transfer consistency and strand surface qualities. An important
consideration is the nature of the casting flux which is used
both for its thermal transfer characteristics and for the
lubricity which it imparts to the solidifying shell as it passes
through the mold system. A lubricity measuring system has been
developed and has proven effective in preventing strand rupture
and breakout[8].

Secondary cooling takes place along the various segments of
the casting machine critical to avoid sensitive ranges at points
of high stress. This can be accomplished by controlling the
strand velocities and spray cooling rates, but to be effective
spray cooling efficiencies must be predictable and reproducible.
A recent innovation has been the development of air-mist cooling
sprays which provide a range of control not possible with conventional sprays.

Electromagnetic stirring is used in one form or another on
many casting machines, and improvement in internal quality as
measured by centre line segregation is often reported. There is
no fundamental understanding of the precise mechanism by which
the metallurgical structure is refined, and how this influences
metallurgical quality in particular products.

INGOT PROCESSING

For some steel producers ingot processing remains a viable
approach and will continue to be used on at least some portion of
the product mix for the foreseeable future. One integrated system was developed to track the movement of ingots and relay information between melt shops and soaking pit operations. A
hierarchical network of computers exchanges information and controls the processing in each area. At the heart of this system
is a real-time imbedded mathematical representation of the be-
haviour of ingots during cooling, solidification and reheating

which is used to dictate the soaking pit charging and firing strategies. Later, it was discovered that special yield, quality, and energy improvements might be realized by drastically modifying the processing of ingots by purposefully producing slabs with substantial liquid centres, Figure 4. Extrordinarily high yields, improved quality, energy savings of more than 80%, and throughput advantages exceeding 3:1 compared with the best of previous practices were realized[1].

The reference model in the control system allowed the process to be moved to a regime which had not been considered when the system was designed. The model provided fundamental insight into the problem, and the opportunity would not have been recognized without the model projections.

Recent efforts have been devoted to closing the loop using rolling forces as feedback. That "rollability" information is used to develop statistical process control parameters which comment directly on the overall system performance, the adherance to standard operating practices, and the accuracy of information available to the system. A secondary objective is to improve yield and throughput by dynamically modifying the drafting procedures to take the largest draft possible, Figure 5.

REHEATING PROCESSES

One of the keys to enhanced competitiveness is energy efficiency within the constraints of adequate quality. One of the main tenets of our philosophy is to recover the energy which would otherwise be lost, and to eliminate cooling and reheating cycles which serve no metallurgical purpose. One process we have been considering is slab reheating, Table 8. Slabs produced by both continuous casting and ingot processing arrive in an assembly yard but scheduling and mechanical and metallurgical constraints often allow the residual thermal energy of these slabs to be dissipated. One way of avoiding this loss and the need for subsequent reheating is to roll ingots directly to strip. This introduces some scheduling problems, which are not insurmountable, but the addition of cold slabs and slabs arriving from the caster complicates matters.

The approach we are developing is to use a process model of the slab during the cooling, reheating, soaking and reduction phases to precisely control the reheating process. There are a number of constraints which must be recognized such as production rates, variable slab thicknesses, widths, and grades. Imperfect information from the slab furnaces, changes in surface conditions, emissivities, surface scaling, etc. add complexity. Our biggest problem is one of monitoring fully the heat transfer mechanisms within the furnace. In order to obtain reliable information on heating rates under operating conditions we use a

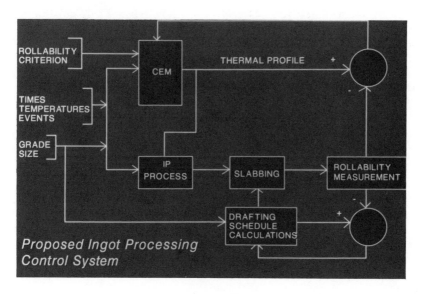

Figure 5.

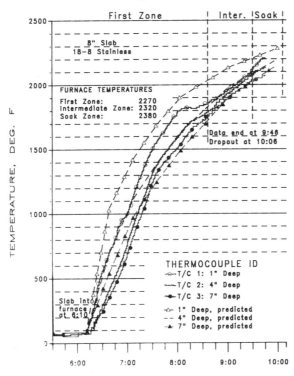

TIME (24—HOUR CLOCK)

Figure 6.   Thermophil Test, Sept. 13, 1984 Multi-Zone Continuous
Reheat Furnace

recently developed device[9] which rides with the slab during the reheating process. This device digitizes and records the voltages generated from thermocouples imbedded at various depths within a sample slab. The information is recovered along with the slab and recording device at the end of the heatup cycle. A typical set of data recovered from a stainless steel slab processed in a multizoned continuous furnace is shown in Figure 6.

HOT AND COLD WORKING

Hot and cold working accomplishes both geometic and metallurgical objectives, Table 9. Not only are the width and coiling temperatures of the strip important, but also the thickness, shape and surface finish. Sensing systems to measure these parameters increasingly are taking advantage of computing capability, multiple sensor inputs, and powerful algorithms to more reliably extract the needed information. One recent device for gauge control relies on a measurement of mill stretch, accommodates changes in incoming strip thickness and rollability, and by means of an adaptive algorithm extracts the actual rolling force from the eccentricity forces generated by external disturbances.

Surface flaws and dimensional control in processed materials are of critical importance in certain product lines. One AISI sensor task group is proposing to develop a method for in-line inspection of surface defects in cold rolled strip[10]. The problem is one of detection and classification of defect type and severity. The approach is to locate all detectable flaws, categorize, record and analyze. The amount of information could be insuperable unless adequate means of discriminating, classifying and differentiating defect classes are developed.

One recently available device for the measurement of strip width is typical of the new breed of smart sensors[11]. This stereoscopic width gauge employs two solid state charge coupled devices to provide information to an on-board computer system. Special processing techniques are used to remove camera imperfections and to compensate for background noise. The system employs an efficient algorithm to detect strip edges with unusually high accuracy, compensates for strip edge flutter, and provides information for real-time control purposes.

CONCLUSIONS

The steel industry will continue to evolve and to adapt new technologies as they become available. Innovative processing methods promising lower costs, better process control and producing materials with equivalent or better, or new and unusual properties are being developed. The full exploitation of these innovations will take many years and depend in large part on the availability of sensing methods and novel control strategies.

Modernization is being aided by an improved understanding of the basic mechanisms and an enhanced ability to characterize the materials and control the processes. The inadequacy of existing sensors for some critical applications has been acknowledged and sensing systems are being developed to meet those needs. Even so, some limitations are fundamental in nature; these may be overcome by model reference and inferential approaches, or by altering the processes to improve the ability to observe and control the important characteristics.

REFERENCES

1. J. R. Cook, T. R. Dishun, and D. F. Ellerbrock. "A Systematic Study of the Factors Influencing Ingot-to-Strip Yield, Energy and Quality". Iron & Steelmaker.
2. R. Whiteley. "Steel Industry Priorities for Process Control and Sensor Development". AISI Research Task Force Report, Jan. 8, (1981).
3. AISI Task Group 3-4; C.D. Rogers Chairman. "Automatic Detection of Pipe and Gross Porosity in Hot Steel Billets, Blooms or Slabs".
4. "Progress in Development of Ultrasonic Sensors for Monitoring Hot Steel Product", M. Linzer, B. Droney, F. Mauer, S. Norton, C. Rogers, R. Rudolph, K. Sandstrom, J. Toth, & H. Wadley, private communication, to be published. Report on some aspects of NBS efforts for AISI Task Groups 3-4 (above), and 5-4, "Rapid Measurement of Temperature Distribution within a Solid or Solidifying Body of Hot Steel". J.R. Cook, Chairman.
5. "Rapid Analysis of Steel using Laser-Based Techniques", D. Cremers, F. Archuleta, & H. Dilworth, private communication, to be published; AISI Task Group; F. Achey Chairman.
6. AISI Task Group 4-2; D Huffman, Chairman. "Detection of Slag in Liquid Steel During Teeming".
7. G. Bryant, J. Swab and T. Hynes. "Development of a Sheath for Sensor Protection in Molten Steel Applications". U.S. Army Materials & Mechanics Research Centre; AMMRC - DOE Agreement #DE-AIOI-82-CE40552.
8. "Mold Lubrication and Oscillation Monitoring for Optimizing Continuous Casting", B. Mairy, D. Ramelot, M. Dutrieux, L. Deliege, M. Nourricier, private communication, to be published.
9. Thermophil STOR Datalogger, Ultrakust-Geratebau, GMBH & Co., Ruhmannsfelden, NDB.
10. AISI Task Group 4-5; W. Wilson, Chairman. "On-Line Inspection for Surface Defects on Hot and Cold Strip".
11. Stereoscopic Width Gauge; European Electronics Systems Limited, Maldon, Essex, United Kingdom CM9 6SW.

* Drawing courtesy of H. Wadley, National Bureau of Standards.

APPLICATION OF CURRENT NMR TECHNOLOGY TO MATERIALS

CHARACTERIZATION

Douglas P. Burum

Bruker Instruments, Inc.
Manning Park
Billerica, MA 01821

INTRODUCTION

Let me begin by clearing up a point of confusion which has arisen fairly often recently: NMR stands for "Nuclear Magnetic Resonance," not "Nuclear Medical Resonance." Many of you, I'm sure, are familiar with the usual application of NMR to the chemical analysis of liquids and solutions. In such cases a well resolved spectrum is produced in which the dominant interactions determining line positions and splittings are the magnetic shieldings of the electron clouds surrounding the nuclei and the indirect couplings of neighboring nuclei via chemical bonds. Integration of the spectral lines yields quantiative information in terms of relative members of nuclei.

Occasionally, someone asks me: "What's new in NMR these days"? My feeling is that solution-state FT-NMR has grown in the recent past mostly in a quantitative way. By that, I refer to continuing improvements in resolution and sensitivity, and in the power and automation of the data system. There have, however, been a number of improvements in experimental technology such as pulsed cross-polarization and decoupling techniques and so-called "2-D" experiments such as COSY, which correlates a spectrum against itself and reveals which nuclei are coupled to which.

SOLID-STATE NMR

In the case of solid samples, on the other hand, a tremendous revolution has occurred in the last few years in a qualitative sense, i.e., in the types of information which can be extracted. An

311

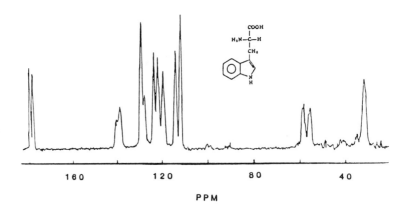

Figure 1. Spectrum obtained using CP/MAS.

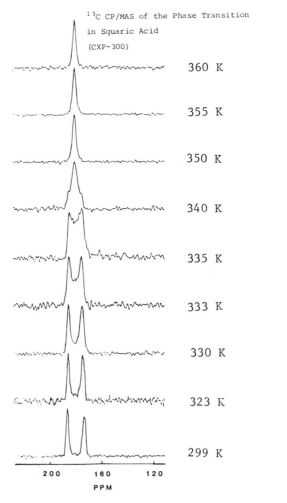

Figure 2. Squaric Acid is characterized as a function of temperature

interesting historical fact which is seldom mentioned is that
solid-state NMR is just as old as solution-state NMR. In fact,
Block and Purcell shared the Nobel Prize for the discovery of NMR
in the late 1940's, with Block having used water as his sample and
Purcell having used paraffin.

Nevertheless, solution-state NMR quickly gained prominence as
an important technique for chemical analysis, while solid-state
NMR was confined to some limited studies of molecular dynamics via
either relaxation times or line shape analysis. This is because
the direct dipole-dipole coupling between nuclei, which is averaged
to zero in solutions by rapid molecular motion, dominated and
obscured the chemical information in solids.

It was not until about 5 years ago, that improved hardware
performance and more sophisticated techniques finally made chemical
analysis of solids via NMR a practical reality. By far the most
successful technique to date is the "Cross-Polarization and Magic
Angle Spinning," or "CP/MAS" experiment. It is applicable to nuclei
such as ^{29}Si, ^{27}Al, ^{31}P, ^{15}N and most importantly, ^{13}C. In this
method a thermodynamic cross-polarization technique is used to
enhance the ^{13}C signal and to make use of the faster proton relaxation time. Subsequently, brute force, in the form of high power,
continuous rf irradiation, is used to decouple the protons from
the carbon spins while the ^{13}C FID is acquired. Throughout the
experiment, the ^{13}C chemical shift anisotropy, which would otherwise
cause the resonance lines to overlap and obscure one another, is
removed by physical rotation of the sample about an angle inclined
at 54.7^0 from the vertical. This is the "Magic Angle," so-called
because most people fail initially to see why this angle should be
significant. However, it is in fact possessed of special properties,
such as being the angle which causes the first Legendre Polynomial
to vanish and the angle which a vector can form simultaneously with
all three co-ordinate axes.

Figure 1 shows a typical spectrum obtained using CP/MAS, namely,
the amino acid, L-tryptophan. Although the lines are somewhat
broader, the spectrum is clearly of the same nature as is obtained
routinely from solutions by NMR. An application of this technique
to materials characterization is illustrated in Figure 2, in which
a phase transition of squaric acid is characterized as a function
of temperature. Molecular motions of most types can be detected by
this technique, such as in the case of Bullvaline in Figure 3, which
undergoes a conformational exchange motion at room temperature that
is suppressed at -50^0C, thus narrowing the NMR lines.

The relative degree of crystallinity of a solid is also readily
determined in most cases by NMR. In Figure 4, two spectra of
aspirin are shown: the first taken directly from the bottle and the
second after sublimation. The change in crystallinity is directly

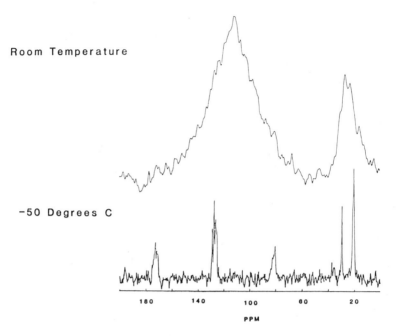

Room Temperature

−50 Degrees C

180 140 100 60 20

PPM

Figure 3. Bullvaline undergoes a conformational exchange motion at
room temperature.

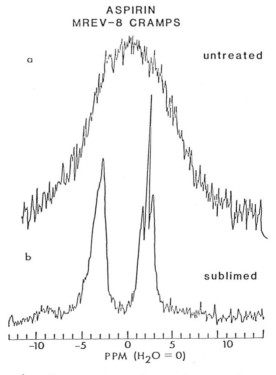

ASPIRIN
MREV-8 CRAMPS

a

untreated

b

sublimed

−10   −5   0   5   10

PPM (H₂O = 0)

Figure 4. Two spectra of asprin are shown.

314

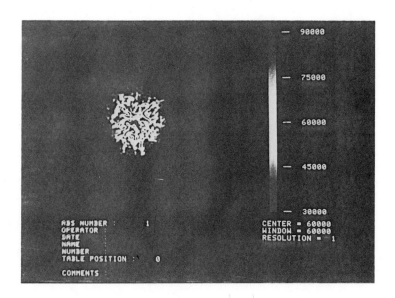

Figure 5. Cured by aeromatic amine.

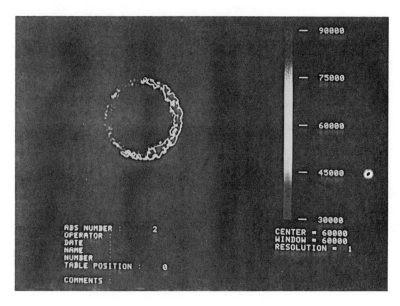

Figure 6. Cured in a anhydrous fashion.

reflected in the reduction of the linewidths in the second spectrum. By the way, the spectra in this figure are proton spectra, which were obtained by combining a pulsed, homo-nuclear decoupling technique with magic angle spinning. A more typical spectrum of, e.g., a polymer would consist of a sharp peak rising out of the center of a broad line. Deconvolution of the two components would then yield the relative crystallinity.

IMAGING

Today, when NMR is mentioned, a mental image is often formed of a huge magnet with someone's legs protruding out of it, and a cross-sectional image of the person's head or chest being formed on the console display. Indeed, NMR imaging of proton density in biological samples such as mice and men has already become an NMR application of overwhelming interest and importance, and will clearly continue to be so into the indefinite future. For small samples, resolution of 0.2 mm or less, can be routinely obtained, prompting some people to become slightly over-enthusiastic and refer to the technique as NMR "microscopy."

While biological imaging receives the majority of the attention, a related application which should be of great interest to materials scientists, namely imaging of solids, is slowly being born. So far, the most successful applications have involved imaging of liquids which have penetrated into a solid, but it would appear that true solid-state imaging is not far in the future. The next five figures, courtesy of Shell Development Company in Houston, TX, illustrate the current state-of-the-art.

Figures 5 and 6 are cross-sectional images of glass reinforced epoxy resin composite rods which were approximately one inch in diameter and 25 cm long. Both were submerged in water at 93^0C for several weeks and, as a result, absorbed about 1-2% water by weight. The difference was that in Figure 5, the sample was cured by aeromatic amine while in Figure 6, the sample was cured in an anhydrous fashion. The result was that small thermal shrinkage cracks in the first sample allowed the water to penetrate to the center of the rod, whereas the water was confined to the outer region in the second case.

Figure 7 presents a time resolved study of the penetration of Toluene into a 1.5 cm x 5 cm polystyrene rod. The progressive absorption and the formation of a large stress relief crack are clearly evident and diffusion constants can be readily determined. The combination of more traditional NMR techniques with imaging is illustrated in Figure 8. Here the proton density image has been

316

weighted more and more heavily by the NMR relaxation time, T_2, with the resulting conclusion that a region of long T_2 exists near the surface of the rod, whereas T_2 is shorter in the interior. The

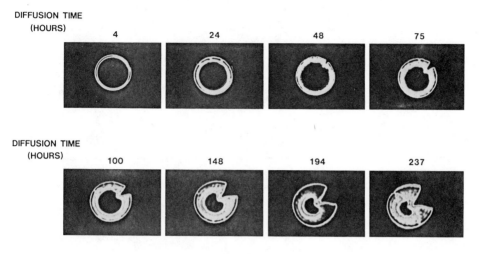

Figure 7. Time resolved study of the penetration of Toluene into 1.5 cm x 5 cm polystyrene rod.

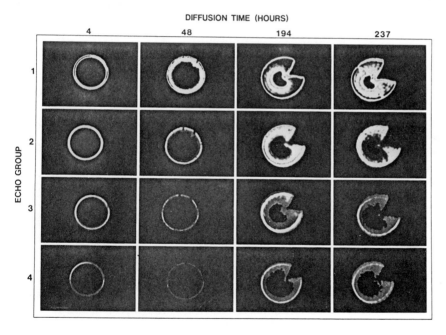

Figure 8. Traditional NMR techniques with imaging

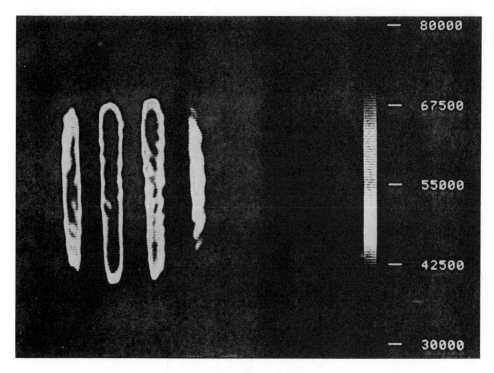

Figure 9.   One further step in the direction of true solid state
            imaging.

authors attributed this difference in relaxation time to strong
boundary interactions between the liquid and solid in the interior,
and this hypothesis was borne out by subsequent observation of
microscopic pores in the surface region, which would allow more
fluid-like behavior.

Figure 9 illustrates one further step in the direction of true
solid-state imaging. The sample consisted of a stack of alternating
plates of glassy polymers at 180^0C. The bright regions arise from
material for which T_g=100^0C, whereas for the dark areas, T_g=180^0C.

I hope that I have given you some idea in the last few minutes
of the variety of new NMR techniques for solids which have appeared
in the last few years which show some exciting possibilities for
materials characterization. The progress has been rapid, and shows
no signs of slowing down in the foreseeable future.

AN ADVANCED TECHNIQUE FOR CHARACTERIZATION OF POLYMER MATERIALS

BY WIDE ANGLE X-RAY SCATTERING

C. R. Desper

Army Materials and Mechanics Research Center
Materials Characterization Division
Watertown, Massachusetts 02172

ABSTRACT

A data acquisition method for characterization of polymer materials by Wide Angle X-ray Scattering (WAXS) has been devised which is capable of obtaining the requisite data at greatly enhanced speed, without sacrificing accuracy, and which allows immediate data analysis by a digital computer. The WAXS system combines a position-sensitive X-ray detector with a 65 kbyte microcomputer capable of operation as a multichannel analyzer. The data acquisition process is speeded up by a factor between 250 and 1000 by measuring that many data points over a range of diffraction angles simultaneously rather than sequentially. The method is demonstrated by two applications, one involving precision lattice parameter determination in polyethylene, the other involving determination of preferred orientation in a polyester film.

Now that a need for more definitive materials characterization is becoming evident, it is incumbent on characterization laboratories to provide an efficient means for gathering and processing the required information. In its entirety, this means that a characterization laboratory must establish a total laboratory data acquisition and processing system, providing the means for interacting with a variety of experiments, archiving data, and for analyzing data by means of both generalized and specialized digital computer programs. Furthermore, each experimental module of such a system must provide the most efficient and accurate means of data acquisition incorporating, where possible, advances in microcomputer technology to facilitate date handling and manipulation. Our purpose here is to describe the modernization of one such module, a facility for polymer characterization by Wide Angle X-ray Scattering

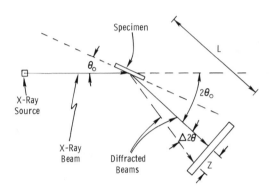

Figure 1. Experimental arrangement for measurement of 2-θ pattern.

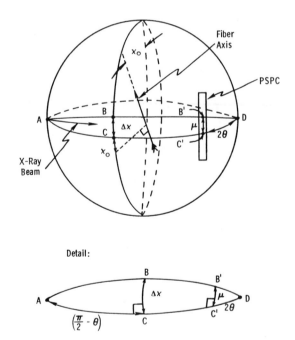

Figure 2. Experimental arrangement for measurement of fiber orientation pattern.

(WAXS), to illustrate its use through specific characterization applications, and to delineate the advantages accruing through the modernization of the laboratory module.

The X-ray diffractomery system in question, a computer-controlled device manufactured in 1967, was considered a state-of-the-art instrument at that time. Although designed and intended for use as a single-crystal diffractometer, the angular motions available proved of value for the study of oriented polymer specimens, and it was adapted[1] for that purpose in this laboratory. The features of the 1967 vintage instrument are outlined in Table 1.

The system has been modernized by replacing the Xray scintillation detector, pulse counting electronics, and computer, while retaining the original X-ray generatory and diffractometer hardware, as summarized in Table 2. Detailed discussion of the position sensitive proportional counter system is outside the scope of this report. The present system operates similar to systems described by Schultz[2] and by Borkowski and Kopp[3]

EXPERIMENTAL

The system has been used for measuring two types of diffraction patterns: 2θ patterns and fiber orientation patterns. To obtain a 2θ pattern, the detector is placed, as shown in Figure 1, with its centerline positioned at a diffraction angle $2\theta_0$. The detector sensitivity axis is positioned parallel to the plane of the diffractometer as defined by the incident beam and the diffracted beam at $2\theta_0$. Each detector pulse provides information on its linear position z (measured from the centerline) which translates into a diffraction angle 2θ according to:

$$2\theta = 2\theta_o + \tan^{-1}(z/L), \tag{1}$$

where L is the perpendicular distance from the center of the diffractometer to the detector axis. In practice, data is accepted over a range of 14° in 2θ which maps into 714 channels in the Lecroy 3500 multiple channel analyzer, thus recording data at 0.02^o intervals. (The angular range may be increased, with a corresponding increase in the angular width of each channel, by reducing the working distance L.)

A fiber orientation pattern is obtained, as shown in Figure 2, by placing the detector sensitivity axis perpendicular to the plane of the diffractometer. Again, the diffraction angle in the diffractometer plane is $2\theta_o$. For generality it is assumed that the

fiber axis makes an arbitrary angle χ_o with the plane of the diffractometer, controlled by the appropriate diffractometer motion.
The position z of a pulse on the detector axis translates into an
elevation angle μ above the equatorial plane of the diffractometer
(see Figure 2) by the equation:

$$\mu = \tan^{-1} (z/L) \tag{2}$$

Viewing intensity at the elevation angle μ is not equivalent,
however, to rotation of the fiber by the same angle. The angle μ
may be translated into an incremental orientation angle $\Delta\chi$ by
spherical trigonometry. As shown in the detail of Figure 2, this
reduces to the problem of relating the indicated small circle angle μ
to the length of the great circle arc $\Delta\chi$. The solution is:

$$\tan \Delta\chi = \tan \mu \ / \ 2 \sin \theta \tag{3}$$

which then gives the true orientation angle χ by:

$$\chi = \chi_o + \tan^{-1} (\tan \mu \ / \ 2 \sin \theta). \tag{4}$$

Substituting (2) into (4), the orientation angle χ is given in
terms of detector coordinates by:

$$\chi = \chi_o + \tan^{-1} [z/(2L \sin \theta)] \tag{5}$$

A very useful way of characterizing polymer orientation is by
means of the orientation function, originated by Hermans[4] and
used extensively by Stein[5]. The orientation function of any
orientation distribution is a second moment of that distribution,
namely the average value of the second spherical harmonic function.
As implemented in this laboratory[6], the orientation function
f_{hkl} is determined by integration of the intensity distribution for
a given diffraction plane (identified by Miller indices h,k,l) as a
function of the orientation angle χ. The limiting values for f_{hkl}
are given in Table 3. A more complete description of the characterization of polymer orientation is given by recent reviews[7-9].
We shall not dwell on details of an established method, but shall
concern ourselves here with its implementation in an advanced
instrument.

CHARACTERIZATION OF BRANCH CONTENT IN POLYETHYLENE RESINS

The first problem to which the rapid WAXS method has been
applied is to the characterization of polyolefin materials for
binary munitions storage. This laboratory has had a continuing

project for determining the materials properties required for storage of DF (methylphosphonic difluoride) over extended periods of time and at elevated temperatures. The materials having the

Figure 3. Branch point (B) incorporated inside a polyethylene folded-chain crystallite. (R): regular folds; (I): intercrystalline links.

best combination of the requisite chemical resistance and long term stability are polyolefins; in particular, copolymers consisting predominantly of ethylene with a small mole fraction of an α olefin comonomer. The comonomer introduces short chain branches into the polyolefin molecule, but the determination of this short chain branch content, which appears to be crucial to the polymer properties required for this application, has been a source of difficulty.

Application of the rapid WAXS method to this problem hinges on the fact that the branch points in the polyethylene can be incorporated as defects within the crystal structure of polyethylene, as shown in Figure 3. This has the effect of expanding the crystal lattice, particularly in the $\underline{a}$ direction, depending upon the amount of branch content in the resin[10-12]. Precision measurement of the $\underline{a}$ lattice parameter is, therefore, potentially a means of characterizing branch content in the polymer molecule. Such a method is needed, especially in the regime of short chain branches and low branch content, where methods based on infrared absorption[13] or molecular dimensions in solution[14] become insensitive.

The need for a better technique for characterizing short chain branching arises from the fact that certain mechanical properties, such as toughness and resistance to environmental stress cracking, are known to be sensitive to short chain branch content for polyethylenes. Indeed, these properties seem to optimize at short chain branch content values on the order of 1% or less, leading for the commercialization of ethylene copolymer grades in which such branches are deliberately introduced by incorporation of a comonomer such as vinyl acetate, butene-1 or hexene-1. As was noted earlier, present methods are inadequate to characterize resins in this regime of branch content. Consequently, the possibility of characterizing such branch content by X-ray diffraction has been explored.

In order to implement the method, calibration samples of known branch content are required. A series of polyethylene resins which have been extensively studied in the past was chosen. These samples, made available to the polymer science community in the 1960's by the Plastics Department of E. I. duPont de Nemours & Co., Inc., are labeled "University Contact Polyethylene" and are numbered PE-75, -76, -78, -79, -80, -81, -82, and -85. Characterization data are available in the work of Bodily and Wunderlich[11]. Except for the last one, these resins has density values in the 0.91 to 0.92 g/cm^3 range and are termed low density polyethylenes. The final sample of this series, PE-85, along with two other resins,

Marlex 6007 (Phillips Petroleum) and Chemplex 6109, served as standards of essentially linear polyethylene content. All of the standard samples were obtained as pellets and were compression

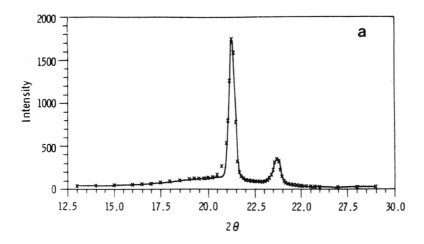

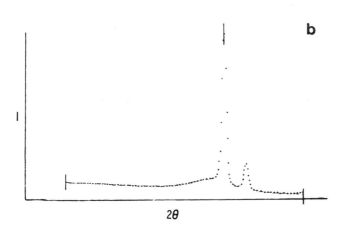

Figure 4. WAXS patterns of polyethylene sample 1981-6B, resin M407MQ. (a) using step-scan method; (b) using PSPC method. In (b), the left and right cursors are at 5° and 30°; the maximum count (marked at 21.580°) is 22,100.

molded at 170° to films of 0.2 mm thickness. Data for the branch content of the low density polyethylenes, as determined by the infrared method[13], were available; the high density resins were assumed from their method of polymerization and from their density values in the 0.95 to 0.96 g/cm³ range to be free of branching. The density, crystallinity, and branch content of the reference samples are given in Table 4.

The DF container samples for which branch content data are needed were available as cylindrical containers of wall thickness 1.5 and 6 mm, and as thick sheets of the same thickness values. These specimens were prepared for X-ray studies by microtoming a section of 0.2 mm thickness, which was subsequently heat-relaxed by melting briefly on a glass microscope slide resting on a hot plate at 170°. These and the standard samples were then further prepared by applying a very thin coating of graphite powder to one surface. To apply the graphite, the surface was first sprayed with a clear lacquer (Illinois Bronze, Federal Stock No. 8010-515-2487), then dusted with graphite powder. The graphite powder provides an internal reference line in the X-ray diffraction pattern, namely the strong and sharp graphite (002) line. Earlier work in this laboratory[15] has shown the value of this internal reference line for improving the accuracy of the lattice parameter determination.

For comparison, diffraction patterns for one of the polyethylenes obtained by the present PSPC method and by the previous step-scanning method are shown in Figure 4. The parameters describing the comparative performance of the two methods of data acquisition are given in Table 5. The step-scanning method has the advantage, in this case, in only one category: the time required to obtain a single datum, 5 minutes compared to 15. However, in the 15 minutes in question, the PSPC method obtained not just one, but all 800 of the data points required, while the step-scan method required a total time of 5 hours. To finish in that 5 hours, a selective data acquisition strategy was required: (a) the range of 2-θ values was smaller, (b) the interval between 2-θ values was coarser, especially in regions of the pattern where peaks were not anticipated, and (c) individual determinations were stopped before the time limit if the count limit of 10,000 occurred first. Thus, the PSPC method is not only faster (by a factor of 20 in this case); it also provided a larger number of data values, more closely spaced, and over a wider range of angle values. Indeed, were the step-scan run in such a way as to replicate the range and number of data values, the time advantage for the PSPC method would be much greater: it would equal the number of data points (800) in the PSPC determination.

The PSPC pattern for one of the graphite-coated standard polyethylenes is shown in Figure 5, along with three regions of interest

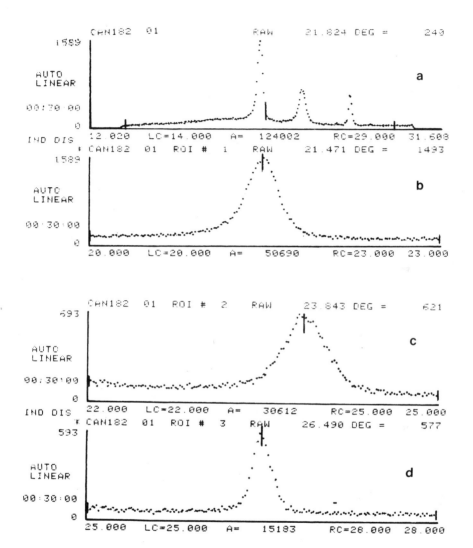

Figure 5. WAXS 2-θ pattern of polyethylene sample from can no. 182
using PSPC method: (a) entire pattern; (b) region of polyethylene
(110) peak; (c) region of polyethylene (200) peak; (d) region of
graphite (002) peak.

from the pattern showing the three diffraction peaks observed. The microcomputer finds these peaks by a peak-search algorithm, then calculates the precise peak position by locating negative-going zero crossings in the first derivative of the data. The microcomputer also applies a background correction and determines the breadth and area of each peak, but these quantities are not of interest in the present case. A precise value for the a lattice parameter of each specimen is obtained using the position of the polyethylene (200) peak with respect to the graphite (002) internal standard, which is assigned a $2-\theta$ value of $26.576°$, corresponding to a d spacing of 3.354 Angstroms. This procedure compensates for any instrumental drift or misalignment, and results in a precision of 0.0005 Angstroms in the polyethylene a lattice parameter value, as determined by replicate measurements on a single specimen. The temperature of the specimen (identical with the room temperature) is the factor limiting the precision of the a determination. The above precision is valid for a temperature range of $25° \pm 0.3°$.

Control of the specimen temperature is essential to obtaining the maximum precision in this method of branch content determination. In instances where the room temperature varied between $23°$ and $28°$, good correlations were obtained between the polyethylene a parameter and temperature, yielding an expansion coefficient da/dT of 2×10^{-3} Angstrom/deg. Sample temperature obviously must be controlled to a fraction of a degree to obtain results insensitive to temperature. In the present work this was accomplished by using the room air conditioning, observing the room temperature to ensure that the specimen temperature is in the required range.

The a lattice parameter values were obtained for the seven reference samples of Table 4 and for eight polyethylene samples of unknown branch content. Using the infrared values for branch content for the reference samples, a linear regression line was obtained (Figure 6) relating branch content to the a lattice parameter. The regression line was then used to evaluate the branch content of the unknown samples, plotted as circles in Figure 6. The branch content of the unknowns may be read from the corresponding ordinate values in Figure 6, or may be calculated from the equation of the regression line:

$$(CH_3 / 100C) = 33.03 (a - 7.4020) \tag{6}$$

The experimental branch content values for the DF container materials, obtained from the a lattice parameter values, are given in Table 6.

The DF container samples are made from three different grades of polyolefin, all products of Phillips Petroleum: resin M407MQ, the constituent of the containers labelled Can 19, Can 182, Can

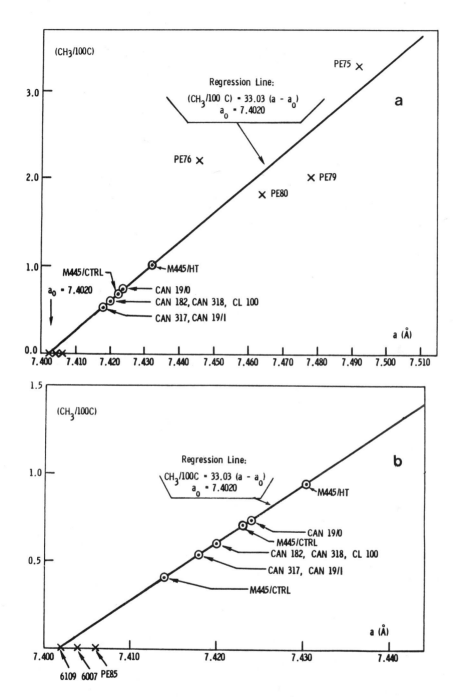

Figure 6. Linear regression correlation: polyethylene branch content versus <u>a</u>; solid line is least squares fit. (a) full range of data including standards (X); (b) detail of data for unknowns (O).

317, and Can 318; resin M445, a replacement for the now discontinued grade M407MQ; and resin CL100, a rotational molding grade. Examining the data in Figure 6 and Table 6, it is seen that, with the

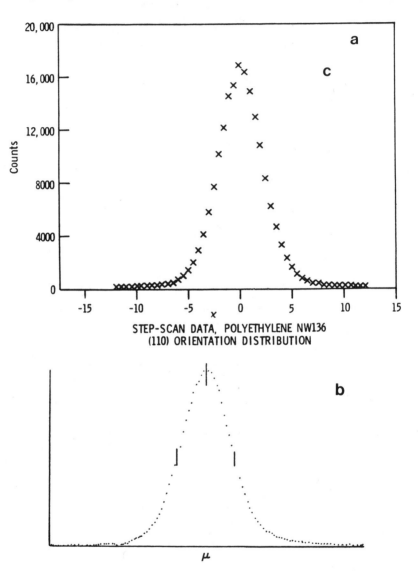

Figure 7. (110) orientation distribution of extruded polyethylene strand #NW136. (a) using step-scan method; (b) using PSPC method. In (b), the left and right cursors are at -0.917° and $+0.917^{\circ}$; the center marker, at 0.000°, registers 21,820 counts.

exception of one sample (M445/HT) which was subjected to prolonged heat treatment, the branch content values for all of the DF container specimens are in a narrow range of 0.50 to 0.75 CH_2/100C. This relatively narrow range of variation indicates a materials consistency over a number of lots, obtained over a period of years from the manufacturer. At present, however, the degree of sensitivity of materials properties to branch content in this range is not known, so the allowable tolerance in this materials characteristic is also unknown.

The higher value for the one heat-treated resin (M445/HT) may arise from a small degree of chain oxidation, leading to the incorporation of carbonyl groups into the polyethylene crystallites. As an alternate hypothesis, the prolonged annealing probably caused thickening of the lamellar polyethylene crystallites, which has been shown by Davis and coworkers[16] to have an effect on the lattice parameters. This hypothesis is discounted for two reasons: (a) all specimens were melted and recrystallized on a glass slide, which should remove all memory of the previous crystalline morphology; and (b) the hypothesized effect is in the wrong direction, since thicker crystallites have smaller lattice parameters.

In terms of the method of characterization, the greatest area for improvement is in the standard samples. The X-ray lattice parameter method has the advantage of speed, precision, and simplicity for work in this area, but does not, of itself, give an absolute measure of branching. Standards are needed over a wide range of branch content values, not just at the low and high ends, to test for curvative in the correlation between branch content and lattice parameter. This has not been possible in the past because a better method was needed for establishing the branch content of the standards. C-13 nuclear magnetic resonance could fill this need, and such instruments are becoming more widely available. When such characterized standards become available, the X-ray lattice parameter method may be put on a more solid absolute basis. Even lacking a wide variety of accurate standards, the lattice parameter method for measuring branch content is valuable in terms of giving precise comparisons of various resins.

CHARACTERIZATION OF PREFERRED ORIENTATION IN FIBERS AND FILMS

A second application of the method is in terms of characterizing preferred orientation in polymer fibers and films. The experimental configuration for the measurement of preferred orientation is described in the experimental section and is shown in Figure 2. The first order of business is to check the validity of the method,

Table 1. Earlier Step-Scan X-ray Scattering System (1967)

1. Picker Model 6238H X-ray Generator
2. Picker Model 3645 Four-Circle Diffractometer
3. Picker Model 2811B Scintillation Detector
4. Picker Model T55-518 Radiation Analyzer
5. Digital Equipment Corporation Model PDP-8/S Computer
    a. Parallel interface to scaler, timer, angle readout
    b. CPU with 8k of core memory
    c. Paper tape I/O
    d. Limited computing capability

Table 2. Wide Angle X-ray Scattering System - 1984

1. Picker Model 6238H X-ray Generator
2. Picker Model 3645 Four-Circle Diffractometer
3. Technology for Energy Corporation Model 210 Position-Sensitive
Proportional Counter
4. Ortec Nim Bin Electronics
    a. Model 446A High Voltage Supply
    b. 2 ea Model 575 Shaping Amplifiers
    c. 3 ea Model 551 Timing Single Channel Analyzers
    d. Model 433A Dual Sum and Invert Amplifier
    e. Model 457 Time to Amplitude Converter
5. Lecroy 3500 Microcomputer
    a. Model 3511 ADC input channel
    b. Multiple Channel Analyzer firmware in ROM
    c. Intel Model 8080 CPU with 65k RAM
    d. Flexible disk I/O
    e. Fortran and Basic programming systems available

Table 3. Limiting Values for the Hermans Orientation Function[*]

Value of f_q	Type of Orientation	Description
+1	Parallel	Crystal direction aligned perfectly parallel to fiber axis
0	Random	Random orientation with respect to fiber axis
-1/2	Perpendicular	Crystal direction aligned perfectly perpendicular to fiber axis

[*] defined by $f_q = (3 \langle \sin^2 \chi_q \rangle - 1) / 2$,
where q is any crystallographic direction (crystal axis or
reciprocal lattice vector) in the unit cell.

Table 4. Characterization of Standard Polyethylene Resins

Sample Designation	Density (g/cm^3)	Crystallinity[*]	Branch Content[**] $(CH_3 / 100C)$
PE-75	0.914_3	0.44_6	3.3
PE-76	0.921_3	0.49_6	2.2
PE-79	0.921_1	0.49_3	2.0
PE-80	0.921_9	0.49_9	1.8
PE-85	0.951_3	0.69_5	0.0
6007	0.953_2	0.70_4	0.0
6109	0.962_0	0.76_3	0.0

[*] from density
[**] by infrared method.

Table 5. Comparison of Step-scan and PSPC WAXS Methods
for Determination of Polymer 2-θ Pattern

[Polyethylene Resin M406MQ, Sample 1981-6B,
Bragg-Brentano Reflection Method]

Parameter	Step-Scan Method	Advantage	PSPC Method
Range	16°	=>	25°
Interval	0.1 to 1°	=>	0.03^o
No. of data pts.	65	=>	800
Time / datum	5 min.	<=	15 min.
Total time	5 hrs.	=>	15 min
Maximum datum count	10,000	=>	22,100

Table 6. Characterization of DF container samples

Sample	Density (g/cm^3)	Crystallinity	a (A^o)	Branch Content $(CH_3 / 100C)$
Can 19/0[*]	0.943_6	0.64_5	7.424	0.73
Can 19/I	$-$	$-$	7.418	0.54
Can 182	$-$	$-$	7.420	0.60
Can 317	0.938_1	0.60_8	7.418	0.54
Can 318	0.940_8	0.62_8	7.420	0.60
M445/CTRL	0.942_8	0.63_6	7.423	0.70
M445/HT[*]	0.947_6	0.66_8	7.431	0.95
CL100	0.940_{21}	0.62_{82}	7.420	0.60

[*]/0 – specimen from outside of can; /I – specimen from interior of
can; /CTRL – control; /HT – heat treated 4 days at 110°.

Table 7. Comparison of Orientation Parameters
as Determined by Step-scan Method and PSPC Method

[Extruded Polyethylene Strand Sample # NW136, (110) Plane]

Parameter	Step-scan Value	PSPC Value
FWHM *	4.86°	4.90°
f_{110}	-0.497_{2}	-0.496_{8}

*
 FWHM: Full Width at Half Maximum in (110) χ orientation
distribution.

Table 8. Comparison of Step-scan and Rapid WAXS Methods
for Determination of Polymer Fiber Orientation

[Extruded Polyethylene Reference Sample #NW136, (110) plane]

Parameter	Step-Scan Method	Advantage	PSPC Method
Range	40°	=>	55°
Interval	0.5°	=>	0.22°
No. of data pts.	81	=>	250
Time/datum	40 sec.	<=	5 min.
Total time	54 min	=>	5 min
Maximum datum count	16,800	=>	21,800

Table 9. Results: Preferred Orientation Determination

[Poly(ethylene Terephthalate) Sample A906
Drawn 6X by Co-extrusion at 90°]

(hkl)	$\langle \sin^{2} \chi_{hkl} \rangle$	f_{hkl}
010	0.017_{5}	-0.473_{8}
-110	0.010_{6}	-0.484_{1}
100	$* 0.009_{7}$	-0.485_{4}
c axis (polymer chain)	0.973_{0}	0.959_{5}

*
 Derived quantities based on (010), (-110), and (100) data.

and in particular, the validity of the angular transformation equation 5. For this purpose, the (110) orientation pattern was redetermined for an extruded polyethylene monofilament specimen which has been carefully characterized by the step-scan method[6]. The patterns obtained by the two methods is shown in Figure 7, while the experimental results are compared in Table 7. This particular type of specimen, an ultraoriented polyethylene strand from the laboratory of Professor Porter at the University of Massachusetts, is a critical test of orientation resolution, since these samples show some of the highest levels of crystalline orientation observed in polymer science. The resulting orientation parameters in Table 7 are in good agreement. Having established the validity of the method, Table 8 compares it in terms of efficiency with the step-scan method. Again, the advantage goes to the PSPC method in all areas except timer per datum, which is unimportant, since all data are obtained simultaneously in the PSPC method.

As an experimental "worst case" test, data obtained for a thin poly(ethylene terephthalate) film shall be presented. This specimen constitutes a critical test of the sensitivity of the method, since the crystallinity is low (30%) and the specimens are less than 25 microns thick. As is often the case in research, the amount of available sample is limited, so it is not possible to laminate multiple layers of sample to improve intensity. Furthermore, such a multiple layering procedure has an inherent source of error: the apparent orientation distribution may be broadened by the misalignment of successive layers with respect to each other. Obtaining an orientation distribution using a single layer of such a low crystallinity specimen is not possible using conventional step-scanning techniques. Thus, the use of the position sensitive detector has been explored as an alternative.

The results on this and other poly(ethylene terephthalate) films shall be discussed in detail elsewhere[17], so only a brief discussion shall be undertaken here. It has been shown by Wilchinsky[18] that for a polymer such as poly(ethylene terephthalate) having a triclinic unit cell, the orientation function of the $\underline{c}$ axis (the polymer chain direction), which may not be measured directly, may be calculated from experimental orientation functions for three (hk0) diffraction planes. For poly(ethylene terephthalate), the planes selected for this purpose are the (010), (-110), and (100) planes. Orientation patterns for these three planes were obtained using the geometry of Figure 2 with a data acquisition time of 24 hrs. for each plane. (N.B. a knife-edge beam stop was placed 1 mm beyond the specimen in the path of the X-ray beam; without this, air scattering led to excessive background). Using the equation of Wilchinsky[18], and the general definition in Table 3 of an orientation function, the $\underline{c}$ axis orientation function in this case is calculated from:

$$f_c = - k_{010} \, f_{010} - k_{-110} \, f_{-110} - k_{010} \, f100, \qquad (7)$$

where

$$k_{010} = 0.8807; \ k_{-110} = 0.7755; \text{ and } k_{100} = 0.3438 \qquad (8)$$

are determined by the parameters of the unit cell. The results for the poly(ethylene terephthalate) film are summarized in Table 9. The $\underline{c}$ axis orientation function is approximately 0.97, which is rather high but not quite in the ultrahigh range of the polyethylene strands[6], where it exceeds 0.99. The point here, however, is that this is the only way in which this orientation function could have been measured; step-scanning is useless in this case because of the poor signal to noise ratio.

SUMMARY

Wide Angle X-ray Scattering (WAXS) provides information which is quite detailed and very useful for the characterization of polymer materials. WAXS would be more widely used for polymer characterization except for one limitation: the speed with which data may be acquired and analyzed. To deal with these limitations, a WAXS system has been devised combining a position sensitive X-ray detector with a 64 Kbyte microcomputer. In one application, the system has been used to measure the branch content in a series of LLDPE polyethylene resins, using the expansion of the $\underline{a}$ unit cell parameter with included branches. Patterns were obtained spanning a 14° 2θ range and recorded in 714 channels of data memory. Using the microcomputer, the centroid of the (200) line was determined to a precision of better than 0.002 Angstroms. Using standards of known branch content, the branch content of unknown resins could be determined to a precision of 0.1 $CH_3/100C$. In a second application, orientation functions were determined for a series of drawn poly(ethylene terephthalate) films, using the intensity distribution perpendicular to the equatorial plane of the instrument at fixed 2θ. In this case, an appropriate spherical construction was used to relate the instrument angle to the crystallite orientation angle. A polyethylene strand of ultrahigh orientation was used to establish the accuracy of the method. The orientation function thus determined agreed with earlier step-scanning data with a discrepancy of only 0.0004. Subsequently, f_c values were determined for co-extruded poly(ethylene terephthalate) films using experimental orientation functions for the 010, -110, and 100 diffraction planes. Because of the low crystallinity and thickness of these films, orientation functions could not be determined by the conventional step-scanning method.

REFERENCES

1. C. R. Desper, Adv. Xray Anal., 12, (C. S. Barrett, G. R. Mallett, and J. B. Newkirk, eds.), Plenum, New York, (1968), 404; AMMRC TR 69-27, (1969).
2. J. M. Schultz, J. Polym. Sci. Polym. Phys. Edn., 14, 2291 (1967).
3. C. J. Borkowski and M. K. Kopp, Rev. Sci. Instr., 46, 951 (1975).
4. J. J. Hermans, P. H. Hermans, D. Vermaas, and A. Weidinger, Rec. trav. chim., 65, 427 (1946).
5. R. Stein, J. Polym. Sci., 31, 327 (1958).
6. C. R. Desper, J. H. Southern, R. D. Ulrich, and R. S. Porter, J. Appl. Phys., 41, 4284 (1970).
7. C. R. Desper, CRC Critical Reviews in Macromol. Sci., 1, 501 (1973); AMMRC TR 73-50, 1973.
8. G. L. Wilkes, Adv. Polym. Sci., 8, 91 (1971).
9. C. R. Desper, "Characterization of Materials in Research, Ceramics and Polymers" (J. J. Burke and V. Weiss, eds.), Syracuse Univ. Press, ch. 16, (1975).
10. B. Wunderlich and D. Poland, J. Polym. Sci. A, 1, 357 (1963).
11. D. Bodily and B. Wunderlich, J. Polym. Sci. A-2, 4, 25 (1966).
12. B. Wunderlich, Macromolecular Physics, Vol. 1, Academic Press, New York, 153-154, (1973).
13. ASTM D233868, Annual Book of ASTM Standards, Vol. 35, Am. Soc. Testing Matls., Philadelphia, 70, (1981).
14. F. W. Billmeyer, Jr., J. Polym. Sci. C, 8, 161 (1965).
15. C. R. Desper, "Characterization and Selection of Polymer Materials for Binary Munitions Storage", AMMRC TR 83-45, (1983); op. cit. "Part 2. Characterization of LLDPE Resins", AMMRC TR 84-28, (1984).
16. G. T. Davis, J. J. Weeks, G. M. Martin, and R. K. Eby, J. Appl. Phys., 45, 4175 (1974).
17. T. Sun, C. R. Desper, and R. S. Porter, submitted to J. Matls. Sci.
18. Z. W. Wilchinsky, Adv. Xray Analysis, 6, 231 (1963).

APPLICATIONS OF THE RAMAN MICROPROBE TO MATERIALS CHARACTERIZATION

Fran Adar

Instruments SA, Inc.
173 Essex Avenue
Metuchen, New Jersey   08840

ABSTRACT

The MOLE[R] Raman microprobe is an analytical tool providing molecular and crystalline information on samples with 1 μm spatial resolution. The instrument consists of a standard microcomputer-controlled Raman monochromator optically and mechanically coupled to an optical microscope. Sample preparation is identical to that for optical microscopy, and actual measurements are performed under ambient conditions (or in corrosion or catalyst cells under controlled conditions with a long working distance objective, when required). Examples of successful applications of the technique are derived from work on polymers, ceramics, corroded samples, semi-conductor materials, and manufacturing contaminants (particularly organic contaminants). In all cases the information derived from the Raman microprobe complements that obtained on other analytical instruments. The unique contribution of the Raman microprobe to our understanding of these materials is always emphasized.

INTRODUCTION

Raman scattering is a vibrational spectroscopic technique that can both fingerprint organic species and identify crystalline polymorphs of organic and inorganic materials. By coupling an optical microscope to a conventional Raman spectrometer, the technique becomes a microprobe with spatial resolution of 1 μm, determined by the wavelength of the radiation (ca.0.5 μm) and the numerical aperture of the microscope objective. Since Raman microprobe (RMP) spectra are identical to spectra acquired in a classical (macro) geometry, literature spectra and spectral files can be used

339

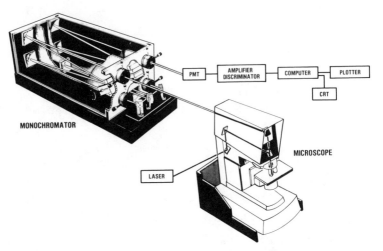

Figure 1.   Instrument schematic of the MOLE[R] Raman microprobe.

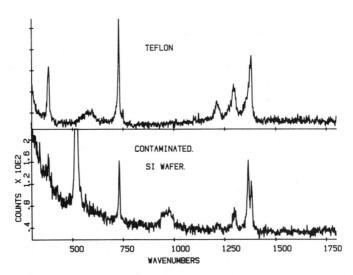

Figure 2.   RMP spectrum of fluorinated hydrocarbon contaminant on
            silicon.

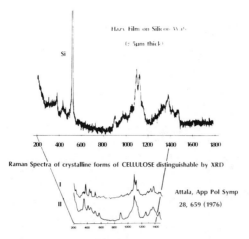

Figure 3.   RMP spectrum of a cellulose film contaminating a silicon
wafer.

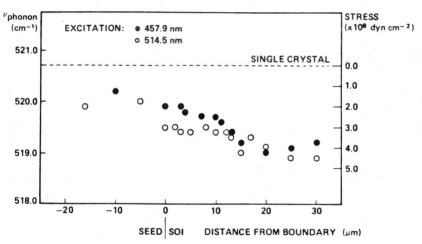

Figure 4.   Raman frequency and stress of silicon on oxide laterally
epitaxially annealed and plotted as a function of
distance from the edge of the oxide islands.   The
original spectra were collected with $0.1cm^{-1}$ increments
and were reproducible to the same value.   Spectra were
generated with argon laser wavelengths 514.5 and
457.9nm.

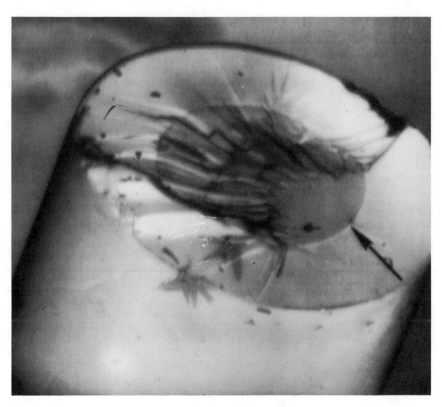

Figure 5.   Photomicrograph of ZrO_2 inclusions precipitating break
in as optical fiber.

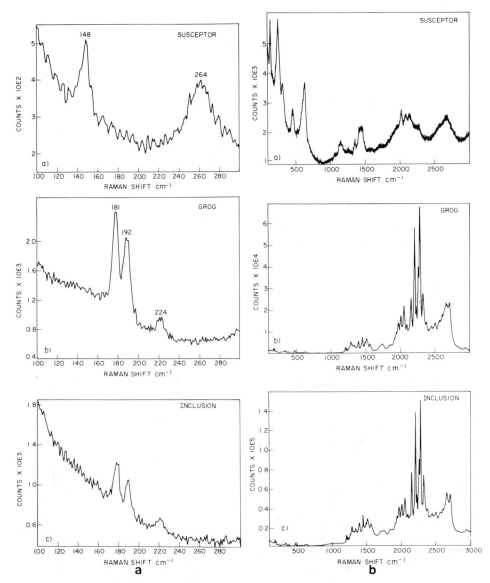

Figure 6a. RMP spectra between 100 and 300cm^{-1} of ZrO$_2$ susceptor and grog materials used in fiber drawing furnace, and of ZrO$_2$ inclusion precipitating break in an optical fiber.

Figure 6b. Microfluorescence spectra between 100 and 300cm^{-1} of ZrO$_2$ samples.

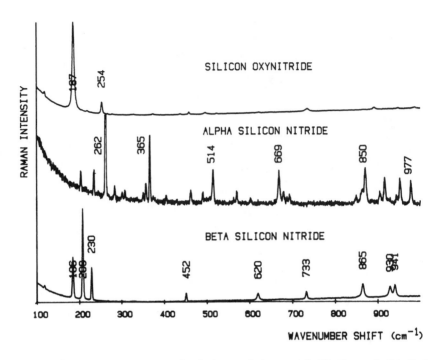

Figure 7. RMP spectra of alpha and beta of Si_3N_4 and Si_2N_2O.

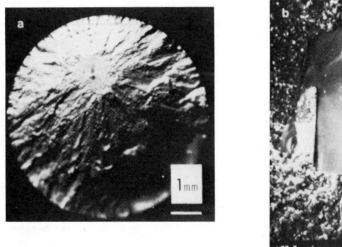

Figure 8a. Low magnification optical micrograph of fracture
surface of Si_3N_4 test bar showing a particl precipitating
precipitating the fracture.

Figure 8b. Electron micrograph of inclusion precipitating fracture
of the Si_3N_4 test bar.

for interpretation. Applications include fingerprint identification of microscopic contaminants and inclusions, and characterization of technologically important new materials.

GENERAL INSTRUMENT DESIGN

The coupling between micro-sampling optics and a Raman spectrometer was accomplished simultaneously in France and in the U.S. Delhaye and co-workers coupled a commercial research grade microscope to a Raman monochromator producing the MOLE[R] (Molecular Optical Laser Examiner)[2,3] while Rosasco coupled an elliptical reflector[4] to a standard Raman monochromator[5,6]. The work described here was accomplished on a second generation instrument of the French (Delhaye/Dhamelincourt) design.

The instrument is shown in Figure 1. The coupling of laser to microscope to sample to monochromator is achieved by the beam splitter in the nose piece of a classical metallographic microscope. The beam splitter enables disentanglement of illuminating and scattered radiation with high numerical aperture at the sample. Standard microscope objectives serve to focus the laser irradiation and to collect the scattered light. Sample preparation is identical to that required for standard optical microscopy. Sample alignment is achieved by positioning the region of interest under the focussed laser beam. Subsequently the focal plane is adjusted to maximize the signal when samples are large and high spatial resolution is not essential. When maximum spatial resolution in the axial and/or radial directions is required, an iris diaphragm behind the microscope (not shown in Figure) can be closed. The iris is mounted at an image plane conjugated to the sample, and has the effect of eliminating radiation originating outside of the laser focal volume[6-8].

INTEGRATED CIRCUIT MANUFACTURING PROBLEMS

ORGANIC CONTAMINANTS – Over the past years we have been asked to identify organic contaminants that appear on silicon wafers during some processing operations. As the scale of integration of the circuits increases, and the size of the smallest features decreases, the size of contaminants that can destroy a device's operation decreases and the ability to identify foreign materials becomes more critical. The amount of material present is often too small for analysis by infrared absorption or X-ray diffraction. The Aujer and electron microprobes are incapable of yielding chemical (i.e., molecular) identification. The Raman microprobe is unique in its ability to identify organic contaminants that appear as particles as small as 1 μm, or as films as thin as 1 μm.

EDTA – A very large particle (>100 micron) on a wafer was identified as EDTA (ethylene diamine tetraacetic acid). Because of

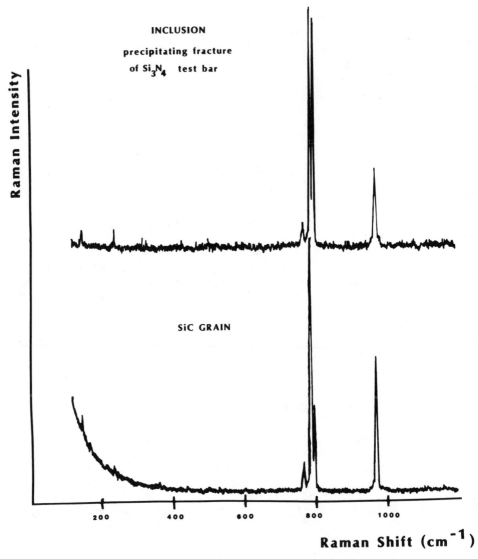

Figure 9. RMP spectra of inclusion in Si_3N_4 test bar and reference
spectrum of SiC.

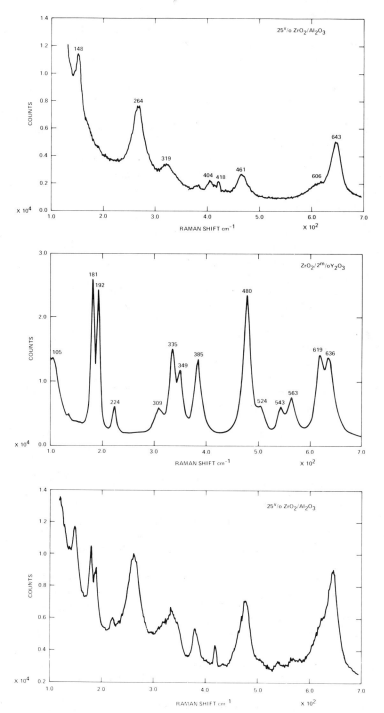

Figure 10.   RMP spectra of the monoclinic and Al_2O_3 - stabilized tetragonal phases of ZrO_2 ceramics.

the size of the particle, it was apparent that the EDTA had not precipitated on this wafer after inadequate rinsing (precipitated crystals would exhibit morphology of growth on a substrate). It was inferred that this particle had been sprinkled on the wafer when clean room gloves had not been appropriately changed.

SILICONE - A silicone film about 5 micron thick was identified on a silicon substrate. An elemental probe would have difficulty detecting a small excess of silicon atoms in the film and would certainly not be capable of identifying the source of the contaminants. However, the Raman spectrum, by fingerprinting the polymer, provides clues to the source of the contaminant. In this case, a degraded silicone-rubber gasket was suspected of introducing material that settled on this sample.

PET Chip - A pollutant chip was fingerprinted as a piece of polyethylene terephthalate. The source of this pollutant was identified as the wafer handling baskets. This result was especially surprising because these baskets were "guaranteed" not to chip and produce debris. Replacement of baskets at this site was estimated to be $150,000.

FLUORINATED HYDROCARBON - The bottom of Figure 2 shows a spectrum of material that had settled on another wafer. The features at 521 and 950cm^{-1} are one- and two phonon bands of the silicon substrate. The other features come from the contaminant. The top of the figure shows a spectrum of teflon (polytetrafluoroethylene). While the two spectra in the figure are not a perfect match, it is significant to note that the 1450cm^{-1} region, where most hydrocarbons have a strong band, is free of any structural features.

Interpretation of this spectrum requires information on the history of the wafer. This specimen had been polished with a slurry containing organic solvents and then plasma etched. It is possible that low molecular weight material was dissolved and carried from a teflon container. A second suggestion involved the deposition of $(CF)_n$ from a $CF_4 + H_2$ plasma containing excess H_2.

CELLULOSE - The upper spectrum of Figure 3 was recorded from a hazy film (<5 micron thick) on a Si wafer. In addition to the silicon phonon at 521cm^{-1}, other bands appeared which could be matched to literature spectra of native cellulose (polymorphy I), whose spectrum is shown in the lower portion of the figure.

The source of this contaminant would be cellulose filter material in one of the processing operations. However, there is a problem in reconciling the assignment to cellulose I and the film morphology. It has usually been observed that cellulose I cannot be recovered from cellulose solutions. Thus one would not expect

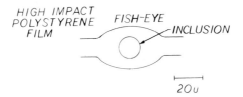

Figure 11a. Schematic of cracks in tetragonal ZrO_2 ceramic parts emanating from the corners of as indent produced by the stressbearing tool. RMP spectra were acquired as a function of distance from the crack.

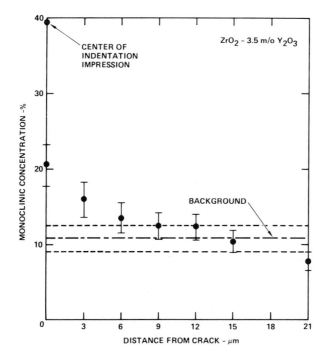

Figure 11b. Concentration of mono-clinic ZrO_2 as a function of distance from the stress-induced crack in a sample that had been prepared as stabilized tetragonal ZrO_2.

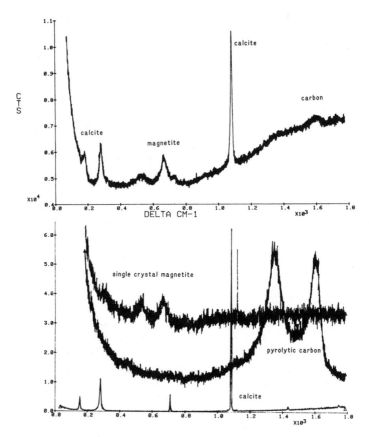

Figure 12. RMP spectrum of corrosion fuel on steel that had been exposed to a brine containing 2610 ppm NaCl, 1065 ppm NaSo4, 885 ppm CaCl$_2$, 100 atm CO$_2$ at a temperature of 375°F.

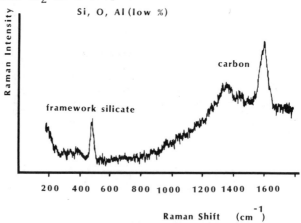

Figure 13. RMP spectrum of white particle removed from heat exchange showing band close to 500cm^{-1} characteristic of zeolites (framework silicates incorporating alumina in the lattice) and cathedral carbon bands at ca. 1360 and 1600cm^{-1}.

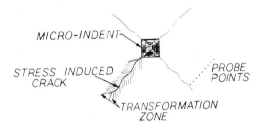

Figure 14a. Schematic cross section of fish-eye in high-impact
polystryrene film showing inclusion in center.

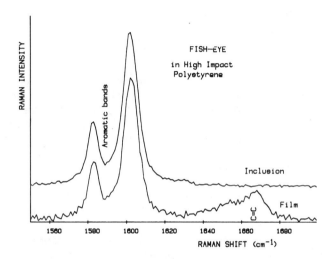

Figure 14b. RMP spectra between 1550 and 1700cm^{-1} of high impact
polystryene film and an inclusion in a fish-eye that
was exposed by cross-sectioning.

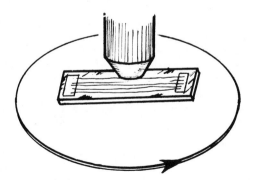

Figure 15. Schematic microscope rotating stage enabling examination
of single point of a fiber in several orientation/
polarizaion combinations.

to observe cellulose I from a film. However, Atalla has been able to show that cellulose I can be recovered from solutions of cellulose by precipitation at elevated temperatures (>150^{0}C)[9]. Thus, this Raman spectrum enables one not only to identify the chemical composition of the contaminant, but also allows determination of the time during processing during which the pollutant precipitated. It is useful to note here that identification of crystalline polymorphs of many other polymers provides information on their thermal and/or stress history.

MICROANALYSIS - BREAK IN WIRE CONNECT - An integrated circuit package that had failed during use was submitted for analysis. The ceramic package had been opened and it was found that some of the aluminum wire connects had broken not far from the gold bonding pads. It was requested that material deposited on the gold around the broken aluminum wire be identified - the material was suspected to be organic.

Repeated attempts to collect evidence of organic deposits yielded no success. However, a band at approximately 519cm^{-1} was observed on every pass - and it was not observed when the instrument was focussed on the gold pad or aluminum wire. We were forced to the conclusion that the "contaminating" material was elemental, semi-crystalline silicon. It was then imperative to explain the origin of the material and the reason for the failure.

From infrared analysis, it was known that significant amounts of silica were present. It was subsequently argued that silica gel had contaminated the package and acted as a carrier for corroding chemicals. The Raman microprobe did not find evidence of silica. We argued that the problem could have originated in the aluminum wires themselves. These often have a small percentage of micro-crystalline silicon added as a hardening agent. If the silicon is inadequately dispersed there might be local heating during use which would promote electro-migration and could end ultimately in a break in the wire.

In order to reconcile the Raman microprobe results with those of infrared, it is important to recognize the limitations of the techniques. The Raman-active phonon mode in silicon is coated with, at least, a thermal oxide to which infrared absorption is very sensitive. While the oxide is also active in the Raman effect, its signal is much weaker than that of the silicon itself. Thus an infrared spectrum of a chip of silicon would show the oxide while the Raman would show the silicon itself.

STRESS IN LASER-ANNEALED SILICON ON OXIDE - In collaboration with Hewlett Packard Laboratories lateral epitaxially regrown films of laser-annealed silicon on oxide were examined[10]. This study is motivated by the attempt to produce high quality single crystal silicon films isolated by oxide from the silicon crystal wafer. Such materials could then be used to manufacture devices with increased density per package. It is known, however, that mismatch at the silicon/oxide interface will usually result in stress in the annealed material that causes dislocations and eventually grain boundaries. The RMP study[10] described here was an effort to combine the ability of the Raman signal to monitor stress[11] with the high spatial resolution of the technique. The measured stress could then be correlated with the geometry of the sample.

The samples were prepared by depositing .55 μm of polysilicon over a (001) wafer patterned with 480 μm square islands of SiO_2 that were 1.4 μm thick. The edges of the islands were oriented along (110) directions. The polysilicon was encapsulated with 60A of silicon nitride. The samples were held at 500^0C during re-crystallization which was accomplished by scanning an 11W, 80 μm diameter Gaussian argon ion laser beam across the sample at a rate of 25cm/sec, raster stepping 10 microns between scans.

The Raman spectra were recorded digitally with .1cm^{-1} resolution and repeatability. Figure 4 shows a plot of the Raman phonon frequency as a function of distance from the edge of the oxide island. The figure also shows the dependence of the stress on the position of the probe.

Two sets of data are illustrated in the figure; Raman spectra were excited at wavelengths 514.5 and 457.9nm. Because the optical penetration depth of silicon is different at these two wavelengths, the information generated reflects different thicknesses of samples. Both sets of data indicate an increase in stress as the distance from the oxide edge increases. A striking difference between the two spectra is the indication of a drop in stress at approximately 25 μm from the edge that is detected in the 457.9nm-excited data. This is attributed to the shallower penetration depth of this radiation; the conclusion to be drawn is that after the sample breaks into polycrystalline grains approximately 11 μm from the edge (as measured by Nomarski contrast micrographs), there is relief of stress at the surface of the recrystallized material.

OPTICAL FIBERS - Several manufacturers of optical fibers have suggested that a Raman microprobe could be a useful tool in characterizing fibers. The literature already shows that it is possible to monitor the concentration of additives to silica down to the 1 mole percent range[12,13]. Polished sections of preforms or drawn fibers have been monitored on the microprobe.

In addition, foreign material in fibers that precipitate breaks have been identified. Recently analysis of ZrO_2 particles causing breaks in AT&T optical fibers enabled identification of the origin of the contamination[14]. Figure 5 is an electron micrograph of one

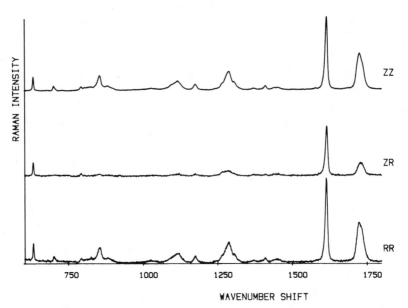

WAVENUMBER SHIFT

Figure 16a. RMP spectra of one spot of a polyethylene terephthalate (PET fiber spun from the melt at 1500 m/min and examined with laser polarization and analyzer combinations ZZ, ZR, and RR where Z refers to the fiber axis and R refers to one of the axes perpendicular to the fiber. The spectra have been displayed so that intensity ratios between the three polarization options can be estimated from the figure.

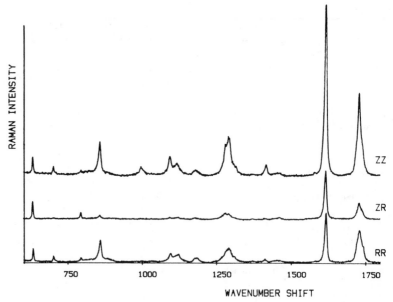

WAVENUMBER SHIFT

Figure 16b. RMP spectra of a PET fiber spun from the melt at 1500m/min and drawn at room temperature. All acuisition conditions from (a) apply.

such break. The design of the furnace used to draw the optical fibers from the preform provides two possible sources of ZrO_2 that can potentially contaminate the fibers. Elemental microanalysis could not identify which ZrO_2 material was the contaminating source. In this case identification of crystalline polymorph enabled identification of the origin of the particles.

The fiber-drawing furnace uses yttria-stabilized tetragonal zirconia as the susceptor material and granular zirconia insulation ("grog") which occurs in the monoclinic phase. RMP spectra between 100 and $300cm^{-1}$ of the susceptor and grog materials and of one of the 2 x 4 micron inclusions are shown in Figure 6a. It is clear from identification of the crystalline phase that the inclusion is a particle of the grog material. Figure 6b shows RMP spectra out to $3000cm^{-1}$. All bands with frequency shifts greater than $1000cm^{-1}$ are fluorencence bands of rare earth metal impurities. The sensitivity of the technique to these impurities is quite high[14] and work is proceeding to identify which rare earth ions in the zirconia matrix are responsible for these lines.

III-V SEMICONDUCTORS - The use of Raman spectroscopy to characterize crystalline films of compound semiconductors has been reviewed[15]. Effects that can be monitored are orientation, carrier concentration, charge carriers' scattering times, mixed-crystal composition[15] and Group V deposits in the native oxides[16].

POLYPHASE CERAMICS

A series of examples of applications of the Raman microprobe to ceramics characterization appeared in a publication by Clarke, et. al.[17]. Differentiation of polymorphs and identification of unknown phases were shown to be useful.

SILICON NITRIDE - Silicon nitride is known to exist in two polymorphs - alpha and beta. The alpha phase occurs as a product of chemical vapor deposition, of the reaction of silane and ammonia, and of the low temperature nitridation of silicon. The beta phase occurs during the liquid phase consolidation occurring during hot-pressing. Raman spectra of the alpha and beta phases of Si_3N_4 are quite different as can be seen by examining Figure 7. These spectra agree well with the published spectra of Kuzuba, et. al.[18] and Wada, et. al.[19]. Figure 7 also shows spectra of silicon oxynitride, another phase which can occur in silicon nitride ceramics.

Figure 8a shows an optical micrograph of the fracture surface of a silicon nitride tensile test bar that was broken as part of the DARPA/AFOSR Quantitative Nondestructive Evaluation Program. Figure 8b shows an electron micrograph of the center of the fracture surface. The fracture markings indicate clearly that an inclusion acted as the origin of the fracture. The inclusion could not readily be examined by X-ray diffraction because of its size and the curvature of the surface. X-ray microanalysis detected only silicon. The Raman spectrum of this inclusion and a reference spectrum of 6H silicon carbide are shown in Figure 9. The reference spectrum is clearly a good match whereas none of the spectra in Figure 7 match that of the inclusion (nor is crystalline silicon with one strong band at 520.6cm^{-1} a possible match).

ZIRCONIA - As mentioned earlier in the example of the contaminant in the fiber optic zirconia exhibits more than one crystalline phase and the Raman spectra of them are good fingerprints[20,17]. Normally zirconia is monoclinic at room temperature, and transforms to tetrogonal and cubic phases at elevated temperatures. However, the higher temperature phases can be stabilized by mixing the starting material with additives such as alumina or yttria. Figure 10 shows spectra of pure monoclinic and room-temperature stabilized tetragonal zirconia reproduced from an early application of the use of the Raman microprobe to measure the size of the transformation zone of a stress-induced crack in tetragonal zirconia[21]. In fact, this is probably the first room temperature spectrum of pure tetragonal zirconia which enables one to clearly define a region diagnostic of polytype. The region between 100 and 300cm^{-1} exhibits bands of one form or the other which do not overlap. A calibration procedure was established in Reference 21 which provided the ability to plot the relative monoclinic concentration as a function of distance from a stress induced crack in a stabilized tetragonal sample. A representative set of data is presented in Figure 11.

Reference 17 shows an example of a particle of ThO_2 (fluorite structure) identified in a ZrO_2/ThO_2 ceramic sample, a material proposed for use as a tough electromagnetic window material. It was pointed out there that there are technical difficulties in characterizing the phases of this system by X-ray diffraction, but that the Raman microprobe provided a quick clear identification of the thoria particle.

CORROSION

During the last several years several laboratories have shown the usefulness of Raman spectroscopy to the study of corrosion films. Figure 12 shows the results of one application in which calcite (one phase of $CaCO_3$), magnetite (Fe_3O_4) and carbon were identified on the surface of a steel coupon exposed to a brine containing 2610 ppm NaCl, 1065 ppm Na_2SO_4, 855 ppm $CaCl_2$, 100atm. CO_2, at a temperature of 375^{0}F. It was confirmed that hematite (Fe_2O_3)1, a more oxidized phase, was not present. Based on the presence of the carbonate and carbon, one can argue that there is more than one redox reaction occurring at the metal surface. (Note that this steel contained neither chrominum to inhibit corrosion,

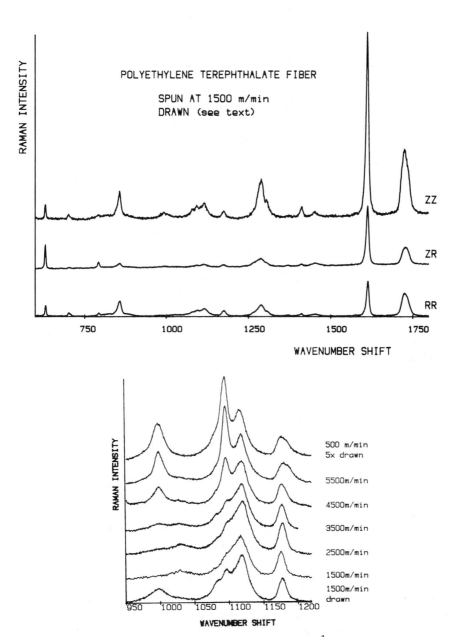

Figure 17 PMP specrta between 950 and 1220cm^{-1} of a series of fibers of
PET spun from the melt at speeds varying from 500m/min to 5500
m/min. The fiber whose spectrum is at the bottom of the figure
was drawn at room temperature. Fibers whose spectra are shown
in the top three traces were partially crystalline, as determined
by x-ray diffraction. This region of the spectrum reveals
information regarding the conformation of the atoms in the glycol
linkage.

nor carbon at levels greater than 0.2%).

Another example was the identification of the composition of a white particle removed from a heat exchanger. From energy dispersive X-ray analysis it was known that this particle was composed of silicon, oxygen, and small amounts of aluminum. Infrared spectroscopy could not, however, identify the phase of the sample because of the intensity of the Si-O band. The Raman spectrum acquired is reproduced in Figure 13. The band at ca.480cm^{-1} is diagnostic of zeolites, framework silicates which incorporate small amounts of aluminum in well-defined crystal sites[22]. In addition, bands of disordered carbon at 1360 and 1600cm^{-1} that are present indicate again the possible involvement of dissolved carbon dioxide in chemical processes that oxidize metal surfaces.

POLYMERS

DEFECT IDENTIFICATION - Films, fibers, or other morphologics of copolymers or polymer blends often exhibit defects that reflect non-uniform dispersal of the component polymers. One sample examined was a film of high impact polystyrene. Defects in the film were caused by inclusions in the film that were easily exposed by cross-sectioning. Figure 14 shows a portion of the spectrum of such an inclusion compared to that of the surrounding film. The doublet at 1600cm^{-1} arises from the aromatic group while the feature at ca.1650cm^{-1} is diagnostic of polybutadiene which is added to provide impact resistance. Clearly the defect is an inclusion of pure polystyrene.

SPIN-ORIENTATION - Polarized Raman spectra fibers of polyethylene terephthalate, spun from the melt with take-up speeds of 1500 to 5500 m/min were recored in order to measure the increasing orientation and crystallinity of the polymer as the take-up speed is increased[23]. Representative spectra are shown in Figure 15. Comparison of spectra from fibers spin oriented at 1500 and 5500 m/min show clearly the effects of orientation; in unoriented fibers the ZZ and RR intensities will be equivalent. When oriented, the ZZ and RR intensities will be quite different because the molecular axis will be aligned along the fiber axis and the Raman tensor is not isotropic in this material. From optical birefringence it was known that the 1500 m/min fiber had little orientation whereas the 5500 m/min fiber was highly oriented. The polarized Raman spectra confirm this orientation. Spectra of a fiber spun at 1500 m/min and drawn at room temperature also indicated a high degree of orientation, which was confirmed by birefringence measurements. In addition, measurements in the region between 950 and 1220cm^{-1}, Figure 16, which is sensitive to the conformation of the glycol part of the polymer, indicate order in this region that increases with the take-up speed. It is useful to realize, however, that the increasing molecular order of this part of the molecule can occur independently of crystallization; the bottom trace of Figure 16 was recorded from the non-crystalline, oriented fiber and shows substantial intensity at 1080 and 1096cm^{-1} in bands that presumably reflect trans conformation of the ester atoms[23]. In contrast, the carbonyl band width which has been the classical indicator of crystallinity, is decreased in fibers that are known to be partially crystalline (take-up speeds of 4500 and 5500 m/min). However, the carbonyl bandwidth in the drawn, non-crystalline fiber spun at 1500 m/min is very similar to that of the non-drawn fiber, (see the 1730cm^{-1} band in Figure 15).

In this polymer, a combination of polarization measurements and inspection of band contours enable separation of the molecular orientation of the polymer from its crystallization. In contrast all previous Raman measurements did not attempt to separate the effects of crystallization from orientation. The other physical measurements used on these samples enabled appropriate interpretation of the Raman data but were not redundant with it. Optical birefringence measures total orientation of fibers but cannot differentiate between amorphous and crystalline phases. X-ray diffraction can measure the degree of orientation of crystalline material only and is not amenable to probing individual fibers. Thus the Raman microprobe data provide additional structural information on these fibers.

SUMMARY

The development of the Raman microprobe MOLE^{R} has enhanced the ease of recording Raman spectra and the power of its application to technologically important materials manufacturing problems. Examples presented here have been derived from work or electronic devices, ceramics, corrosion scales and polymers. In all cases the information derived from the RMP could not have been extracted by other techniques because of size limitations or because of the molecular/crystalline content of the Raman spectra. The examples presented were meant to be representative of a large field of materials characterization where RMP spectra will certainly have increasing impact in the coming years.

REFERENCES

1. F. Adar, "Microbeam Analysis - 1981," R. H. Geiss, Editor; San Francisco Press, pp. 67-72 (1981).
2. M. Delhaye, P. Dhamelincourt, J. Raman Spectrosc. 3, 33 (1975).
3. P. Dhamelincourt, "Microbeam Analysis - 1982," K. F. J. Heinrich, Editor; San Francisco Press, pp. 261-269 (1982).
4. T. Hirschfeld, J. Opt. Soc. Am. 63, 476-7 (1973).
5. G. J. Rosasco, E. S. Etz, W. A. Cassatt, Appli. Spectrosc. 29, 396 (1975).

6. G. J. Rosasco, "Advances in Infrared and Raman Spectroscopy Volume 7," R. J. H. Clark, R. J. Hester, Editors, Heyden, London, Chapter 4 (1980).

7. P. Dhamelincourt, Phd. Thesis, L'Universite des Sciences et. Techniques de Lille, Lille (1978).

8. F. Adar, D. R. Clarke, "Microbeam Analysis - 1982," K. F. J. Heinrich, Editor, San Francisco Press, pp.307-310 (1982).

9. R. H. Atalla, B. E. Dimick, S. C. Nagel, Jr., Editors, ACS Symposium Series No. 48, American Chemical Society.

10. P. Zorabedian and F. Adar, Appl. Phys. Lett. 43, 177-179 (1983).

11. E. Anastassakis, A. Pinczuk, E. Burstein, F. H. Pollack, M. Cardona, Solid State Commun. 8, 133 (1970).

12. W. A. Sproson, K. B. Lyons, J. W. Fleming, J. Non-Crystal. Solid 45, 69-81 (1981).

13. U. Shibata, M. Horiguchi, T. Edahiro, J. Non-Crystal. Solids 45, 116-126 (1981).

14. L. Soto and F. Adar, "Microbeam Analysis - 1984," A. D. Romig and J. I. Goldstein, Editors, San Francisco Press, pp. 121-124 (1984).

15. G. Abstreiter, E. Bauser, A. Fischer, K. Ploog, Appl. Phys. (Springer Verlag) 16, 345-352 (1978).

16. G. P. Schwartz, G. J. Gualtieri, J. E. Griffiths, C. D. Thurmond B. Schwartz, J. Electrochem. Soc. 127, 2488 (1980).

17. D. R. Clarke and F. Adar, "Advances in Materials Characterization, Materials Science Research Volume 15," D. R. Rossingto R. A. Condrate and R. L. Snyder, Editors, Plenum, New York pp. 199-214 (1983).

18. T. Kuzuba, K. Kijima and Y. Bando, J. Chem. Phys. 69, 40 (1978).

19. N. Wada, S. A. Solin, J. Wong and S. Prochazka, J. Noncryst. Solid, 43, 7 (1981).

20. G. M. Phillippi and K. S. Mazdiyasni, J. Am. Ceram. Soc., 54, 254 (1971).

21. D. R. Clarke and F. Adar, J. Am. Ceram. Soc., 65, 284 (1982).

22. C. L. Angell, J. Phys. Chem. 77, 222 (1973).

23. F. Adar and H. Noether, submitted for publication.

X-RAY STRESS MEASUREMENT OF STRUCTURAL CERAMICS

C.O. Ruud[1], C.P. Gazzara[2], D.J. Snoha[1], D.P. Ivkovich[1], and D.P.H. Hasselman[3]

[1]The Pennsylvania State University, University Park, PA
[2]Army Materials and Mechanics Research Center
   Watertown, MA
[3]Virginia Polytechnic Institute and State University
   Blacksburg, VA

INTRODUCTION

This paper* describes the application of a unique x-ray diffraction stress instrument to the measurement of residual stresses in three structural ceramic components.

The inherent strength, high-temperature performance, and corrosion and erosion resistance of structural ceramics have promised marked improvement over certain metallic components in service performance. The military applications where this promise has been investigated include gun barrel tubes and turbine engines. Unfortunately, ceramics not only are inherently much stronger than metals, they are also much less ductile, i.e., more brittle. The fact that metals are ductile, i.e., will undergo a significant amount of plastic deformation before fracturing, provides for relief of unexpected high-stress concentration in components prior to their failure. However, the virtual absence of plastic deformation that might relieve stress concentrations, means that the stress profile must be accurately known in all of the critical areas of ceramic structural parts. It is therefore important that the residual stresses resulting from a ceramic component's fabrication and processing be known, especially in regions of high service stress concentration. A few attempts have been made to measure or estimate residual stresses in ceramic bodies. Kirchner[1] in studying the beneficial effects of compressive residual stresses attempted to estimate them through

*The work described herein was partially funded by the U.S. Department of the Army through the Army Materials and Mechanics Research Center under contract number DAAG46-83-K-0036.

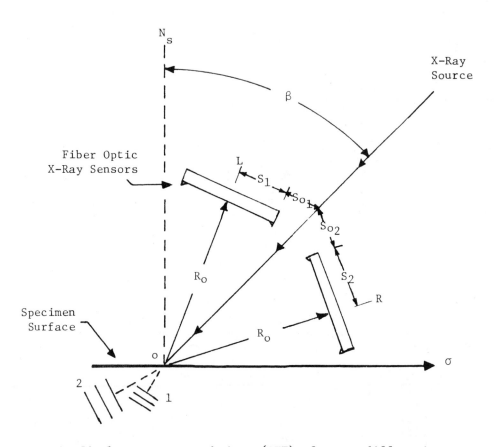

Fig. 1. Single-exposure technique (SET) of x-ray diffraction
residual stress measurement. The numbers 1 and 2 denote
certain crystallographic planes immediately below the
specimen surface. These planes are diffracting the
incident x-ray beam (note the arrows pointing toward the
specimen) backward to the fiber optic x-ray sensors (note
the arrows pointing toward S_1 and S_2). The angles between
the incident and diffracted beams are related to $S_1 + S_{o1}$
and $S_1 + S_{o2}$ and the stress at the surface of the specimen
at point o is directly proportional to $(S_2 + S_{o2})$ –
$(S_1 + S_{o1})$. The angle between the specimen surface normal
(N_s) and the incident x-ray beam is beta (β).

tests relating the flexural strength and fracture mirror radius. Most researchers, however, have used rather crude mechanical techniques, where the deflection or strain change in a component is measured as the result of material removal. The mechanical techniques are by nature destructive and provide only a coarse measurement of stress, so that in situations where high-stress gradients exist, the true magnitude of the existing stresses cannot be determined. Semple[2] proposed that X-ray diffraction (XRD) techniques be used to measure residual stress in alumina bodies. The XRD techniques are nondestructive for surface stress mapping and when used in conjunction with material removal techniques can provide excellent three-dimensional stress pattern determination. Unfortunately, until now the XRD techniques have been time consuming, ten or more minutes per reading under ideal conditions, and in many cases awkward to apply. In spite of this fact, portable x-ray diffractometers, for easier laboratory application and for field measurements, are still being produced.

The measurement of residual strains in ceramics has imposed more stringent conditions on the XRD method, and this study deals with these demands.

BACKGROUND

The Materials Research Laboratory of The Pennsylvania State University has developed an advanced XRD stress measuring instrument which provides for unprecedented stress measurement speed consistent with excellent accuracy. The instrument was developed in order to facilitate a more thorough understanding of residual stress patterns in crystalline materials. Since its installation, studies of nickel and copper base alloys as well as ferritic and stainless steels have been conducted providing a detailed understanding of the stress patterns in the subject components. This unique XRD instrument is based upon the use of a position sensitive scintillation detector (PSSD), the principle of which is based upon the coherent conversion of the diffracted x-ray pattern into an optical signal, i.e., light; the conduction of this light signal, over several linear centimeters of coherent, flexible fiber optic bundles; the amplification of this signal by electro optical image intensification; the electronic conversion of the signal; and the transfer of the electronic signal to a computer for refinement and interpretation[3]. In the present PSSD detector design a film of scintillation material has been adhered to one end of each of two flexible fiber optic bundles, 3 by 13 mm in cross-section. This scintillation-coated end of the fiber bundles is the x-ray sensitive area which is located in the vicinity of the x-ray source and specimen. The length of the bundles vary from tens of centimeters to two meters, depending upon the x-ray diffraction application. The other, non x-ray

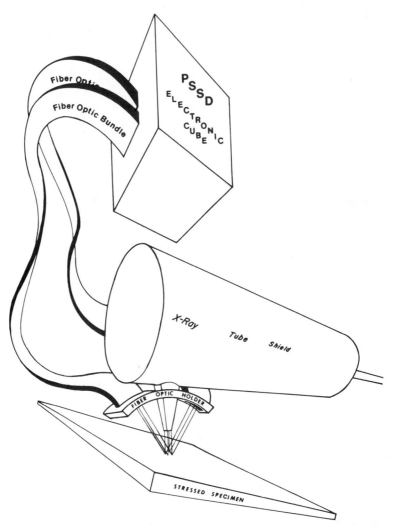

Fig. 2. Residual stress measurement application of the PSSD. This
sketch shows the x-ray source in the form of a shielded
x-ray tube, and the combination fiber optic holder/incident
x-ray beam collimator which provides the single-exposure
technique x-ray optics. The flexible fiber optic bundles
are shown as providing an optical coupling between the
scintillation coating and the PSSD electronic cube.

sensitive cross-sectional end of the flexible fiber optic bundles are optically coupled to a proximity focussed image intensifier which amplifies the visible light version of the x-ray pattern. The light signal then having been amplified several thousand times is transferred, via a rigid fiber optic couple, to two parallel self-scanning, linear silicon diode arrays where the light pattern is converted to an analogue electronic signal. This signal is then sent to an analogue-to-digital converter, then on to a PDP 11/03 computer where it is stored, refined, and interpreted. The image intensifier and the diode arrays are enclosed in a cube which is approximately 50 x 50 x 50 mm. This enclosure is hermetically sealed and the temperature-sensitive diode arrays are cooled by a solid-state cooler also contained in the electronics cube.

One advantage of this PSSD-based stress analyzer is that its pinhole focusing XRD optics can be applied to give a very small irradiated area on the specimen so as to provide for high spatial resolution of stress readings. Also, the geometry of its x-ray optics allows for stress readings in very confined areas such as gear teeth, holes, and turbine vane bases. The rapidity of stress measurement provides for data collection over a period of a few seconds which is often a two order of magnitude improvement over conventional XRD instrumentation. Furthermore, the speed and accuracy of this equipment has been demonstrated in tests on stainless steel where a precision of ±4 KSI (28 MPa) has been obtained in less than 30 seconds[4].

The high speed, consistent with accuracy, of this unique instrument is due to the two independent x-ray sensitive surfaces possessed by the detector[3,5]. This feature is especially useful for residual stress measurements in ceramics, where long counting times are required, due to the high elastic moduli, low Poisson's ratios, and the usually low peak-to-background ratios encountered in the back-reflecting diffraction peaks. These two independent x-ray channels allow this instrument to apply all three of the most commonly used residual stress measurement techniques, i.e., single-exposure, double-exposure, and sin-square-psi techniques[6]. In fact, when this PSSD is applied in the $\sin^2\psi$ mode, two $\sin^2\psi$ data points are collected simultaneously, thus reducing the standard deviation of the residual strain/stress, within the limited data collection time. All conventional laboratory diffractometer instruments and most demountable types are incapable of applying the single-exposure technique (SET) since it requires two separate detection surfaces as shown in Figure 1. Figure 2 shows a schematic drawing of the Ruud-Barrett PSSD (R-B PSSD) as it is usually applied to residual stress measurement.

Table 1

Characteristic Radiation and Crystallographic
Plane Combinations Tested for Alpha-Alumina

| X-RADIATION | MILLER INDICES | | | TWO-THETA |
	h	k	ℓ	(degrees)
Cu K–alpha	1	1	15	142.7
Cu K–alpha	4	0	10	145.5
Cu K–alpha	0	5	4	149.6
Cu K–alpha	1	0	16	150.6
Cu K–alpha	3	3	0	152.8
Fe K–alpha	1	2	10	152.3

Table 2

Characteristic Radiation and Crystallographic
Plane Combinations Tested for Alpha Silicon Carbide

| X-RADIATION | MILLER INDICES* | | | | TWO-THETA |
		h	k	ℓ	(degrees)
Cu K–alpha	(15R)	0	2	37	144.4
Cu K–alpha	(15R)	2	0	38	150.7
Fe K–alpha	(12H)	1	2	1	149.5
Fe K–alpha	(15R)	0	1	35	151.2
Fe K–alpha	(15R)	2	1	7	156.1
Cr K–alpha	(15R)	0	0	30	131.4
Cr K–alpha	(15R)	0	2	10	
Cr K–alpha	(12H)	0	0	12	
Cr K–alpha	(12H)	2	0	4	
Cr K–beta	(12H)	2	0	8	147.1
Cr K–beta	(15R)	0	2	22	163.7

*Note: The designations (15R) and (12H) refer to polytypes and
may be found in the JCPDS powder diffraction file as 22-1301 and
29-1131, respectively.

PROCEDURE AND RESULTS

Three types of ceramic specimens were selected for study: two were alumina (alpha Al$_2$O$_3$) and one was silicon carbide (alpha SiC). The first type of alumina specimen was represented by forty broken halves of rods, twenty of which had been thermally treated to provide compressive surface residual stresses. The second type of alumina specimen was a ceramic tile for composite armor. The third type of specimen was a silicon carbide gun tube insert, placed in a compressive stress state by a steel jacket.

X-Radiation Selection

Often in the XRD measurement of residual stress, a number of x-radiations might be used to provide the measurement. Table 1 shows the most promising combinations of radiation and crystallographic planes for alumina using two characteristic radiations, i.e., Cu and Fe. XRD patterns in the far back-reflection range, i.e., 130 to 160° two-theta were collected using the R-B PSSD instrument and a fine saphire (-400 mesh Al$_2$O$_3$) powder, as well as the specimens described herein. Figures 3 and 4 show CRT photos of the XRD patterns from Cu K-alpha and Fe K-alpha for the alumina armor specimen, respectively, and are typical of the patterns from the saphire powder. The data collection times for these two patterns were 30 and 45 seconds, respectively, with 45 KV and 12 Ma from a full-wave rectified x-ray power supply. The (1 0 16) peak using Cu K-alpha radiation was selected for the residual stress investigation of the alumina specimens. It should be noted that James et al.[10] mentioned use of a (6 0 2) peak in error; they used the (1 1 15) peak.

Table 2 shows the most promising crystallographic plane and x-radiation combinations for alpha silicon carbide using three radiations. XRD patterns in the far back-reflection range, i.e., 130 to 160° two-theta, were collected using the R-B PSSD instrument and a fine alpha silicon carbide (-400 mesh) powder obtained from the manufacturer of the gun tube insert material. Figures 5 and 6 show CRT photos of the XRD patterns from Cu K-alpha and Fe K-alpha for the gun insert (gun tube) specimen and are typical of patterns from the powder. The data collection times for these two patterns were 120 seconds at 35 KV and 13 Ma, and 240 seconds at 44 KV and 12 Ma, respectively, with a full-wave rectified x-ray power supply. The Cu and Fe K-alpha radiations were both used for residual stress measurement on the silicon carbide gun insert.

Calibration and Confirmation

Calibration of the R-B PSSD instrument was performed using the two -400 mesh powders mentioned previously and·setting the x-ray beam so that it was normal to the powder specimen surface,

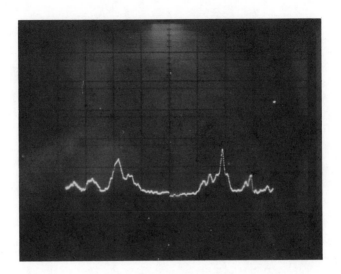

Fig. 3. CRT photos of the XRD pattern from the alumina armor block using Cu K-alpha x-radiation. The pattern shown is actually two patterns, one from each detector channel and are mirror images of each other. The center peak in each channel is approximately 150.6° two-theta.

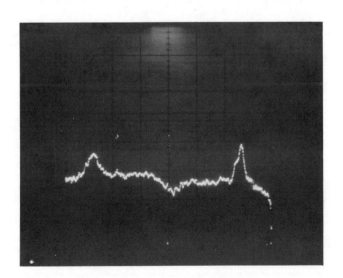

Fig. 4. CRT photos of the XRD pattern from the alumina armor block using Fe K-alpha x-radiation. The pattern shown is actually two patterns, one from each detector channel and are mirror images of each other. The peak shown in each channel is approximately 152.3° two-theta.

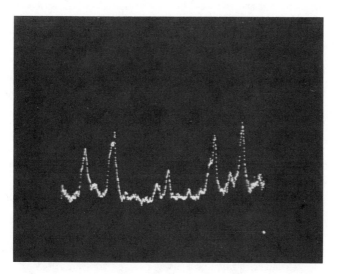

Fig. 5. CRT photos of the XRD pattern from the silicon carbide gun
tube insert using Cu K-alpha radiation. The pattern shown
is actually two patterns, one from each detector channel
and are mirror images of each other. The two peaks shown
on each channel are at approximately 144.4 and 150.7° for
the outer and inner, respectively.

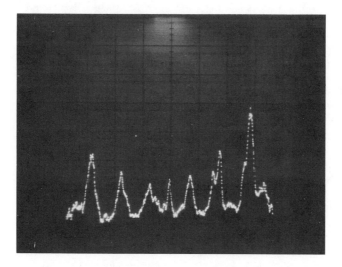

Fig. 6. CRT photos of the XRD pattern from the silicon carbide gun
tube insert using Fe K-alpha radiation. The pattern shown
is actually two patterns, one from each detector channel
and are mirror images of each other. The three peaks
shown on each channel are at approximately 149.5, 151.2,
and 156.1°, respectively, from the outside edge of the
photo to the center.

i.e., $\beta = 0°$. Confirmation as to the validity of the calibration was performed on each powder with the x-ray beam incident at thirty degrees, i.e., $\beta = 30°$. The incident x-ray beam in each case was collimated to provide the same shape and size of irradiated area that was used for stress measurement on the various specimens. The specimen-to-detector distance (R_O, in Figure 1) was allowed to vary 40 ± 2 mm for the powder readings to confirm the relative insensitivity of the R-B PSSD to errors in R_O[5]. Several readings were taken at various R_O distances between 38 and 42 mm and the mean and standard deviation of this value was obtained. Table 3 shows the mean and standard deviation of nine readings obtained under the above stated conditions for the two ceramic powders and the x-radiations subsequently used for residual stress measurement. The mean stress should be nearly zero if the calibration is valid, since a ceramic powder will not sustain a macro residual stress between the powder particles. Such a zero stress standard has been accepted by ASTM for metal, and data for an iron powder is included in Table 3, since this report also shows data from the steel sleeve and jacket constraining the silicon carbide gun tube insert.

Elastic Constant Determination

It is recommended for XRD residual stress measurement that empirical derivations of the elastic constants E and ν used to calculate stress be performed[7]. This derivation is recommended because the constants usually found in the literature are bulk values representing the average elastic properties for all crystallographic directions. However, in the XRD stress measurement technique, strain measurements in a single selected crystallographic direction, dependent upon the Miller indices of the diffracting planes, are used. The difference between the bulk values and the empirical XRD values can be substantial; however, for the ceramic stress measurements in this study, bulk constants were used because specimens suitable for experimental determination of E and ν were not available. Table 4 shows the elastic constants used for the two ceramics and the steel.

Residual Stresses in Thermally Treated Al_2O_3 Rods

The forty alumina rods in which residual stresses were measured were actually the broken halves of twenty Al-300 alumina rods, 0.20 inches (5 mm) in diameter and are described by Tree et al.[8]. Ten of these rods had been annealed at 1000°C for 8 hours and slow cooled so as to provide as stress-free a material as possible, and ten had been annealed at 1000°C for 8 hours, equilibrated at 1450°C, then force air-quenched to provide a compressive surface residual stress.

Table 3

Results of the Calibration Confirmation on Zero Stress Powders

X-RADIATION	POWDER	RESIDUAL STRESS IN KSI (MPa)	
		Mean	S.D.
CuKα	α-Al₂O₃	2.2 (15)	±3.5 (±25)
CuKα	α-SiC	3.9 (27)	±3.8 (±27)
FeKα	α-SiC	-5.0 (-35)	±6.7 (±47)
CrKα	Fe	-1.3 (-9)	±2.0 (±14)

Table 4

Elastic Constants Used for XRD Stress Calculation

MATERIAL	YOUNG'S MODULUS E	POISSON'S RATIO ν
α-Al₂O₃	46.0 x 10^6 psi (324 x 10^3 MPa)	0.21
α-SiC	59.4 x 10^6 psi (418 x 10^3 MPa)	0.142
Steel	30.0 x 10^6 psi (211 x 10^3 MPa)	0.30

Tangential residual stress measurements were performed upon the broken halves of the twenty specimens by locating a line-shaped irradiated area, 0.04 x 0.51 inches (1 x 13 mm) of copper K-alpha radiation, such that the long dimension was parallel to the rod axis and near the center of the length of the rod piece. Measurements were performed with the rods in a horizontal position and the incident beam at a sixty-degree angle to the tangent of the convex surface at the irradiated area so as to provide a beta angle of 30 degrees (see Figure 1). The rods were rotated at o.15 rps during stress measurement so as to obtain the average of the stress around the circumference. The time of data collection from each rod was 30 seconds with a copper tube at 40 KV and 13 Ma from a full-wave rectified x-ray power supply. All of the measurements were performed in less than six hours, including three repeats of data accumulation and stress calculation, and the placement of the specimens in the rotation device for x-ray data collection.

The XRD measured tangential residual stress is plotted versus the breaking strength data in Figure 7. A linear regression fit to all the data from both types of specimens shows a correlation of 90%. This would indicate that the XRD residual stress measurements performed herein could be used to predict the breaking strength within a standard deviation of ±9.7 KSI (±68 MPa).

The XRD data indicates that the quenched rods possessed an average tangential stress of 12 KSI (83 MPa) more compressive than the annealed rods. However, these XRD stresses were calculated using bulk elastic constants, a practice that can produce errors greater than ten percent in reasonably isotropic metals, and possibly much larger errors in an anisotropic material, such as alumina. However, such an error would affect the calculated residual stresses proportionally. Also, it is not known how well the zero residual stress saphire (alumina) powder represents the crystallographic, elastic anisotropic condition of the rods. This could cause an absolute error in the stresses determined; thus the average stress value from the annealed specimens was assumed to represent zero stress, and all other readings from the rods normalized against that average. Further, a slight position error due to defocussing of the x-ray beam by the convex surface of the rod is predicted, but calculations indicate that that error would be on the order of 1 KSI (7 MPa) and therefore negligible[6]. The difference in the XRD results from the annealed versus the quenched rods is in good agreement with Semple[2] who measured surface stresses in AD-94 and Lucalux alumina 0.25 x 0.25 x 4.0 inches (6.4 x 6.4 x 101.6 mm) bars by XRD and noted that the sum of the axial and transverse stresses were about 15-20 KSI (105-140 MPa) more compressive than annealed bars.

The precise relation between the residual tangential surface stress and the breaking strength can not be calculated unless the

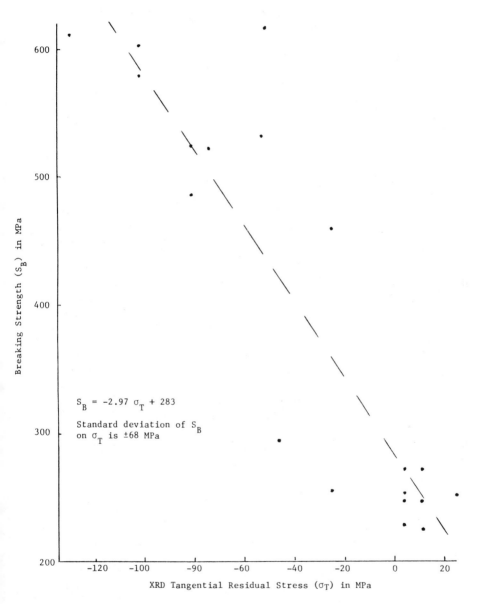

Fig. 7. Breaking strength plotted vs XRD determined surface
tangential stress for annealed and quenched Al₂O₃ rods.

axial stress, as well as the gradients for the stresses into the rod interior, are known. However, qualitatively, the XRD tangential results compare well with the breaking strength, as illustrated by the plot in Figure 7 and the correlation of 0.90 for the linear regression fit of this plot. The standard deviation of the line fit implies that the breaking strength of the rods, which varied between about 29 and 86 KSI (200 and 600 MPa), can be predicted by XRD tangential stress measurements to within ±10 KSI (70 MPa).

Residual Stresses in Alumina Tile

The second type of alpha alumina specimen in which residual stresses were measured in this study was a 0.5 x 5.6 x 5.7 inch (13 x 142 x 145 mm) Coors AD-94 sintered alumina armor tile. Measurement was performed in only three areas of the tile using the same size of irradiated area as used for the alumina rods, i.e., 0.04 x 0.51 inches (1 x 13 mm). Also, the same data accumulation times and x-ray tube power settings were used. Figure 8 shows a sketch of the tile and indicates the position, direction, and value of the stress readings. These measurements were made holding the specimen-to-detector distance within ±0.004 inches (±0.1 mm); therefore, the precision was better than that indicated in Table 3 and was less than ±3 KSI. The absolute accuracy of the residual stress measurements on the tile is not known; however, comparison of the readings is valid and indicates a trend toward more tensile residual stresses in the center of the tile. Further, the residual stresses perpendicular to the edges seems to be more tensile than those parallel to the edges. The aforementioned trends in residual stresses on the tile could be rationalized as due to cooling since the corner would cool the fastest, thereby tending to develop compressive residual stress, and the center cooling slowest, thereby tending toward more tensive stress. However, caution must be observed in drawing conclusions from this cursory examination of an armor tile.

Residual Stresses in a Silicon Carbide Liner

The third type of specimen in which residual stresses were measured in this study was a silicon carbide 50 calibre gun tube liner. The liner was the inner tube of a composite gun tube with two outer steel tubes. The two steel tubes were designated the sleeve and jacket; with the sleeve shrunk fit into the jacket with a 0.002 inch (0.05 mm) interference. The silicon carbide liner was shrunk fit into the sleeve-jacket assembly with a 0.003 inch (0.08 mm) interference. The result of these interferences was to produce compressive radial and tangential stresses on the ceramic liner. A paper by Bunning et al.[9] discusses the procedure in some detail and the predicted stress patterns shown in Figure 9 were obtained from that reference.

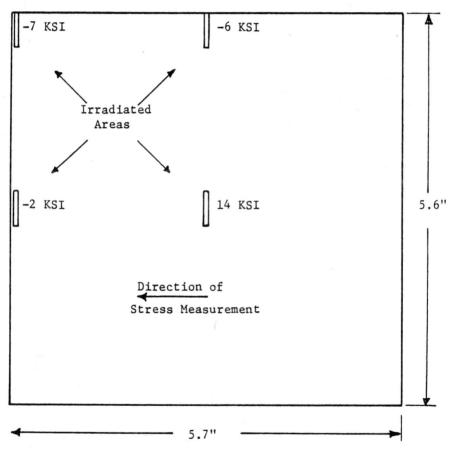

Fig. 8. Residual stress measurement results from a 0.5 inch (13 mm) thick Coors AD-94 sintered alumina armor tile.

The specimen examined was an approximately 2.1 inch (53 mm) long segment of the composite gun tube that had been sectioned

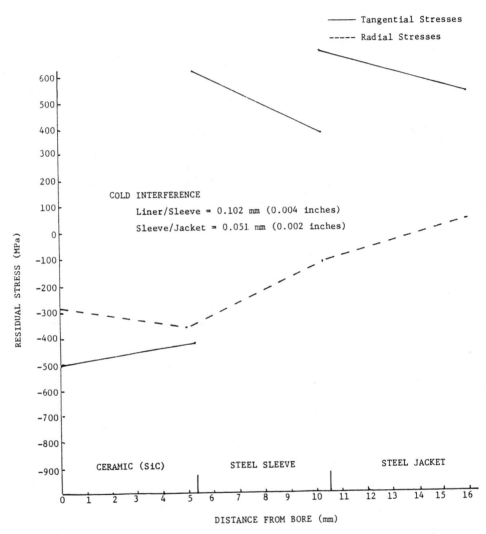

Fig. 9 Predicted radial and tangential stress patterns in the steel and ceramic tubes of a shrink-fit composite 50 calibre gun tube, assuming cold interferences of 0.004 inches (0.10 mm) and 0.002 inches (0.05 mm) for the liner--sieve and steel--jacket interferences, respectively.

through a plane normal to the axis from the longer original tube. The sawed surface had been subsequently mechanically polished and stresses in the ceramic were measured in that condition. However, subsequently, the steel cross-section of the specimen was electro-polished so as to remove the stresses induced by mechanical polishing prior to stress measurement. The tangential and radial stresses induced by the shrink fits in the ceramic liner were measured using both copper and iron K-alpha radiation. These same stress directions were measured in the steel cross-section using Cr K-alpha radiation. The stresses were measured along three radial traverses approximately óne hundred and twenty degrees apart.

The XRD parameters for stress measurement in the silicon carbide liner and steel sleeve and jacket are listed in Table 5 for the various radiations used. The specimen-to-detector distance, i.e., R_0 in Figure 1, was maintained at 1.57 ± 0.004 inches (40 ± 0.1 mm) throughout the measurements and the incident beam inclination to the surface normal, i.e., β in Figure 1, at thirty degrees. The R_0 distance is automatically calculated by the R-B PSSD software so that it is convenient to maintain a close tolerance in R_0 on a flat, polished surface. However, the tolerance in R_0 could have been increased tenfold without significantly increasing the error in stress measurement.

Figures 10 and 11 show plots of the induced stress patterns measured on the cross-section of the composite gun barrel from one of the radial traverses, i.e., traverse number one. Figure 10 shows the tangential stress distributions, as determined using two x-radiations for the ceramic and chromium K-alpha for the steel. Data from the various radiations are plotted as symbols and the key to those symbols indicate the estimated precision of the readings as bars. The stresses in the steel are lower than the calculated (predicted) stresses as might be expected with the smaller shrink fit, i.e., 0.003 for the experimental, instead of 0.004 as used for the calculated stresses. The estimated precision and accuracy of the stress measurement on the steel is 2 KSI (14 MPa). The measured stresses on the ceramic liner are less compressive than predicted, which is logical since the steel sleeve and jacket stresses were lower. The measured stresses from the two types of radiation are in good agreement; however, only one measurement was possible in the tangential direction for the Fe K-alpha radiation since this placed the long dimension of the irradiated area along the radius of the liner. As is shown in Figure 10, the estimated precision of measurements on the silicon carbide using copper radiation is better than for iron radiation.

Figure 11 shows the radial stress distribution using iron radiation for the ceramic and chromium for the steel. No data was

Table 5

X-Ray Diffraction Stress Measurement Conditions

for the Composite 50 Calibre Gun Tube

($R_o \simeq 1.6$ inches (40 mm) and $\beta = 30°$)

MATERIAL	RADIATION	X-RAY POWER		DATA ACCUMU- LATION TIME	SIZE OF IRRADIATED AREA	
		KV	Ma	Sec	Inches	(mm)
SiC	Fe K-α	44	10	240	0.04 x 0.16	(1 x 4)
SiC	Cu K-α	35	20	120	0.04 x 0.08	(1 x 2)
Steel	Cr K-α	40	10	75	0.04 x 0.06	(1 x 1.5)

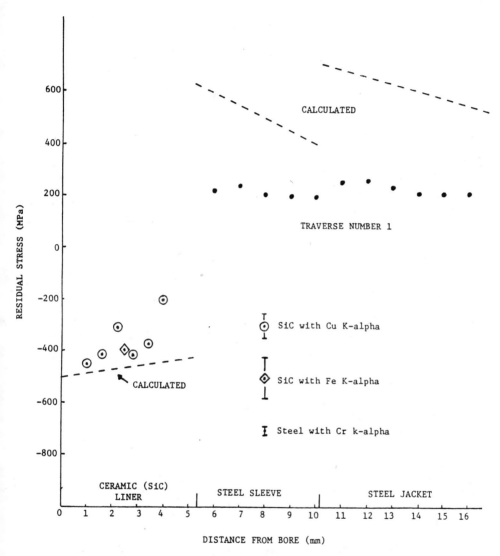

Fig. 10. Measured and calculated (predicted) stresses induced by shrink-fit assembly of a composite 50 calibre gun tube.

available for the ceramic with copper K-alpha radiation due to the inability of the R–B PSSD software to adequately fit the (2 0 37)

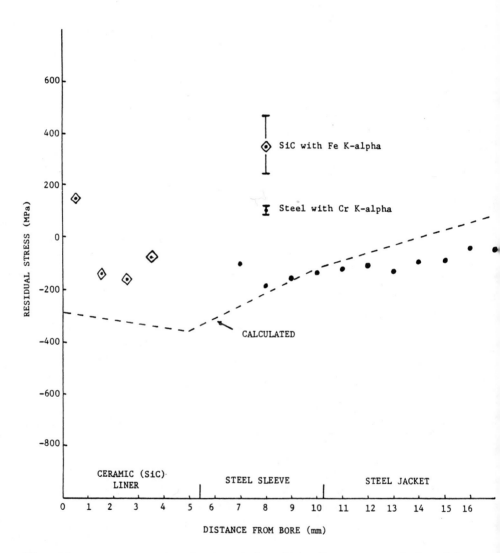

Fig. 11. Measured and calculated (predicted) stresses induced by shrink-fit assembly of a composite 50 calibre gun tube.

peak of silicon carbide. The stresses in the steel are lower, i.e., more compressive, at the outside radius than predicted and show a smaller gradient from the outside to the inside radius of the steel components. The radial stresses in the ceramic are less compressive than predicted which is again consistent with the lower shrink-fit interference.

It should be noted that Bunning et al.[9] predicted tensive stresses in the axial direction in the steel portion of the composite gun tube and compressive in the ceramic. Since the specimen studied was a cross-section of the original tube, it is reasonable to expect that the cutting of the specimen produced some stress relief in the axial direction due to the creation of the free surface studied. In other words, the axial stresses would have to be reduced to zero at the surface upon which the tangential and radial stress measurements were made. This stress relief would cause the stresses in the plane of the cross-section to become more tensile in the steel and more compressive in the ceramic. No correction for this free surface stress relief was made to the data shown in Figures 10 and 11.

SUMMARY AND RECOMMENDATIONS

Application of the XRD stress instrument developed at The Pennsylvania State University to residual stress measurements in alpha alumina and alpha silicon carbide structural ceramic components showed that it provided improved speed, accuracy, resolution, and ease of measurements over previous work described in the literature. Further, the data described in this paper represents a significant contribution to the total number of XRD stress measurements on ceramics reported in the literature. The precision reported herein is at least as good as previously reported, i.e., approximately 3 KSI (21 MPa) for alumina and 4 KSI (28 MPa) for silicon carbide with copper radiation. Finally, the rapidity of XRD stress measurement is one to two orders of magnitude faster than with other types of instrumentation, and the superiority of the R-B PSSD for residual stress measurements in ceramics has been demonstrated.

However, more work needs to be performed in the development of software to provide accurate location of the narrow ceramic x-ray peaks. Also, software to allow use of all of the several XRD peaks available in the back-reflection pattern, see Figures 3-6, in residual stress determination would most certainly improve the accuracy and precision of stress measurement. Finally, it is most important that the elastic constants used in the calculation of residual stress be derived empirically and suitable specimens and strain applying devices must be fabricated and/or developed.

Further investigation needs to be performed on the thermally treated AL-300 rods to provide surface axial residual stress data which should provide a better correlation to breaking strength than the XRD measured tangential stresses reported herein. Also, a more thorough study of the composite gun tube should be performed by examining several cross-sections at more than three radial traverses.

CONCLUSION

1. Of the three types of characteristic radiations investigated for XRD stress measurement of alpha alumina and silicon carbide, copper K-alpha provides the best precision.
2. The XRD tangential residual stress data in the thermally treated AL-300 alumina rods correlated well with breaking strengths.
3. The XRD residual stress data on the sintered alumina armor tile indicated a more tensile surface stress at its center.
4. The XRD stress distribution on the alpha silicon carbide gun tube insert was compatible with the predicted stress pattern.
5. The precision of XRD stress measurement on alpha alumina was approximately 3 KSI (21 MPa) and on alpha silicon carbide was 4 KSI (28 MPa), using copper K-alpha radiation.

ACKNOWLEDGMENT

The investigation described herein was in part made possible by funding from the U.S. Department of the Army through the Army Materials and Mechanics Research Center under contract number DAAG46-83-K-0036.

The authors wish to thank P. Wong and G. Quinn, of AMMRC, for the SiC gun tube section and the ceramic armor plate.

REFERENCES

1. H.P. Kirchner, Strengthening of Ceramics, Treatments, Tests, and Design Applications, Marcel Dekker Inc., N.Y. & Basel, (1979).
2. C.W. Semple, "Residual Stress Determinations in Alumina Bodies," Report No. AMMRC TR70-14, June (1970).
3. C.O. Ruud, "A Unique Position Sensitive Detector for X-Ray Powder Diffraction," Ind. Res. and Dev. Magazine, December (1982).

4. C.O. Ruud, P.S. DiMascio, and D.M. Melcher, "Application of a Position Sensitive Scintillation Detector for Nondestructive Residual Stress Measurement Inside Stainless Steel Piping," Adv. in X-Ray Anal., Plenum Press, Vol. 29, (1983).

5. C.O. Ruud and D.J. Snoha, "Displacement Errors in the Application of Portable X-Ray Diffraction Stress Measurement Instrumentation," J. of Met., Vol. 36, No. 2, February, pp. 32-38, (1984).

6. SAE, "Residual Stress Measurement by X-Ray Diffraction-J784a," Soc. of Auto. Eng., Warrendale, PA, (1971).

7. P.S. Prevey, "A Method of Determining the Elastic Properties of Alloys in Selected Crystallographic Directions for X-Ray Diffraction Residual Stress Measurement," Adv. in X-Ray Anal., Vol. 20, Plenum Press, pp. 345-354, (1977).

8. Y. Tree, A. Venkateswaran, and D.P.H. Hasselman, "Observations on the Fracture and Deformation Behavior During Annealing of Residually Stressed Polycrystalline Alumina Oxides," J. Mat. Sci., Vol. 18, pp. 2135-2148, (1983).

9. E.J. Bunning, D.R. Claxton, and R.A. Giles, "Ceramic Liners for Gun Tubes: A Feasibility Study," Presented at the Ceramic-Metal Systems Division, Fifth Annual Conf., Merrit Island, FL, January 20, (1981).

10. M.R. James, D.J. Green, and F.F. Lange, "Determination of Residual Stresses in Transformation Toughened Ceramics," Adv. in X-Ray Anal., Vol. 27, pp. 221-228, (1984).

A RENAISSANCE IN NEUTRON RADIOGRAPHY VIA ACCELERATOR NEUTRON SOURCES

John J. Antal

Army Materials and Mechanics Research Center
Materials Characterization Division
Boston, MA 02172

ABSTRACT

As a nondestructive inspection tool, neutron radiography could complement the application of x-ray radiography to in-process inspection. However, the best source of neutrons for this purpose is the nuclear reactor which is both uneconomical and inconvenient to apply to the task. Small accelerator sources of neutrons seem to be a way of bringing neutron radiography out of the reactor and into the workplace. The recent successful demonstration of the usefulness of a mobile small accelerator-based system by AMMRC has awakened many to consider new applications for neutron radiography. The paper includes a review of neutron radiography technology, particularly the problem of finding an efficient and economical source of neutrons.

INTRODUCTION

We know radiography as the imaging of internal components of bodies by employing penetrating radiation in a nondestructive manner. Any radiation may be used, but x-rays, gamma-rays and neutrons are the only radiations which have found wide application. Radiography has generally been applied after-the-fact. After a weld is made, it is inspected for soundness by radiography; after a crash landing, an aircraft landing gear is inspected for signs of defects formed earlier. Radiography is an inspection technique which needs to be reevaluated for a change from its traditional role of final inspection to the evaluation of materials in the process of becoming final products. Neutron radiography would seem to be particularly far from this sort of application, but recent advances in hardware demand a new look at the emerging possibilities.

Historically, neutron radiography has grown step-wise with a
spurt of development followed by a long period of relative idleness.
The technique of neutron radiography was well-founded in the 1960's[1],

Fig. 1. X-ray Radiography Setup. (A) Recording film; (B) Intensi-
fying Screen (when necessary); (C) Object to be radiographed; (D)
Aperture to reduce stray radiation; (F) X-ray source.

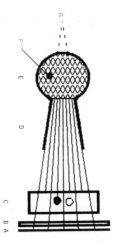

Fig. 2. Neutron Radiography Setup. (A) Recording film; (B) Converter
Screen; (C) Object to be radiographed; (D) Collimator; (E) Moderator
assembly; (F) Source of high-energy neutrons; (G) Location of
virtual point source.

but there were essentially no problems for it to solve. By 1969, neutron radiography capabilities had become available at many reactor facilities and one author was encouraged to write: "There is strong evidence that we are on the threshold of a new era for neutron radiography"[2]. Then the following years saw a multitude of research reactors shut down and their neutron radiographic capabilities dismantled. In the mid 1970's the third flurry of activity took place when the then Atomic Energy Commission in the U.S.A. began a search for uses of the man-made neutron-emitting isotope californium-252[3]. For the first time neutron radiographers could hope for a simple, even portable source of neutrons which would allow them to apply their trade beyond the confines of the nuclear reactor. But this source's constant need for shielding and surveillance discouraged many who might have had a use for the technique, and activities in neutron radiography leveled off once more. Now it is the mid-1980's and a fourth reawakening to the promise of neutron radiography is taking place following advances in hardware sponsored by the Army Materials and Mechanics Research Center. The remainder of this paper will explore the origin and development of these advances.

NEUTRON RADIOGRAPHY

Neutron radiography should not be segregated in the mind from other forms of radiography, particularly x-radiography. This highly-developed technology of tremendous versatility will remain king of radiographies for a long time and is the measure of all other radiographic processes. In a way this has been unfortunate for neutron radiography, because its application to many problems has been held back, I believe, by the desire always to obtain the perfection of image in the neutron case that is easily attained in the x-ray case.

In addition to personnel protection provided by shielding, radiography is accomplished with three items:

1. A source of radiation
2. A sample to be examined
3. An imaging detector

Figure 1 shows the arrangement of these items for x-ray radiography with the image reproduced on film. The detector in the case of either x-rays or neutrons can be x-ray film. Sometimes a lead screen or a fluorescent screen is placed on either side of a film to enhance the x-ray image. In the case of neutron radiation, an additional screen is absolutely necessary to convert the neutrons to a type of radiation-light, beta-rays or gamma-rays - to which the film is sensitive. A nuclear reaction in the screen material produces the desired radiation when a neutron is absorbed by the nuclei of special elements in the screen such as gadolinium.

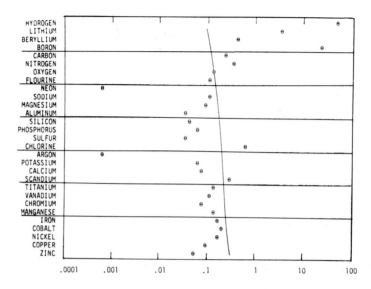

Fig. 3. Mass Attenuation Coefficients for a number of light
elements. The continuous line denotes values for 125 KeV X-rays.
The points denote values for 0.025 eV neutrons interacting with
elements having their natural isotopic abundance.

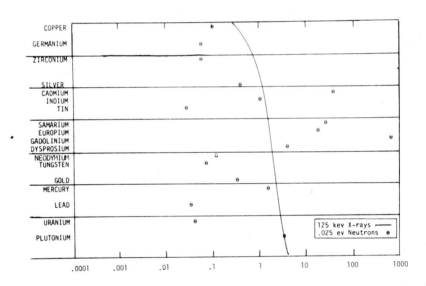

Fig. 4. Mass Attenuation Coefficients for a number of heavier
elements. The continuous line denotes values for 125 KeV X-rays.
The points denote values for the interaction of 0.025 eV neutrons
with elements having their natural isotopic abundance.

The detecting system can also be a fluorescent screen which is viewed by an electronic imaging device which presents the image information to a viewer on the face of a cathode-ray tube. The image information may also be deposited into an electronic memory at the same time for later processing. All of this technology except the fluorescent screen upon which the radiation first falls can be borrowed directly from other fields[4] such as astronomy. When images are produced in rapid succession the process is called "real-time" imaging.

The sample or item to be evaluated must have a density or thickness and interaction characteristics which are compatible with the radiation used, particularly the radiation energy. As the energy increases, the sample thickness can increase, but the degree of interaction with the radiation may also change.

The source of radiation should be geometrically as close to a point-source as possible. Since within an x-ray tube electrons are easily focussed onto the target, x-ray source diameters are generally good point sources 2 to 3 millimeters in diameter. Neutrons, being uncharged, cannot be focussed by magnetic or electric fields and thus point sources are not possible for neutrons in the most desirable energy range. Neutrons are the product of high-energy nuclear processes which impart high kinetic energies to the emitted neutrons. It is necessary to reduce neutron energies some 8 decades to about 0.025 electron volts before directing them toward the sample.

The process of moderation in energy is illustrated in Figure 2. Neutrons with their high initial energies are injected into a material such as a plastic or an organic fluid which contains large amounts of the element hydrogen. While making billiard ball-like collisions with the nuclei of the hydrogen atoms, the neutrons lose kinetic energy and quickly equilibrate near 0.025 electron volts[5]. Unfortunately, they also lose their original direction in these collisions and all that can be done is to direct them toward the sample with a collimating horn. This collimator provides a "virtual" point source indicated by the dashed lines in Figure 2 and works quite well, but it is not perfect. Even in the smallest sources, the volume of moderator providing the slowed-down neutrons (now called "thermal" neutrons) is about 4 cm in diameter, about 13X the size of the usual x-ray source size.

The legacy of an extended source rather than a point source in neutron radiographic systems is that high resolution imaging is not attained efficiently. The resolution of a collimator is measured by what is spoken of as the "L over D ratio":

L/D = (length of collimator)/(diameter of entrance aperture).

Fig. 5. Neutron radiograph of a variety of small-arms ammunition. Reading counterclockwise from the lower left corner, Two 30mm APDS-T projectiles, two 25mm 792 projectiles, 3 12-gauge shotgun shells, 3 45-caliber ball ammunition, 2 clips of tracer ammunition and 3 38-caliber special rounds. In all cases the powder charge is well detailed although least well in the shotgun shells where the plastic-paper outer case attenuated the neutrons. Hydrogen is the primary imaging element here.

Table I. Summary of Neutron Radiography Uses

CATEGORY	FOR
Detection of Hydrogen-Containing Materials	Imaging of: Water and Moisture Adhesives Corrosion Products Plastics Rubbers Explosives and Pyrotechnics Composite-Metal Assemblies
Determining Presence of Light-Weight Components in Heavy Casings	Inspection of: Explosively-Activated Items Ammunition Rounds Hydraulic Assemblies Absorbers in Nuclear Components
Differentiation of Materials with Similar Density and Atomic Number	Inspection of: Brazed Joints Electronic Devices Nuclear Fuels Lightweight Composite Materials
Examination of Radioactive Materials and Components	Inspection of: Nuclear Fuels Nuclear Reactor Structures
Where X-ray Radiography May Fail	

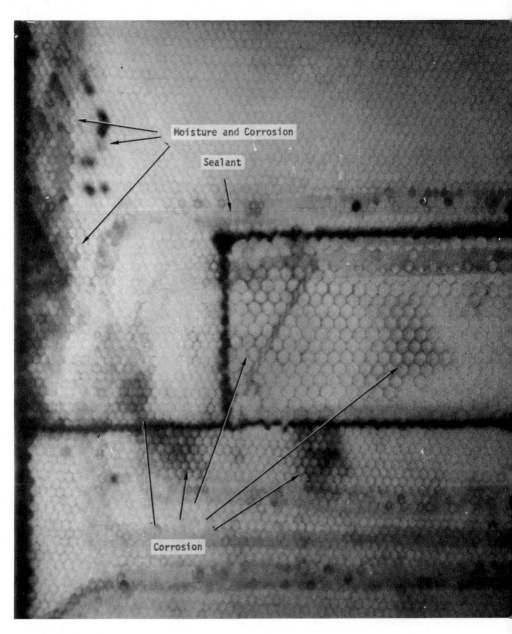

Fig. 6. Neutron Radiography of an aircraft honeycomb-filled airframe panel. This repaired panel shows a newer honeycomb insert of larger cell size and many areas of corrosion from retained moisture. This aluminum honeycomb retains water in its corrosion products which image readily because of the presence of hydrogen.

As the L/D ratio increases, the radiation becomes less divergent, and the resolution possible at the image plane increases. Typically, L/D may vary from 20/1 to 100/1. In order to increase resolution, the collimator is made longer or the entrance port is reduced in size, either procedure leading to a decreased intensity at the sample position. Decreased intensities require lengthened exposures and thus there exists a trade-off between resolution and radiographic exposure time. This is the major consideration in evaluating the usefulness of a small neutron source such as radioactive californium-252.

In order to record an image of a part of an object being radiographed, it must interact with the radiation in a sufficiently different manner from the surrounding material that a contrast is produced on the recording film. To produce contrast the part must be either of a different material than its surroundings or be of a different thickness. Success at imaging a part of an object can often be judged in advance by studying the variation of the mass attenuation coefficients for the elements which make up the object. These are plotted for many elements in Figures 3 and 4. From this plotted data, one can quickly note three areas where neutron radiography might be particularly useful: The region of the lightest elements, the rare earths, and the heavy metals. The mass attenuation coefficients vary rapidly by large amounts for the lightest elements and thus provide good contrast sensitivity in neutron radiographs. The very large neutron interaction with hydrogen provides the major reason for the usefulness of neutron radiography in its ability to image readily water and organic compounds. It is in this light element area that x-rays fail to image well and thereby neutron radiography is often seen as a complementary technique to x-ray radiography.

The rare earth elements and the metal cadmium absorb neutrons readily and are relatively easily detected in radiographed objects. Amongst the heavy elements, which are very absorbing of x-rays, bismuth, uranium, tungsten, lead and tin are relatively transparent to neutrons. Again neutron radiography provides a service which is complementary to x-ray radiography for these materials. Most ceramic materials are quite transparent to neutrons, although boron carbide is exceptionally opaque.

Figures 5 and 6 are radiographs of items of interest to the U.S. Army which illustrate some of the features discussed for both x-ray and neutron radiation. Table I lists areas where neutron radiography has been useful.

NEUTRON SOURCE SYSTEMS

With the obvious complementarity of these two techniques, one might ask why all NDE facilities do not have neutron and x-ray

Table II. Approximate Neutron Source Characteristics

Source	Volume (cm^3)	Heat Generated (WATTS)	Neutron Radiography Flux* (N/cm^2/sec)
Reactor	2×10^8	10,000	1×10^7
Van de Graff	1×10^7		1×10^6
Kaman A-711	1.5×10^3	300	3×10^4
Pu-Be	67	200	7×10^2
Cf-252	2	1.6	2×10^4

* Ave energy = 0.025 eV, Ave wavelength = 1.8 Å

Fig. 7. The AMMRC mobile accelerator neutron radiography system demonstrating radiography of an Army helicopter in the field. The exposure was controlled from the white trailer at left center of the photograph. The location is the Army's Yuma Proving Ground in Arizona.

systems working side by side. The answer is that neutron sources of physical characteristics which provide mobility and exposure capabilities similar to x-rays are not to be found[6]. The nuclear reactor has been and remains the best source of neutrons for radiography today. But the reactor is such a complex facility to own and operate that outside the reactor itself it has not found a home in the workplaces which could benefit from neutron radiography. It requires that objects to be radiographed be limited in size and that they be brought to the source. Too often these are requirements which are very restrictive.

The Department of Defense has requirements for nondestructive methods which can evaluate the integrity of aging airframe structures in its rework and maintenance shops. The required examinations should detect corrosion and moisture entrapment. Neutron radiography could provide this handily if a suitable mobile source of neutrons can be found. Other than the nuclear reactor, accelerator and radioactive sources are available. A listing of the characteristics of representative sources is given in Table II. The limitations of the reactor have been discussed. The large Cockroft-Walton and Van deGraff accelerators are being used as radiography sources in fixed locations, but they are inefficient and require more maintenance than is desirable for the radiation flux obtained. Of those listed in Table II, only the radioactive sources and the Kaman A711 accelerator source are physically as small as x-ray sources.

The U. S. Navy, in conjunction with the IRT Corporation, explore the possibility of using californium-252 as a mobile neutron source for radiography. In operation, a neutron camera unit consisting of a shielded collimator assembly which could be manipulated by a crane was connected to a shielded transport and storage cask located in an accompanying vehicle by a flexible tube. This system was small enough to be transported to and operate around an aircraft at a rework center. After the collimator assembly was positioned above the part to be examined and a film cassette placed below it, the exposure was begun by moving the californium source from the storage cask through the tube to the collimator assembly. Returning the source to storage terminated the exposure.

This californium system produced adequate radiographs for evaluating the integrity of airframes, but the exposure times were deemed to be too long to be practical. Increasing the size of the californium source could reduce the exposure time, but that increases the amount of shielding required to burdensome quantities and further aggravates the ability of the system to meet safety and licensing requirements[7].

The Army Materials and Mechanics Research Center, in 1978 joined the Vought Corporation in an effort which would explore the use of the small Kaman A711 accelerator as a mobile source of neutrons

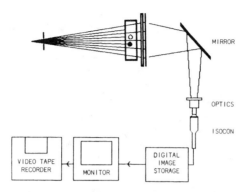

Fig. 8. Sketch of a real-time electronic imaging system. The Isocon is a "TV" type of tube which converts light to electronic signals which are stored digitally and can be viewed on a monitor or recorded on a video tape recorder.

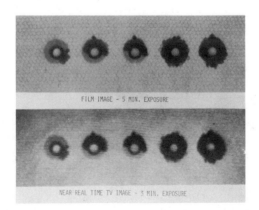

Fig. 9. Neutron radiographs of inspection for adhesive distribution under steel fasteners. The circular steel inserts are penetrated, revealing various degrees of adhesive application within this honeycomb airframe panel. The AMMRC mobile neutron radiography system produced both radiographs. The upper image is that produced normally on X-ray film, the lower image was produced electronically by a "near-real-time" system.

arranged in a system suited to examining Army aircraft or other material in the field. Such a system would eliminate the hazard associated with the transport of a large amount of radioactive material since the radiation would be machine generated and could be turned "off" until required. The machine constructed is shown in Figure 7. It was built as an engineering model, simply to see if the small accelerator source which was designed for chemistry laboratory use could be turned into a field-like system and to explore the limits of its durability and acceptance in field settings. The machine functioned far better than one could predict. It was hauled over 5000 miles in an ordinary enclosed truck trailer, and produced over 700 useful radiographs during two field tests. During the field tests the machine was used by radiographers who had not dealt with neutron radiation before and with them it found ready acceptance as a tool. However it was thought that the exposure times were too long for routine use.

The physical size of the machine, its configuration and "on-off" radiation capability were the major reasons for its acceptance in the field. It is the first neutron machine to look a good bit like an x-ray machine. Acceptance of the machine's physical size and configuration paved the way to acceptance of another factor which allows it to be successful. As mentioned earlier, there is a trade-off between resolution and beam intensity. Because the smallest artifacts being sought in an examination of airframes are not small (relative to a crack in a casting, for example), the tradeoff was taken in favor of intensity at the expense of resolution. The machine is used with an L/D ratio of 13:1 to 22:1, which allows exposures to be made in 10 to 30 minutes. The radiographs produced are not as sharp as those radiographers are accustomed to from x-ray machines, but they are more than adequate for the problems they are expected to solve. Forcing acceptance of radiographs which were not as detailed in all respects as are equivalent x-radiographs was a psychological barrier which had to be overcome for the system to be accepted. All of the radiographs in this paper were taken with the AMMRC system.

Near real-time radiography has been accomplished with this machine, also. Figure 8 shows how real-time imaging systems can be used with neutron radiography sources. The procedure is very similar to that employed with x-rays, where for low intensities the image can be built up in time and stored in an electronic memory until it is of adequate contrast. A real-time image photographed from the viewing screen of a cathode ray tube is shown in Figure 9.

This image was collected in 3 minutes and is quite adequate for the assessment of adhesive spreading problems. This type of imaging capability allows neutron radiography to move into the arena of in-process evaluation and control.

Table III.   Some In-Process Neutron Fluoroscopy Studies Using
             Real - Or Near Real-Time Imaging

DESCRIPTION	LOCATION	REACTOR SOURCE
Two-Phase flow studies of air and steam in water	Oregon State University	TRIGA
Gas dynamics and density gradient studies in optically opaque tubing	Sandia Lab	ACPR
Firings of 7.62 mm rounds in rifle barrels	Lawrence Livermore Lab	Pulsed TRIGA
NAK flow in insulated steel piping	Dept of Energy	for LMFBR
Lubricant flow in helicopter and auto engines, and deposit buildup in turbochargers	Rolls Royce	DIDO

IN-PROCESS APPLICATION

Actually, many have gone to the expense and trouble of doing in-process neutron radiography on an experimental basis using special reactor facilities. A number of these studies are listed in Table III. Rolls Royce, the turbine engine manufacturer, has successfully tracked down a lubricating oil fluid flow problem by getting real-time neutron images at a research reactor in England with a beam of neutrons brought outside the reactor building to a special turbine engine test stand[8].

The next generation prototype of the AMMRC machine, now being designed under a combined Navy and Air Force contractual effort, should be able to provide real-time radiography suited to the solution of the types of problems listed in Table III. Its size will allow it to be brought to the aircraft shop to accomplish what needs to be done. At the present time, three companies are designing new accelerators which will produce higher intensities without proportionately increased sizes or weights. These machines are being designed specifically for application to mobile neutron radiography systems based on the AMMRC engineering model. A major facility for the neutron radiography of complete aircraft is also under construction by the Air Force to supplement x-ray radiography at its McClellan Air Logistics Center, where the AMMRC machine was first field tested. This flurry of activity is an expression of a new approach to neutron radiography, and presages a period of renaissance in neutron radiography application.

REFERENCES

1. H. Berger, "Neutron Radiography," Elsivier Publishing Co., Chapter I, Amsterdam, (1965).
2. J. W. Ray, "Neutron Radiography, A Solution in Search of Problems," Research/Development, V. 20, No. 7, p. 18, July (1969).
3. "Neutron Sources and Applications Proceedings of the American Nuclear Society National Topical Meeting, Augusta, GA, April 1971; CONF-710402, National Technical Information Service, Springfield, VA (1971).
4. L. M. Bieberman and S. Nudelman, Editors, "Photoelectronic Imaging Devices," V. I and II, Plenum Press, NY (1971).
5. L. C. Woods, "Introduction to Neutron Distribution Theory," John Wiley & Sons, NY, p. 11, (1964).
6. J. J. Antal, "Justification for an Innovative Neutron Radiography Device," Paper Summaries, National Fall Conference of ASNT, p. 198, October (1980).
7. W. E. Dance, "N-Ray Inspection of Aircraft Structures Using Mobile Sources: A Compendium of Radiographic Results," Report NAEC-92-116, Naval Air Engineering Center, Lakehurst, NJ, April (1979).

8. P. A. E. Stewart, "Advances in Radiography and Fluoroscopy," Brit. J. NDT, pp. 27-32, January (1982).

9. "GCA Does Testing Robot," MASS HIGH TECH, Mass Tech Times, Inc., October 15-18, p. 24, Burlington, MA, (1984).

10. W. E. Dance and S. F. Carollo, "Demonstration and Evaluation of Mobile Accelerator Neutron Radiography for Inspection of Aircraft Structures at SM-ALC/McClellan AFB," ATC Report No. R-92200/2CR-4, San Antonio Air Logistics Center (SA-ALC/MMEI), Kelly AFB, Texas 78241, January (1982).

ANALYTICAL APPLICATIONS IN THE ARMY OIL ANALYSIS PROGRAM

Cyril M. Brown and Lisa E. Boley

USAMC Materiel Readiness Support Activity
AMXMD-MO, Lexington, KY 40511-5101

Improved equipment readiness and enhanced flight safety are maintenance and equipment-related terms that are not normally associated with the chemical laboratory. There is, however, an Army program that takes advantage of various analytical chemical techniques to supplement the maintenance efforts of the Army in the field. The program, specifically the Army Oil Analysis Program, or AOAP as it is referred to, is part of a Department of Defense effort to detect impending equipment component failures and determine lubricant condition through on-line and laboratory evaluation of oil samples. Through proper application of spectrometric and physical property analysis techniques, flight safety is improved, equipment downtime and maintenance costs are reduced, and oil life is extended resulting in lubricant conservation. Responsibility for program management lies with the United States Army Materiel Command (USAMC) Materiel Readiness Support Activity (MRSA), which has been designated the Program Manager for the Army Oil Analysis Program (PM-AOAP).

The Army's program began in 1961 after several defective Army helicopter transmissions were detected through oil analysis by a Navy laboratory. Recognizing that oil analysis was a valuable tool, the Army opened a laboratory at Fort Rucker, Alabama, in September 1961 to provide support to its expanding fleet. On the first day of operation, a defective aircraft engine was identified through spectrometric analysis. That was the beginning of what was to become an Army-wide program.

The item of laboratory equipment used by the Fort Rucker laboratory was a Jarrell-Ash direct reading spectrometer (3.0 meter Wadsworth), and the analysis technique, which measured wear metals in the lubricant, was based on one that had been used by the railroad industry since 1941. Notwithstanding the age of the technique, it proved highly effective, and numerous internal aircraft component defects were found well in advance of failure. Owing to its immediate success, oil analysis became mandatory for all aircraft components. During the Vietnam conflict, oil analysis was routinely applied to all Army aircraft involved in combat operations.

Between 1961 and 1975, oil analysis was used exclusively to monitor aircraft components. In 1975, however, a significant change took place in the Army Oil Analysis Program: Ground equipment was introduced and with it the requirement for analytical tests capable of determining lubricant condition.

Prior to the addition of ground equipment to the oil analysis program, emission spectrometry had been the Army's only medium for oil analysis, and it is still the one primarily used for that purpose. The spectrometer currently used, the Baird Corporation FAS-2, which was built to government specifications, measures wear metals over a range of 1 to 1000 parts per million. The concentration of wear metal found in used oil samples is indicative of component wear and is used by the laboratory technician to reach decisions about the condition of the components from which the oil is taken. For example, certain concentrations of iron, copper, and silver in a T-53 aircraft engine point to bearing wear. High concentrations of chromium and molybdenum in an AVDS 1790 engine (M-60 tank) point to excessive ring wear, and the presence of silicon in the same engine is a giveaway that air induction problems exist.

As previously mentioned, ground equipment, specifically, combat vehicles, were brought into the Army Oil Analysis Program in 1975. In 1977, tactical and wheeled vehicles were introduced and construction equipment was added in 1979.

With the introduction of ground equipment came the requirement for physical property test procedures. Acceptable procedures had to be considered in light of an average 300-400 sample per laboratory per day workload. The standard American Society for Testing Materials (ASTM) procedures for determining lubricant physical properties were found to be too time-consuming for the average AOAP laboratory. Recognizing that what the technicians in Army laboratories needed were physical property procedures capable of revealing subtle changes from sample to sample rather than absolute values, personnel charged with the development of the Army tests turned to commercial laboratories performing similar work. As a result, the Army Oil Analysis Program came up with the following tests, which have become standard in all laboratories:

1. <u>Viscosity Determination</u>. An oscillating sphere viscometer is used for this test and measurements are made at room temperature (adjusted through temperature charts). The measurement takes about 15 seconds, and although the result is given as Centipoise x g/cm^3, established standards make this a very useful test. Identification of fuel dilution problems has made this a particularly useful analysis. For example, an inherent engine design problem (fuel leaking into lube oil) was detected by means of viscosity measurement, which ultimately resulted in an Army-wide message from the equipment manager directing maintenance actions to correct the deficiency.

2. <u>Crackle Test</u>. Free water is determined by placing one or two drops of used oil onto a laboratory hot plate heated to 150°-175° C. A positive test is indicated by splattering and an audible crackling of the oil. As with the viscosity test, the speed at which this can be accomplished affords the technician an excellent screening test that requires little time to accomplish. This technique has proven very effective. Through early detection, the laboratory technician is able to recommend corrective actions to the user. Common findings involve cracked engine heads, defective seals, oil cooler problems, and improper operation. Early detection of water and notification to the unit that a problem exists provides the Army in the field with a tool to improve their overall maintenance.

3. <u>Blotter Spot Test</u>. Another quick test, the blotter spot test covers the determination of insoluble contaminants and dispersant ability. After vigorous shaking, one drop of the used oil is placed in the center of an oil print filter circle (typically a Whatman 40) and allowed to disperse for about 15 minutes. The resulting spot is evaluated for total contaminants, and lubricant dispersive effectiveness.

4. <u>Insolubles</u>. This comes closest to the standard ASTM tests. It is a modification of ASTM method D893 and provides the technician with a means of determining insolubles. Usually this test is employed when the blotter spot and viscosity point to the need. Improper engine operation and adjustment, oxidation, and extraneous and wear debris all contribute to the level of insolubles found in the oil. The technician uses the information obtained, in conjunction with other test data, to formulate maintenance recommendations.

At approximately the same time that the Army began using physical property tests to determine lubricant condition, a second technique for detecting the presence of metals and identifying potential component failures was adopted. This is microscopic analysis and its primary function involves inspection of debris from aircraft engines, transmissions, and gearboxes. Because spectrometric analysis is limited insofar as the size of particles that can be detected, microscopic analysis provides a good overlap procedure. Microscopic analysis in its traditional form has certain disadvantages that seem to have been overcome by a new instrument, the Ferrograph, which may eventually supplant the current microscopic technique.

Ferrographic analysis is an advanced procedure that may prove highly effective in detecting incipient component results through the microscopic examination of wear debris removed from aircraft and, possibly, ground equipment components. One Army laboratory is already using it to study debris from aircraft components and several defective engines and transmissions have been found as a result. The Ferrograph provides an extended range of particle size detection--the spectrometer is most effective in measuring particles over the 0.8-8.0 micrometer range. The Ferrograph will allow the technician to look at larger particles, e.g., 25 to 200 micrometers, that may be indicative of abnormal wear. The Ferrograph makes use of a special bichromatic microscope for observation of particles deposited on a substrate by means of a magnetic gradient. In addition to its superior ability to capture particles, it also provides the technician with a means of determining severity and mode of wear, e.g., rubbing, cutting, and fatigue. This is something that cannot be done with the microscopic procedure currently employed. Current plans call for locating five Ferrographs at area laboratories to support aviation assets. Future plans for use of the Ferrograph will be based on the findings at these five laboratories.

Built-in test equipment for aircraft is the current trend and is being considered to improve the chances of detecting interval component defects more effectively and faster than can currently be done by means of laboratory analysis. One of these systems being considered for the UH-1 and AH-1 helicopter fleets makes use of an ultra-fine filtration (3 micrometers absolute) combined with a full-flow chip detector superior to those now in use. The advantages of this item of test equipment are as follows:

1. Wear is reduced.

2. Large metal particles invisible to the spectrometer can be detected.

A field test of this built-in test equipment will probably take place in fiscal year 1985. Installation of the UH-1/AH-1 aircraft fleet is scheduled to follow.

Although this filter-chip detector combination seems to offer promise, it does have certain drawbacks insofar as the scope of the Army Oil Analysis Program is concerned: It will not detect small particles that may be indicative of beginning failure and it removes all particles "visible" to the spectrometer. In addition, the high efficiency filter-chip detector combination cannot be used on diesel engines or other "dirty" components enrolled in the Army Oil Analysis Program.

Although each of the tests mentioned above has its limitations, collectively they provide the Army Oil Analysis Program with a means of detecting impending equipment failures and determining lubricant condition. They clearly demonstrate that laboratory bench tests can be adapted to a real-world maintenance situation with a high degree of efficiency.

To ensure adequate monitoring of equipment, oil samples are taken from components, e.g., engines and transmissions, at specific intervals. Typical intervals, which may vary depending on whether equipment is assigned to active Army, Army Reserve or National Guard, are shown below:

1. Rotary Wing Aircraft

   Engines - 12.5 hours
   Transmissions - 25 hours.

2. Combat Vehicles

   Engines and Transmissions - 25 hours or 30 days, whichever comes first.

3. Construction Equipment

   Engines - 50 hours or 60 days, whichever comes first.

4. Tactical Wheeled Vehicles

   Engines and Transmissions - 100 hours or 60 days, whichever comes first.

By analyzing samples taken at specified intervals, unusual changes in wear-metal concentration and changes in the physical properties of the oil are easily detected by means of spectrometric and physical property analysis.

Samples are removed by several means: tubes, pumps, valves. For aircraft, a plastic tube is used. It is inserted into the oil reservoir and the oil is removed while holding a finger over the end of the tube. Only a small amount of oil can be removed in this manner and several repetitions may be required to obtain the amount needed for analysis (about 5/8 ounce). This method is preferred for aircraft, however, since only a small amount of oil is needed for sample analysis. The use of pumps or valves might result in too much oil being removed from a reservoir that holds only a small amount of oil.

Samples from ground equipment are either removed by a vacuum pump fitted with the sample bottle used for sending the oil to the laboratory or by means of a valve mounted on the component being sampled. The latter method is preferable because it makes it easier for the soldier in the field to take a sample, and it ensures that representative oil samples are removed from the component. Samples from ground equipment (3 ounces) are larger than those from aircraft because ground equipment samples must have several physical property tests run on them.

Army equipment enrolled in the Army Oil Analysis Program is being fitted with sampling valves, and new equipment entering the program will have valves installed prior to fielding.

Once the soldier in the field takes an oil sample, he completes a DD Form 2026, Oil Analysis Request. The completion of this form, which accompanies the sample to the laboratory, is critical to the success of the analysis, since all data relating to the component from which the oil was taken is entered on the form. Component serial number, hours since last oil change, and hours since last overhaul are examples of data required by the technician for accurate assessment of component wear or changes to the lubricating oil's physical properties.

Once a sample has been analyzed, the technician reviews the analyses, both spectrometric and physical, if applicable, and makes a decision concerning the component from which the sample was taken. The majority of analyses will have normal results; however, some will require action on the part of the technician and the owner of the equipment. When a sample has abnormal results and indicates a possible problem, the technician acts promptly to notify the owner. Typical recommendations by the laboratory will be for oil changes, inspection of air induction and cooling systems, and removal and repair of components. Once the maintenance recommended is performed, the laboratory and the Army Oil Analysis Program Management Office are notified in writing.

Maintenance data returned are used in computing Army Oil Analysis Program cost avoidance, which is the money saved as a result of timely laboratory and maintenance action. During fiscal year 1983, 1.3 million samples were analyzed by 21 Army Oil Analysis Program laboratories at a total program cost of $6.6 million. Maintenance actions initiated by the laboratory resulted in a cost avoidance of $73.6 million. This is a cost avoidance to cost ratio of 11 to 1.

The analysis of used oils for physical properties has made it possible to identify another area where man-hours and lubricant can be saved. Army equipment entered in the Army Oil Analysis Program is now subject to on-condition oil changes. That is, oil is only changed at the recommendation of a laboratory and not on the basis of a designated calendar interval. During fiscal year 1983, two Army installations reported a 74 percent reduction in the number of oil changes with a combined cost avoidance of $544,403. This computation was carried out at only two installations in fiscal year 83 because it had to be done manually. It was not done in fiscal year 84 for that same reason. However, the new computer system installed in all AOAP laboratories during fiscal year 84 will enable installations to compute lubricant savings.

The number of samples processed by Army laboratories makes it imperative that each laboratory be equipped with a computer system capable of maintaining component history, analyzing oil sample data, and producing maintenance management reports required by maintenance officers and commanders at the division and installation level. To provide this capability, the Army Oil Analysis Program Manager developed and fielded the Army Oil Analysis Program Standard Data System, which consists of a Wang 2200 MVP minicomputer interfaced to the Baird spectrometer (spectrometric results are entered in the computer automatically; physical property results are entered manually).

With this system, laboratory personnel maintain all required records on the computer, which reduces the clerical burden while greatly increasing the efficiency and productivity of the laboratory. The computer, in fact, performs a preliminary evaluation of the analytical results and provides the technician with a maintenance recommendation related to the component from which the oil sample was taken. The technician can override the computer recommendation, but this feature of the computer has greatly increased each Army laboratory's sample-processing capability.

The AOAP laboratories now have the capability of providing routine management reports on a monthly basis, or more frequently if the need arises. These reports, which were developed in coordination with Army maintenance personnel, are designed to help the users in the field manage their equipment to lessen the administrative burden on the soldier, and to provide a means by which the effectiveness of the program can be assessed at unit level. Management reports, which were designed for simplicity of interpretation and use, can be provided on any combination of oil analysis data elements.

The AOAP Standard Data System not only satisfies AOAP processing needs at the installation level, it also provides the medium for collecting Army-wide data in a central Joint Oil Analysis Program data bank which is maintained by the Air Force at Kelly Air Force Base (AFB), San Antonio, Texas. The AOAP data processed at each laboratory will be forwarded to the Materiel Readiness Support Activity, PM-AOAP, for consolidation. Once consolidated, the data will be transmitted electronically to the data bank at Kelly AFB. These data can then be used in the preparation of national level management reports. Plans are underway to provide on-line access to the central data bank for the US Army Materiel Command and its major subordinate commands, as well as other major commands. Data required for computing AOAP cost avoidance and determining AOAP effectiveness will be extracted at the Materiel Readiness Support Activity.

Identifying and putting new laboratory instrumentation into use is one of the ways to improve the effectiveness of oil analysis. To this end, the PM-AOAP, MRSA, is reviewing other types of instrumentation for possible use in the Army laboratories. The equipment currently under consideration is as follows:

1. Ferrograph. This item of laboratory equipment was discussed earlier in this paper.

2. Infrared Spectrophotometer. Although used in industry for oil analysis, infrared spectrophotometry has never been used on the scale required by the Army. However, recent tests conducted at the Joint Oil Analysis Program Technical Support Center, which provides support to the Army, Navy and Air Force oil analysis programs on technical matters, indicate that infrared may be an analytical technique capable of providing all the physical property test data needed to determine lubricant condition. Current plans call for further testing in an AOAP laboratory environment.

3. Water Analyzer. The presence of water in aircraft and ground equipment oils is a cause for concern. This contaminant is usually indicated by the crackle test for ground equipment oils--the percentage is usually high. For aircraft, the presence of water, which may be low, is often revealed through corrosive action on, for example, ferrous metals (gears and bearings) and aluminum-magnesium alloys (transmissions and gearbox cases).

Since the concentration of water in aircraft components may be too low for detection by the crackle test used for ground equipment oils, another method--preferably quantitative--is required. To this end, MRSA, PM-ACAP is currently reviewing the application of a procedure developed by the U. S. Navy. The ultimate goal of quantitative water detection is to verify that metals detected by spectrometric analysis are not corrosion products.

4. Flash Point Tester. Fuel dilution of high concentrations is not uncommon in some diesel engines, and the use of flash point testers to determine this type of contamination is an accepted practice. However, because of the large workload flowing through Army laboratories, standard ASTM procedures for flash point have been found to be unacceptable, primarily because of the time and amount of oil required for analysis. A flash point tester requiring only 4 milliliters of oil and capable of providing a flash point determination in less than 15 minutes is currently being investigated and plans call for a test in a high production ACAP laboratory.

The instrumentation described above will be for use in fixed-base laboratories. Although the current ACAP laboratories have been geographically located to provide maximum support to Army activities in the field, the Army even in peacetime, is a dynamic organization, requiring flexibility in all phases of operation. Oil analysis support is not excluded; therefore, various items of portable equipment for use outside the laboratory are being considered. These are briefly discussed below:

1. Portable Wear-Metal Analyzer. A portable wear-metal analyzer is currently under development by the Air Force and testing is due to begin in the third quarter of fiscal year 85. It consists of two 40 pound "suitcases" and has a nine element wear-metal capability. The analytical technique employed is atomic absorption spectrophotometry (graphite furnace), and the current wear-metal capability was designed with only the analysis of aircraft oils in mind. Air Force findings concerning the portable analyzer's applicability to aircraft analysis will be reviewed by the Army for similar application. In addition, further testing by the developer will be considered to determine if this instrument can be used for the analysis of oils from ground equipment, an important requirement for the Army.

2. Portable Physical Property Test Monitor. Another requirement is that of lubricant physical property determination. The procedures currently used in the AOAP fixed-base laboratories are adequate, but the same requirement governing the need for portable wear-metal analysis governs the need for portable lubricant

411

physical property analysis. Current efforts by the Navy Research Laboratory, in coordination with the Army, are directed toward the use of micro-sensor technology to monitor lubricants for fuel dilution, water/coolant contamination, total solids, and acidity and alkalinity.

Investigation of other techniques having possible applications to the detection of abnormal wear in aircraft and ground equipment components will continue as will the search for better methods for lubricant physical property determination. The Army in the field is a dynamic force and oil analysis support requires a dynamic effort from the scientific community. Army activities involved in research and development can contribute significantly to this effort.

Current analytical procedures used in the Army Oil Analysis Program are based on sound laboratory techniques. They have, however, been modified to provide a maintenance tool to the Army in the field. Improvements in equipment readiness, maintenance effectiveness, and aircraft safety have been realized through the application of analytical laboratory techniques originally developed with other things in mind. The Army Oil Analysis Program has shown that significant benefits accrue when standard laboratory techniques are looked at in light of a maintenance need, and efforts will be continued to sort out and put into use those laboratory methods that offer potential for the Army in the field.

CHARACTERIZING BRITTLE-FRACTURE RESISTANCE OF STEEL

D. M. Kindel,* R. G. Hoagland,* J. P. Hirth and
A. R. Rosenfield**

*The Ohio State University
**Battelle Columbus Laboratories

ABSTRACT

The description of fracture resistance of steel in the ductile/
brittle transition region is treated as a multi-parameter problem.
Some metallurgical factors influencing two of the parameters (lower-
shelf toughness and sharpness of the transition) are discussed.
For a series of HSLA steels which are variants on ASTM A710A, removal
of Cu increased the lower-shelf toughness and tended to sharpen the
transition. Ni removal also tended to sharpen the transition, while
C and Nb additions tended to smooth it.

The two other parameters describing toughness reflect uncertainty
and can lead to severe data scatter: uncertainty in translating the
transition temperature derived from impact energy into an equation
relating fracture toughness to temperature and scatter in fracture
toughness itself apparently arising from microstructural non-
uniformity. Some implications of these findings with regard to
materials characterization are discussed.

* present affiliation: Columbus Auto Parts, Columbus, Ohio

INTRODUCTION

Cleavage fracture resistance of steel presents a particularly difficult challenge to materials characterization. There is the problem that there is no agreed-upon method for measuring fracture toughness within the ductile/brittle transition region[1]. There is also the problem that, with the exception of the transition tempera- ture, there is practically no information on the physical, chemical and structural factors affecting the variation of toughness with temperature. This paper discusses both of these difficulties and some recent research aimed at overcoming them.

RELATING IMPACT ENERGY AND FRACTURE TOUGHNESS

Despite great advances in fracture mechanics technology, the Charpy Impact Test is still the most widely used measure of fractur resistance of steels. The success of this test stems, in a large part, from pioneering advances by the US Army during World War II[2]. However there are several features of the test that cause concern[3]. Among these are the use of a notch instead of a crack, the differ- ence in thickness between test specimens and real structures, and the difficulty in relating transition temperature to stress and crack length. Progress has been made by introduction of the pre- cracked Charpy specimen and instrumented testing. In addition, understanding of the effects of thickness on toughness is beginning to be developed[4,5]. Despite this, Charpy data are not completely satisfactory since they give no insight into safe design stresses.

One approach for utilizing Charpy data in estimates of safe operating loads of flawed structures is contained in the ASME Boile and Pressure Vessel Code, which uses Charpy data in association wit empirically-developed "lower-bound" toughness relations. These relations are of the form:

$$K_{Ic} = K_o + K_1 \exp(A(T-T_o)) \tag{1}$$

where K_{Ic} is plane-strain fracture toughness, T is test temperature T_o is the ductile/brittle transition temperature, and the other quantities are empirical parameters. The transition temperature is determined using Charpy and Drop-Weight Tear Test (ASTM A 208) results. This approach has been quite successful in preventing service failures. However, it has several problems. One is that the lower bound was conceived as being a deterministic property. I reality it should be evaluated statistically. In addition, the constants of Equation (1), and perhaps its form, may very well diff for different steels.

The statistical problem is a serious one. Wallin, et. al.[6] have developed strong evidence that variability in K_{Ic} within the transition region is described by a Weibull distribution with a modulus of four, which corresponds to a coefficient of variation of about 30%. In terms of MIL HDBK Design Allowables, which decree use of small percentiles of the cumulative failure distribution, such a spread in data will result in only a small fraction of the median toughness being safely utilized.

As a more general approach to evaluating scatter, Bishop, et. al.[7] have developed a new statistical technique for providing cumulative failure probabiliities, such as fracture toughness, within the ductile/brittle transition region. The method employs the Weibull distribution and is particularly well-suited for analysis of cleavage of steel, where fracture initiation is believed to be controlled by local brittle regions[1]. Briefly, the method consists of expressing the logarithms of the Weibull parameters as power series in temperature with unknown coefficients. In solving for the coefficients, errors in both central tendency and variance are minimized. Percentiles of the data and of the associated confidence intervals are then calculated.

Figure 1, from Cheverton and Ball[8] shows an application of Bishop's method to the rapid-crack-propagation resistance of steel, as estimated using the stress intensity at crack arrest. Here the dashed lines represent the fifth and ninety-fifth percentiles of small-laboratory-specimen crack-arrest data. The points describe crack-arrest results using very large specimens. As can be seen from the figure, the statistical analysis is representative of the spread of the points, providing support for the approach.

Data such as in Figure 1 have great significance for character-ization research. The large scatter in Figure 1 is representative of a situation where behavior is believed to be controlled by the tail of the size distribution of carbide particles[6]. Usually-reported parameters, such as average particle size and spacing, are thus important only if they are indicative of the largest particle size. To the extent that this example is typical, it would appear that advances in producing steels of more uniform structure may be at least as important technologically as producing steels with higher average toughnesses.

EFFECT OF METALLURGICAL VARIABLES ON FRACTURE TOUGHNESS

While Equation (1) is empirical, it does point out the different factors contributing to the brittle fracture resistance of steel: the lower-shelf fracture toughness (K_0), the steepness of the transition (A), and the level of the ductile/brittle transition temperature (T_0).** As is well-known, the last alternative is the

** The K_1 parameter accounts for the difference in transition temperature between fracture-toughness specimens and Charpy specimens.

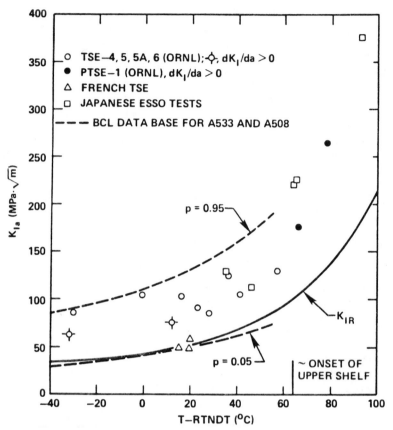

Figure 1. Comparison of Small–Specimen and Large–Specimen Crack–
Arrest Toughness. Small specimen failure percentiles are
represented by the dashed line large–specimen data are
given by the points (from Cheverton and Ball[8]).

only one studied extensively. While recent research at Battelle–
Columbus and The Ohio State University has examined all of the
elements of Equation (1), this paper will concentrate on the two
aspects least investigated: the lower–shelf fracture toughness and
the steepness of the impact–energy curve.

The research program involves the use of six different steels
with compositions that are variants on ASTM A710A, which was
studied earlier[9]. The data reported below were obtained either in
the as–rolled condition or after quenching and tempering according
to Heat–Treatment C1 in Reference 9. Further details are reported
in Reference 10.

Lower-Shelf Fracture Toughness

Values for K_{Ic} obtained for the heat-treated project steels are plotted in Figure 2. The figure shows two trends:

(1) The data can be separated into two categories, those steels containing slightly more than one-percent Cu falling on the lower curve and those that are Cu-free falling on the upper curve.

(2) Within each category, toughness increases with increasing strength. This trend is obtained for cleavage, and is opposite to the well-known trend of decreasing toughness with increasing strength, which is found for the dimpled-rupture mechanism.

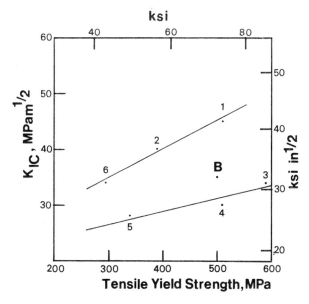

Figure 2. Relation Between Fracture Toughness at -196C and Room-Temperature Yield Strength. The upper curve represents Cu-free steels while the lower curve represents Cu-containing steels. Symbols next to points indicate individual project steels.

While the reasons for this behavior are not yet clear, an
indication of the cause is given in Figure 3, where it is seen that
increasing fracture toughness is associated with increasing roughnes
of the fracture surface. The mechanism by which Cu acts to promote
planarity of the cleavage crack is under investigation.

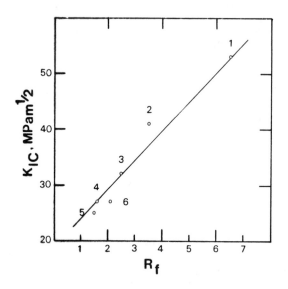

Figure 3. Relation Between Fracture Toughness At -196C and
Fracture- Surface Roughness. The data are for peak-aged
steels.

Sharpness of the Ductile/Brittle Transition

The criteria for defining the ductile/brittle transition
temperature differ from one application to another. If the transition curve is sharp, there is no problem of definition. More often
the transition is gradual and the choice may be somewhat arbitrary.

Surprisingly there is little information on the basic causes
of the steepness of the transition. Reinbolt and Harris[11] have
shown that increasing carbon levels cause the Charpy curve of steel

to become less steep and to exhibit lower plateau-energy values. A somewhat different effect was found for increases in Sulfur content: a lowering of the upper plateau with no change in steepness[12]. This observation is not consistent with the recent suggestion of Sandstrom and Bergstrom[13] that there is an inverse relation between steepness and transition temperature. A more satisfactory explanation of both sets of data may derive from the results of Frazier, et. al.[14]. They found that the transition curve for a ferrite/pearlite steel was essentially bimodal, with the relative percentage of low- and high-toughness specimens varying with test temperature. They also found that the breadths of the energy-distribution functions for both mechanisms were comparable for the steel that they investigated. In general, the breadths may not be comparable and the mean energies associated with the different mechanisms may differ by varying amounts. This possibility provides the scope for a variety of curves for different steels and heat treatments, ranging from extremely steep to extremely gentle.

The data presented in Figure 4 are for as-rolled steels and provide partial support for the Frazier, et. al.[14] view. The curve exhibiting the very sharp transition and very high upper-plateau energy is consistent with the existence of two failure mechanisms with very different and distinct associated energies. In our experiments this behavior is only found in those steels that are very low in Cu and/or Ni, and only for certain heat treatments. The opposite tendency of broad transitions and low upper-plateau energies was more common for as-rolled steels than for heat-treated steels. The microstructures of the as-rolled steels also tend to be less uniform than those of the quenched-and-tempered steels (two different degrees of non-uniformity are shown in Figure 5).

Apparently non-uniformity can lower the energy of the ductile-fracture upper-shelf mechanism and cause a decrease in the temperature dependence of energy absorption. The cause is not clear, but is suggestive of a requirement for a statistically variable probability of a stress concentration coinciding with a region containing large defects; increased inhomogeneity would produce increased spread in fracture probability as a function of test variables in such a case.

DISCUSSION

While cleavage fracture of steel has been a well-recognized problem for a long time, its understanding remains elusive. One reason is that steel research and development have focused (quite successfully) on one measure of the phenomenon, the ductile/brittle transition temperature. The research cited in this paper suggests that there are three other characteristics to consider: the lower-shelf toughness, the sharpness of the transition, and the scatter in fracture resistance. Our results indicate that the micro-

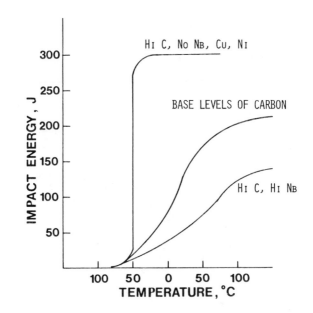

Figure 4. Impact Transition Curves For As-Rolled Experimental Steels

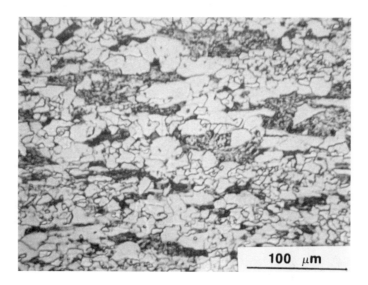

Figure 5. Light Micrographs Of Two As-Rolled Steels. The steel of
the top micrograph exhibited a sharp ductile/brittle
transition while the steel of the bottom micrograph
exhibited a broad transition.

421

structural factors influencing these characteristics are subtle and, to some extent, unexpected. The finding that Cu can promote planarity of a cleavage crack and, in doing so, lower base-line toughness, presents an interesting problem requiring further research. We do note that modelling results[15] indicate that the presence of an extensive array of voids enhances the tendency for plastic instability along Mode I trajectories, i.e. Mode I cracking. To the extent that the small (200 nm) sized Cu particles have lower flow stresses than the matrix, they would have the same qualitative effect.

On the other hand, the effect of Nb on the transition is closer to explanation. Nb causes formation of fine C- and N- bearing precipitates[9], which lower the upper-shelf toughness much as C and S additions to pure Fe do. In contrast, when C is precipitated as Fe_3C the carbides are less-uniformly dispersed. Possibly this microstructure results in a higher upper-plateau energy by inhibiting the final, void coalescensce, stage of ductile fracture.

The as-rolled micrographs (Figure 5) suggest that the role of Ni and Cu in smoothing out the transition may be indirect, resulting from a less-homogeneous microstructure being produced by the rolling schedule used to fabricate these alloys.

With respect to the Conference theme of characterization, the data scatter deserves particular consideration. As indicated above, the scatter is most likely connected to the microstructural inhomogeneity of steels. It appears that the fracture-controlling parameter is the largest particle in a broad size distribution, with the probability of a stress concentration at such a particle having an effect. This probability is another manifestation of the Weibull approach originally developed to analyze the strength of ceramics. Doubtless there are other structure-sensitive properties which obey extreme-value statistics and need to be characterized by the tails of structural distributions and not the means.

CONCLUSIONS

1. Cleavage fracture resistance of steels needs to be described by a many-parameter equation, which includes scatter as an inherent property.

2. Low-temperature fracture toughness is best described by means of a Weibull distribution.

3. Copper can lower base-line toughness by promoting planarity of cleavage cracks.

4. Homogenization on a grain-size scale promotes a sharp ductile/brittle transition.

ACKNOWLEDGMENTS

The authors are greatful to the Army Research Office for support of this research under Grant P18399-MS and to Mr. P. R. Held for expert experimental assistance. We also wish to thank Mr. R. D. Cheverton of Oak Ridge National Laboratory for permission to reproduce Figure 1.

REFERENCES

1. A. R. Rosenfield and D. K. Shetty, ASTM Symposium on Users Experience with Elastic-Plastic Fracture Toughness Test Methods. (in press)
2. A. Hurlich, Appendix to paper by A. F. Landry in What Does the Charpy Test Really Tell Us?, A. R. Rosenfield, et. al., Editors, ASM Metals Park, OH, p. 7 (1978).
3. National Materials Advisory Board, Publication NMAB-328 (1976).
4. K. Wallin, "The Size Effect in K_{Ic}-Results", submitted to Eng. Fracture Mech.
5. J. G. Merkle, Report NUREG/CR-3672 (ORNL/TM-9088), Oak Ridge National Laboratory (1984).
6. K. Wallin, T. Saario and K. Torronen, Metal Science, Vol. 18, p. 13 (1984).
7. T. A. Bishop, A. J. Markworth, and A. R. Rosenfield, Metall. Trans. A, Vol. 14A, p. 687 (1983).
8. R. D. Cheverton and D. G. Ball, "The Role of Crack Arrest in the Evaluation of PWR Pressure Vessel Integrity During PTS Transients," presented at the CNSI Meeting on Crack Arrest, Freibourg, Germany (1984).
9. M. T. Miglin, J. P. Hirth, and A. R. Rosenfield, Metall. Trans. A Vol. 14A, p. 2055 (1983).
10. D. M. Kindel, "Effect of Microstructure on the Fracture Behavior of High-Strength Low-alloy Steels," MS Thesis, Ohio State University (1984).
11. J. A. Reinbolt and W. J. Harris, Jr., Trans ASM, Vol. 43, p. 1175 (1951).
12. J. M. Hodge, R. H. Frazier and F. W. Boulger, Trans. Met. Soc. AIME, Vol. 215, p. 745 (1959).
13. R. Sandstrom and Y. Bergstrom, Metal Sci., Vol. 18, p. 177 (1984).
14. R. H. Frazier, J. W. Spretnak, and F. W. Boulger, ASTM STP 158, p. 286 (1953).
15. V. Tvergaard, J. Mech. Phys. Solids, Vol. 30, p. 265 and p. 399 (1982).

MATERIALS CHARACTERIZATION--VITAL AND OFTEN SUCCESSFUL,

YET STILL A CRITICAL PROBLEM

Don Groves* and John B. Wachtman, Jr.**

*National Materials Advisory Board
National Academy of Sciences-National Research Council
Washington, D. C.
**Center for Ceramics Research, Rutgers University
Piscataway, N.J.

ABSTRACT

A 1967 report of the National Materials Advisory Board (NMAB) on "The Characterization of Materials" gives the definition of characterization as follows: "Characterization describes those features of the composition and structure (including defects) of a material that are significant for a particular preparation, study of properties, or use, and suffice for the reproduction of the material."

While it has now become evident that such characterization action as this is essential to build a stronger, more effective materials capability in this country, it is not yet as evident as to how best accomplish such work in the degree demanded.

This paper reviews some of the past history of materials characterization endeavors, the progress made toward the objective, the relevance of current and future scientific and engineering problems to continued and increased efforts, and the opportunities and roadblocks to progress in characterization.

INTRODUCTION

In today's world, ideally, the manager of a materials characterization laboratory should be able to open an unlabeled container, remove whatever material is within it, and make a series of measurements, which would tell what the material is. Given a second unlabeled container, he should be able to determine whether the

425

material within it is the same material in the sense that it would have the same behavior for some set of properties or processes of interest. For gases or liquids the manager of a good laboratory is likely to be quite successful. For solid materials, complete success is still not always possible. With such materials a combination of knowledge about how the material was made and the best available characterization measurements on the material give more reliable results, but even this combination sometimes fails.

Are such failures due to the lack of available analytical techniques and instrumentation for performing the required characterization of the material? Or, have characterization methods only advanced, in general, to the point where they still merely give expedient and partial descriptions of the preparation method and property measurements? Have we procrastinated on bringing about the improvements in our abilities to describe adequately "those features of the composition and structure (including defects) of a material that are significant for a particular preparation, study of properties, or use, and suffice for the reproduction of the material?' Has characterization knowledge been understood and properly applied in engineering materials?

In an attempt to answer such questions, a review of some of the past history of materials work, various studies on materials characterization and the interdisciplinary nature of the problem appear in order as a basis for an assessment of the current and future status of the field.

HISTORICAL REVIEW

At an early period in history, man discovered that a variety of materials--stone, fired clay, cement, the common biological materials and various metals--were available for a wide range of uses. For example, the production of ceramics (the earliest inorganic material to be structurally modified by man) and copper beads date back to the 9th millenium B.C.; the smelting of minerals to 5000 B.C.; many metal alloys were made by the 3rd millenium B.C.; steel was made (in Iran) as early as 1200 B.C.; and cast iron was first produced (in China) around 500 B.C.

In the several centuries that followed such pioneering developments as these, materials were pretty much taken for granted and it was thought that they required no investigations for their wider range of uses--a situation that, unfortunately, still exists today in some circles.

The first real manifestation of a progressive upsurge of appreciation and knowledge about materials began in the 17th century. The scientific revolution that took place at that time formed the basis and triggered off many subsequent, significant discoveries and developments. Cyril Stanley Smith [1] has reviewed some of these. Professor Smith states:

"...In 1772 Rene Ferchault de Reaumur published an outstanding work on iron, based on observed and hypothetical changes of structure on the level that today we associate with the micro-structure. In the best scientific tradition he designed laboratory experiments aimed at checking and improving the theory and from these he developed an important industrial material, malleable cast iron. His work came, however, at the very end of the period during which Cartesian corpuscular theories could be taken seriously by scientists. Newtonian rigor displaced this kind of structural speculation; micro-crystalline grains came back into science only at the end of the 19th century following the discoveries of microstructure of steel by Henry Clifton Sorby in 1864.

...In 1912 X-ray diffraction was discovered and soon applied to the study of the structure of solids by Lawrence Bragg and his followers. It at once gave a measurable physical meaning to structure on an atomic scale, and made this as real as the larger-scale structures that had been revealed by Sorby's microscopic methods half a century earlier....

For a time the X-ray-diffraction results led to the construction of too idealized a picture. Then the role of imperfections was perceived, first chemical, then electrical, then mechanical errors in the building of crystals. The last served to explain the deformability of metals as well as the nature of the interface between crystal grains, the old grain boundary about which practical metallurgists had long speculated because of its' great practical importance.

Although still dominant, metals thereafter lost their unique position in scientific studies of materials. Ceramics combined all the interesting crystalline complexity of metals with the electrical interest of semiconductors. Organic chemistry had been developing rapidly in the 19th century as analytical methods became available. The awareness that many compounds with the same composition have different properties engendered the organic chemist's particularly fertile concept of structure. Molecular architecture began almost as a notational device but soon became a central part of organic chemistry and was ready to join with X-ray crystallography in guiding the development of the complicated structures that endow synthetic polymers with their properties..."

STUDIES ON MATERIALS CHARACTERIZATION

In the 1930s basic work on materials gathered momentum. Such
work, including progress on characterization, influenced and
enriched the empirical development of new and improved materials.
By 1964, many new alloys and nonmetallic materials had been so
developed and new analytical techniques for the scientific investigation of the materials had come into existence. However, even though
variations in the composition, structure, and defects--features of
the atomic world that can effect the properties and behavior of the
materials--could be then better appreciated, they were not always
understood. As a result, empiricism and empirical predictive
testing of materials for most engineering applications was still
prevalent in the early 1960s. Such techniques were necessary and
valuable, but their limitations became increasingly troublesome.
In short, characterization became more important to scientists as
the understanding of properties became more sophisticated. At the
same time, characterization became more important to technology as
engineering materials became more complex and possessed higher
performance. A small variation in the character of a relatively
low performance material usually is unimportant because such materials are used with a considerable factor of safety. Use of a
large factor of safety with a high performance material removes
much of its advantage. Modern designs use materials closer to their
limits and leave less room for variability in materials properties.
Good quality control based on good characterization is very important
Much progress in characterization and quality control has since
occurred but progress has been and still is impeded by several
critical types of ignorance. One type is ignorance of how to measure
subtle aspects of character such as microcracks of a few microns in
a microstructure with other features of similar or larger size.
Another type is ignorance of the many features of character that are
critical to some properties or processes. For example, just which
of many trace components are critical to sintering behavior in a
particular ceramic?

The seriousness of this lack of understanding was recognized
in the 60s, but a major stimulus to do something about the situation
arose when it became increasingly apparent that there was a very
definite requirement to find improved methods for somehow tailor-
making materials having reliable, uniform and reproducible properties
This need constituted a problem of great national importance since
serious impediments to progress in a variety of areas such as atomic
energy and a spectrum of defense technologies, in particular, were
deemed to be materials limited. In short, the reliable performance
of devices and systems and the development of new devices and systems
were directly dependent on progress in the materials area.

With this background, the Materials Advisory Board (now the
National Materials Advisory Board) of the National Academy of

Sciences--National Research Council was requested by the Department of Defense in February 1964 to form a study committee on the Characterization of Materials. The findings of this landmark study were reported in March 1967.

Basically, the major conclusion of the committee was the confirmation that there existed an urgent need in this country to find ways to better characterize materials. By "better" it was meant that the significant internal or atomistic features of a material (structure, composition, and defects) must be identified, quantified, and these correlated to the physical or behavioral properties that the material exhibits. (An extended abstract of this report, "Characterization of Materials," MAB-229-M, is in Appendix A). Some 16 subsequent NMAB studies conducted to date since 1967 and notably those on electronic device materials, massive glasses for structural applications, IR laser window materials, IR transmitting materials, ceramic processing, rapid solidification processing, cobalt conservation, amorphous semiconductors, structural ceramics, organic polymers, dynamic compaction of metal and ceramic powders, etc. have all endorsed such a national materials need in research and development. Specific recommendations on the direction of such R&D and other suggestions for the implementation of a viable approach to the characterization of materials are summarized in Appendix B. These are taken from some of the various aforementioned NMAB reports.

THE INTERPLAY OF SCIENCE AND ENGINEERING IN CHARACTERIZATION AND THE INTERDISCIPLINARY NATURE OF THE PROBLEM

At least two things are quite apparent in this matter. One, the kind of materials characterization suggested in the first MAB report (MAB-229-M) on the subject is not an easy task. Second, and partly because the job is not easy, materials characterization tends to mean something quite different to people of different backgrounds. To the solid state physicist the interpretation may be quite different from that of the materials scientist or that of the engineer. And, this phenomenon is also quite interesting since in the fulfillment of the objectives of each of these types--materials act as the common denominator in scientific and engineering achievement. Moreover, as Walter Kohl observes [2]:

"...We should now ask in what sense material science differs from solid state physics, which had come into its own as a discipline in the 1930s after the revolutionary concepts of wave mechanics and quantum mechanics had been introduced by De Broglie, Schrodinger, and Heisenberg and applied to the study of atomic systems. It is indeed difficult to make a sharp distinction. If solid-state physics is concerned with the study of electrical, optical and magnetic properties of crystalline solids, materials science embraces the study of all properties of all types of material-crystalline or noncrystal-

line... If solid-state physics is a discipline, materials
science is many disciplines; indeed, its interdisciplinary
nature is one of it's main characteristics... One may also
say that solid-state physics applies existing knowledge to
existing materials in an attempt to understand their properties
Materials science does that too, but it reaches out farther and
attempts to apply this knowledge to the creation of new materials in which desirable properties of several components are
combined with beneficial results... Many such innovations
require the introduction of new techniques for fabrication and
processing. That also is a special domain of materials science
although the materials engineer will be more particularly concerned with applications..."

As long as we are quoting here in this paper, we may as well
add still another interesting quotation on the subject. This one
comes from Sir Peter Hirsch of Oxford University. Professor Hirsch
recently stated [3]:

"...In the 40s, 50s, and 60s we lived through a period in which
the development of solid state physics led to a revolution in
understanding of crystalline solids. In the field of mechanical properties of solids dislocation theory developed rapidly
and in the same period electron microscopy and microanalytical
techniques became available, which allowed materials to be
characterized in unprecedented detail and on a fine scale,
and which helped, inter alia, to establish dislocation theory
on a firm basis. The general advances in electron theory of
solids led to the revolution in semiconductor device technology
while the development of new polymers and plastics has led to
impressive growth and diversity in application of these mate--
rials. The science of composite materials has been largely
worked out and composites are likely to become of increasing
importance in the future...

Over the last ten years or so, there has been a growing real-
ization that in the universities in the U.K. the interface
between materials science and engineering has been neglected:
the motivation for much of the advances in materials science
and physical metallurgy had been to achieve a better understanding of basic mechanisms controlling microstructure-propert
relationships and work aimed at solving engineering problems,
particularly relating to manufacturing technology, had not been
emphasized sufficiently. In the case of microelectronics research this problem has not arisen; the development of new
devices requires sophisticated processing and methods and
monitoring by advanced, often electron optical techniques, area
in which the engineering interface is at the frontier of knowledge. Consequently in this area the universities and industry
collaborate closely together, and the materials "science"

fulfills its proper function of an enabling technology...

Inevitably in a period of financial constraint...it will be more difficult to find support for research projects aimed at furthering our basic understanding of some property if this is not clearly related to achieving some engineering objective, or for developing some new material if there is not recognized need for it. While a shift in emphasis is undoubtedly necessary, it must not go too far...."

CURRENT STATUS OF MATERIALS CHARACTERIZATION

The landmark report of 1967 on Characterization of Materials can still be read with profit today. Its concepts still are sound and many of the recommendations of the report remain to be fully carried out.

On the other hand, certainly significant progress has been made in characterization. Every laboratory manager knows that whatever expensive and sophisticated piece of characterization equipment he buys will satisfy his staff only briefly, and that within a few years he will begin to hear how outmoded it is and at what a disadvantage his people are working. Advances in surface analysis and in electron microscopy alone are dazzling examples of the pairing of advances in science and in engineering of instruments. The availability of powerful, inexpensive microcomputers and minicomputers has revolutionized data collection and analysis. Access to national facilities for synchrotron radiation, neutron scattering, ion implantation, etc. is now vital to progress in many aspects of materials. Analytical chemistry has advanced in many ways and the advance shows no sign of slowing down.

Considering these developments, one might conclude that all is well and that steady progress toward the goals of the characterization report is being made. However, several recent reports seem to reflect a general view that some serious problems remain. For example, reports of the National Materials Advisory Board identify specific characterization problems in the fields of metal and ceramic powders, HgCdTe materials, high purity silicon, organic polymers, and composites. This list of characterization problem areas is certainly not exhaustive.

We would like to enlist your help in assessing the nature and extent of such characterization problems. To this end, we ask that you fill out the questionnaire that has been distributed to you. Please return your completed questionnaire to the Conference Director, Dr. James McCauley, as soon as possible so that the results of this poll can be given at the Workshop Panel Session on Friday, August 17, 1984. To help "prime the pump" on your thinking we offer the following classification of types of problems. However, we emphasize

431

that we want to have your thoughts rather than a reflection of ours. Also, we desire specific examples rather than general statements. Our suggested general framework for characterization problems is as follows:

1. Inadequate knowledge of which features of character are important to the properties of interest. For example, on which types of point defect in a given material should characterization development be centered if the interest is in lower optical absorption? In longer carrier life? In improved sintering behavior? In reduced long-term creep?

2. Inadequate ability in fundamental scientific terms to measure the aspect of character, which is needed. For example, how should one determine which green (i.e., shaped but unfired) ceramics contain defects that will persist through firing and cause unacceptably low strength in final parts?

3. Inadequate use of existing techniques. For example, they may simply be too costly. Or, they may require adaptation that is clear in principle but the field of application may be too small to motivate instrumentation firms to adapt their equipment and procedures.

4. Inadequate knowledge of available techniques. The list of modern techniques is so long, their individual strengths and limitations are so complex, and the field is so compartmented into different specialist groups that many investigators may be lagging seriously behind in their knowledge of what can be done.

We have structured our questionnaire with these thoughts in mind, but we have also left openings for your own viewpoints. Please let us have them. We believe you will find the exercise interesting and the cumulative results of this poll quite revealing and useful.

REFERENCES

1. C. S. Smith, Materials, Scientific American, Vol. 217, No. 3, September, NY (1967).
2. W. Kohl, Personal Communication.
3. P. B. Hirsch, An Enabling Technology. MRS Bulletin of the Materials Research Society, Vol. VIII, No. 6, November-December, Pittsburgh, (1983).

QUESTIONNAIRE

1. Please list your own fields of scientific endeavor and/or
    engineering interest (e.g., precipitation hardening, superalloys
    for gas turbines, etc.):

    _____
    _____
    _____
    _____
    _____
    _____
    _____

2. Is inadequate characterization a major problem limiting either
    scientific progress or engineering applications in your fields
    of interest? If the answer is YES, please give one or more
    examples, and also answer the next question.

    _____
    _____
    _____
    _____
    _____
    _____
    _____

3. What percent of characterization inadequacy is due to:

    o Lack of sufficiently powerful techniques. _____

    o Lack of use of existing and adequate
       techniques. _____

    Please give examples.

    _____
    _____
    _____
    _____
    _____
    _____
    _____

4. Is inadequate knowledge of characterization techniques a problem? If the answer is <u>YES</u>, please answer the next two questions.

           <u>   </u>   <u>   </u>

          (YES)   (NO)

5. What percent of the characterization knowledge problem arises from a lack of broad knowledge of the whole range of techniques including their capabilities, limitations and costs? Is there a need for good survey articles for the whole field?

6. What percent of the characterization knowledge problems is specific to techniques, and is of the nature of "how to get the job done" rather than "what techniques shall we use?"

7. What, if any, additional national or regional characterization facilities are needed (e.g., more synchrotrons, high-voltage electron microscopes, high-field NMR, etc.)?

8. What areas appear to be the most scientifically promising for improved characterization (i.e., those that offer opportunity)?

_____
_____
_____
_____
_____
_____
_____

9. Do you believe that foreign countries, i.e., Japan, Russia, Germany, have carried out significant work in materials characterization? If so, give a few specific examples.

_____
_____
_____
_____
_____
_____
_____

10. Several well-known scientists have recently stated that today materials science, engineering and technology together represent a unified, coherent field. Is this an idealized view considering the real world of current work in materials? If you feel that we have only scratched the surface of the opportunities for a convergence or unification of the field, the role that more emphasis on the characterization of materials can play is vital to this end. Do you agree or disagree? Comment briefly.

_____
_____
_____
_____
_____
_____
_____

NOTE: Please turn in your completed questionnaire to Dr. James McCauley, the conference director. The results of this survey will be discussed on the final day of the conference and documented in the proceedings.

APPENDIX A

A landmark study (NMAB-229M) was the first to outline the
guidelines for the development of a science of materials that would
afford predictable and reliable results in devising new materials
for high performance applications. The cornerstone for such a
science is characterization, defined as describing, "those features
of the composition and structure (including defects) of a material
that are significant for a particular preparation, study of properties, or use, and sufficient for the reproduction of the materials."
In the execution of the study effort, five panels were set up on
composition, structure, defects, polycrystals, and polymers. The
first three covered methods used to improve characterization
generally, while the last two were specific to the unique problems
of two materials classes. The study assessed the situation surrounding some of the greatest needs for characterization, i.e.,
better techniques and instruments, or more and better use of existing
techniques and instruments; better characterization for improved
preparation of materials, or improved study of materials, or
improved use of materials; and more accurate and detailed characterization of materials in general.

A summary of the various recommendations made in this study are:

Technical

1. Composition
    Greatly enhance capability for determination of major
    element stoichemistry. Improve analysis techniques for
    determination of O, N, C, S, B, and other anions. Improve
    valence state determinations. Develop methods for the
    location and analysis of inhomogeneities at the micron
    level. Develop survey techniques for the $\leq$ 1 ppm range.

2. Structural
    Fund greatly increased activity in optical methods
    (especially those utilizing coherent radiation) of struc-
    tural characterization. Maximize the utility of x-ray
    diffraction by increasing the quality of the powder data
    file and extracting the maximum structural information from
    such data. Develop new high pressure and high temperature
    x-ray apparatus. The rapid utilization of the scanning
    electron microscope should be sponsored. Megavolt-range
    electron microscopes and pulsed-neutron spectroscopy like-
    wise offer promise and deserve support.

3. Defects
    Absolute point defect determination (concentration and
    structure) needs substantial support. Methods for surface
    defects need development.

4. Polycrystals

For characterizing polycrystalline systems, methods and theories of measuring internal stresses (micro and macro) are most important. Research on determining homogeneity and structure on the finest scale (< 1000 Å) should be supported. Characterization of dislocation structures in heavily cold-worked and shock-hardened metals is needed. Quantitative metallography is needed for surface and transmission microscopy. Improve thinning techniques for transmission microscopy.

5. Polymers

In the polymer field, research should be supported in: rapid methods of molecular weight distribution; determination of supermolecular order in amorphous polymers, and in semicrystalline polymers; analysis of network structuration; methods to separate polyblends into their components; studies of nonpolymeric analogs (low molecular weight) of polymers; methods for characterization at the molecular level in presently intractable polymers.

Finally, there is an urgent need for immediate attention to all of these recommendations and action on as many as can be initiated in line with current requirements.

General Recommendations

1. The term: characterization, should be used as defined herein.

2. A substantially larger fraction of the funds available for materials research should be allocated and used for characterization.

3. Government agencies concerned with materials work should take positive steps to ensure that characterization is given greater emphasis and the continuity of support that are required to advance materials science.

4. Greater awareness of the basic need for better characterization (of more and better materials) should be promoted by sponsors, faculty, supervisors, and participants in work on materials research, development, and engineering.

5. Editors, referees, and policymakers of technical societies should insist on characterization of materials whose measured properties are submitted for publication.

6. A strong and sustained effort should be made to increase the effectiveness and status of those who work on characterizing materials.

7. Government agencies, such as the National Bureau of Standards, should be encouraged to exhibit stronger leadership in advancing characterization and its beneficial uses, especially in providing characterized reference materials.

8. Government agencies should encourage and support the growth of several strong centers of excellence in characterization of materials.

APPENDIX B

In Appendix A an extended abstract of the Materials Advisory Board report (MAB-229-M) on "Materials Characterization" is given. This abstract contains the recommendations of the study, which was published in 1967.

Since 1967, the National Materials Advisory Board (the successor of the Materials Advisory Board) has conducted a number of committee studies on various materials systems wherein the need for better characterization of materials was stressed in the findings. A sampling of some such studies include the following:

NMAB-223, "Ceramic Processing," Feb. 1968.

NMAB-284, "Fundamentals of Amorphous Semiconductors," Sept. 1971.

NMAB-332, "Organic Polymer Characterization," 1977.

NMAB-362, "Preparation and Characterization of Silicon for Infrared Detectors," Oct. 1981.

NMAB-368, "Rapidly Solidified (RS) Aluminum Alloys--Status and Prospects," 1981.

NMAB-377, "Assessment of Mercury-Cadmium Telluride Materials Technology," Sept. 1982.

NMAB-394, "Dynamic Compaction of Metal and Ceramic Powders," Mar. 1983.

A brief synopsis of each of the seven above-mentioned studies is as follows:

Ceramic Processing (Report NMAB-223)

In the field of ceramics, the NMAB has conducted several major studies. The first concerned the Processing of Ceramics (NMAB-223) and emphasized that a detailed examination of ceramic processing was a necessary step toward obtaining reliable high-integrity ceramic materials with superior properties. Technical recommendations given in the report are that (1) starting materials should be fully characterized as should each step in processing, (2) new tools and techniques should be provided to characterize material in process and the final product, (3) particular attention should be paid to the character of the ceramic surface, (4) standardized lots of starting materials and standard test methods should be made available, (5) the scientific approach should be used to overcome limitations in size without sacrificing reliability, and (6) improved understanding of character-property relationships must be developed. The report states that these essentials should be brought forcefully to the attention of all concerned, with interdisciplinary programs developed including consortia among universities, research laboratories and industry; to address the problems in a pragmatic manner.

Fundamentals of Amorphous Semiconductors (Report NMAB-284)

The study of glasses has been important historically because of their usefulness. Members of a comparatively new class of these materials, the amorphous semiconductors, have evoked interest in the last few years because they exhibit certain unique properties (semiconductivity, photoconductivity, low sensitivity to high-energy radiation, and ease of undergoing phase changes). Such properties are of considerable technological significance. In the report of this amorphous semiconductor study (NMAB-284), it was recommended that increased efforts be made in the gathering of data on physically realized glass structures, development of better methods of material preparation and characterization, investigations leading to better understanding of structure control and radiation hardness, and research aimed at the technological exploitation of unique properties.

Organic Polymer Characterization (Report NMAB-332)

This report attempts to define those properties of organic polymers that are critical to their use in current and advanced structural applications. It discusses and evaluates the characterization methodology that is available to measure and control those properties. It suggests some specific areas in which this technology can be employed to achieve improved performance and reliability through its application to procurement and quality control procedures. Case studies are presented to illustrate the utilization of characterization. Conclusions and recommendations are presented. A list of more than a hundred useful methods of characterization and commentary on use and limitations is given in an appendix.

The Preparation and Characterization of Silicon for Infrared
Detectors (Report NMAB-362)

In this report materials and processing requirements for IR-
type silicon were analyzed and defined. The status of the related
processing technology was reviewed and deficiencies were identified.
The major subjects addressed are:

> Device needs.
> Materials characterization.
> Preparation of polycrystalline silicon.
> Preparation of single-crystal silicon.
> Device process-induced contamination.

Materials requirements for high-speed, high-sensitivity IR
detectors are significantly beyond the present capability of
crystal growth technology. The preparation of ultra-high-purity
polysilicon can be achieved in principle by upgrading or modifying
present purification procedures. However, the preparation of both
ultra-high-purity and homogeneous uncompensated In-doped single-
crystal silicon, either by Czochralski growth or by float-zoning,
with the established procedures is impossible. Substantive modifications of conventional crystal-growth procedures and the development of appropriate alternative approaches to silicon crystal growth,
now in the research stage, are mandatory to meet the materials
requirements.

Included in the recommendations of this study is the following:

> "All sponsored work pertaining to IR-device development
> and fabrication should include a strong materials charac-
> terization component. This procedure could insure the
> advancement of pertinent characterization techniques and
> contribute to the establishment of as-yet-unknown cause
> and effect relationships between materials deficiencies
> and device yield and performance. Such knowledge ulti-
> mately could remove much of the empirical element in
> materials processing..."

Rapidly Solidified (RS) Aluminum Alloys--Status and Prospects
(Report NMAB-368)

This study was conducted to evaluate the potential of particulate (rapidly solidified) aluminum alloys for a broad range of
structural applications. The study included analysis of current
experimental and near-term production alloys; selection of representative target properties and analysis of structural performance
in representative aircraft systems; evaluation of alternative
methods for producing sheet, plate, extrusion and forging mill products with emphasis on approaches for processing particulate directly

to mill products; assessment of structural fabrication and assembly processes and potential associated problems; review of the metallurgical state of the art of these alloy systems; and extensive examination of potential applications in aircraft, military, and space systems and commercial products. Significant conclusions and recommendations are presented that identify the future work required to support adequately the continued development of particulate aluminum alloys and to ensure the eventual availability of large-scale production quantities of these alloys. Among these recommendations was the following:

"Present knowledge concerning phase relationships, metastability of alloy microstructures, and microstructure-property relationships in RS aluminum alloys is inadequate. This lack of knowledge extends to the relative importance of particulate cooling rates, particulate sizes, grain and dendrite sizes, solid solution decomposition kinetics, and alloy composition. Recommendation: A continuing, long-range basic research program should be undertaken to provide adequate support for current developmental and application activities. This program should stress the generation of fundamental structure-property relationships and the understanding of alloy systems and behavior rather than the development of specific RS alloys..."

Assessment of Mercury-Cadmium Telluride Materials Technology (Report NMAB-377)

This report surveys the material requirements and existing material limitations for HgCdTe in its varied applications as a photovoltaic detector. This primary emphasis throughout this report has been the status of the material used for detection of infrared radiation in the 3- to 12-µm wavelength band. The status of the knowledge of the basic semiconductor properties of HgCdTe relevant to the operation of photovoltaic detectors is reviewed and related to device and focal plane performance and future needs. The material preparation aspects of HgCdTe are given primary consideration in this report. This includes a review of the phase relations in HgCdTe required for crystal growth, and a discussion of the defect chemistry of this material system. The crystal growth covers all aspects from derivation of the raw materials to the existing crystal growth techniques. With the current emphasis on epitaxial growth for HgCdTe, the status of substrate growth is also reviewed. The characterization techniques most commonly used in conjunction with the growth are reviewed and critiqued in detail.

Dynamic Compaction of Metal and Ceramic Powders (Report NMAB-394)

In this study on Dynamic Compaction of Metal and Ceramic Powders the state of the art and the technological potential for the dynamic consolidation of metal and ceramic powders was assessed. The

fundamental consideration of dynamic consolidation, consolidation
phenomena during dynamic compaction, dynamic compaction and conditioning of metal and ceramic powders, characterization of dynamically consolidated metal and ceramic powders, computer codes
applicable to dynamic compaction, practical and potential applications, problem areas, and the current position of the United States
in dynamic compaction were examined.

In the findings of the study it was recommended that a systematic study of the dynamic compaction process should be conducted;
existing techniques should be improved and new ones developed to
permit the monitoring of the dynamic events as close to the micro-
scale as possible for temperatures, shock velocities, pressures,
and particle motion; data and information from the systematic
experiments recommended above should be utilized to form data information for the modeling codes; coordination among those investigating dynamic compaction should be maintained; a sufficiently
funded, sustained, coordinated, and concentrated research and development effort should be initiated to strengthen the United Sates
position in the dynamic compaction field. Such a R&D effort includes
the recommendation that at least four types of characterization are
needed to understand the details of dynamic compaction of metal
and ceramic powders:

1. Characterization of the starting powder (including chemical,
    particle and crystallite dimension, X-ray lattice measure-
    ments, surface area, density of particles, shape distribu-
    tion and distributions, etc.).

2. Characterization of the initial pressed powder contained
    in the die fixture (including green density, porosity, and
    texture details).

3. Characterization of the experiment in terms of the pressure-
    time-temperature relationship (in real time) of the pro-
    jectile or explosive on the pressed powders.

4. Characterization of the resulting compact both axially and
    radially (including density versus position and the grain
    size data and shape observations based on detailed metal-
    lographic as well as X-ray TEM studies).

In addition to the aforementioned seven studies, a current
(1984) study in progress is:

Nondestructive Examination for Characterization and Quality
Assurance During Manufacturing and Processing

This study is being conducted to critically assess the
current and future role of characterization and evaluation

442

techniques in materials processing and manufacturing. Due to the broad nature of the topic, the scope has been focused by studying one or two model systems, and where possible, drawing generic conclusions.

Metal and ceramic powder production and consolidation are used as model systems since these are undergoing revolutionary changes primarily through new processing techniques that result in vastly improved properties. For instance, rapid solidification of aluminum, iron, and superalloy powders has received a great deal of attention in the past several years. Major efforts are underway by government and industry in the development of quality components for airframes, engines, spacecraft and missile structures, and other applications. Ceramics have also come to the forefront in recent years and have been used in critical applications primarily due to advances in processing. High technology applications are being explored and already exist in electronics, cutting tools, and automotive engines, among others.

In summary, this study will (1) define the state of the art of powder characterization and evaluation techniques, their applications and limitations as applied to metal and ceramic powder production and consolidation; (2) define current and future application needs and concomitant research and development; (3) examine federal roles and mechanisms for effective coordination among federal agencies; and (4) assess the technology transfer and educational requirements.

RECENT ISSUES AND POLICIES CONCERNING QUALITY ASSURANCE

Harry L. Light

U.S. Army Material Development and Readiness Command
5001 Eisenhower Avenue
Alexandria, VA   22333

Throughout the Army there is increasing command emphasis on quality and reliability and the role that all scientists and engineers must play so that DARCOM fulfills its goal of providing "Quality Equipment and Support for an Excellent Army." The purpose of this article is to explain the issues behind this increase in command emphasis and the policies recently promulgated which all DARCOM engineers and scientists must follow to achieve the DARCOM goal.

THE ISSUES

In an interview with the New York Times, the Army Chief of Staff stated "There are some things that I can get emotional about and quality control is one of them."

More recently, the DARCOM deputy commander for Research, Development and Acquisition conveyed to the Army Chief of Staff the DARCOM approach to quality: "We are stressing that we design for performance, perform producibility engineering and manufacturing methods and technology early to insure repeatability in volume production and adequately test for component qualification, as well as system performance. Quality is not accomplished through inspection. Quality must be designed into the initial system and we must and will hold scientists and engineers responsible for quality and costs. Inspections verify conformance to design. Quality is a mindset and must be achieved through active participation by everyone from the corporate management to the worker on the production line. We must motivate and discipline. We shall do that. Our soldiers deserve no less."

This high-level command emphasis on quality stems from what quality assurance is all about - soldier satisfaction. Poor quality control results in a loss of confidence by the soldier in the field, unsatisfactory reliability performance, and increased costs of weapon systems. The credibility of DARCOM depends on the quality and reliability performance of its equipment and support.

Dissatisfaction with the quality, cost and reliability, availability and maintainability performance of several major Army systems and the process by which they evolved have led to the recognition that improvements to the process need to be made immediately.

Because of difficulties experienced with the development and production of certain newer ammunition and other items, the DARCOM commander appointed a Product Assurance and Test Review Board, chaired by GEN Walter T. Kerwin (USA retired), to evaluate the adequacy of the DARCOM Product Assurance and Test Program.

The Board's objectives were to review the adequacy of quality assurance and field procedures, assess whether deficiencies exist in the interface between DARCOM and the Defense Logistics Agency, and determine whether management expertise and the quality of production-line workmanship are declining.

Although the Kerwin Board initially focused on traditional quality problems - contractors' negligence during production, failure of quality assurance personnel in procuring activities to fulfill their responsibilities, and the negligence or incompetence of Defense Contract Administrative Services inspectors - the Board quickly realized that lapses in these areas contributed only in small part to the quality problems of the Army. The real problem the Board realized was errors in the design and development process, prior to production.

After recognizing that improvements in the design and development process afforded the high-leverage needed to improve the quality of DARCOM equipment and support, the Board refocused its efforts. Inquiries were made to examine the DARCOM/TRADOC interface, the technical performance of new items, the extent to which ASARC or DSARC decision points control the development process, the suitability of test procedures, the accuracy and completeness of the technical data package and the role of the project manager.

This refocus resulted in the findings that problems found during design and development stages were not satisfactorily resolved prior to transition into production, that quality assurance is considered only after cost and schedule, and that lack of up-front quality assurance guarantees problems downstream.

Although the Kerwin Board dealt primarily with ammunition, the issues and recommendations have validity for most DARCOM commodities as is evidenced by the findings of Contractor Assessment Reviews. These joint HQ DARCOM and major subordinate command reviews were initiated by the Deputy Commanding General for Research, Development and Acquisition last fall as a result of costs and quality problems of major systems.

The purpose of these reviews is to identify and make recommendations concerning productivity, cost and quality control. Typical problems identified which result in loss of control of costs and quality of systems entering production include:

- Systems entering production with unresolved design issues and test failures.

- Long duration between identification of a problem, completion of failure analysis and implementation design changes and corrective action.

- Quality and producibility considered only after completion of design and redesign efforts.

- Inadequate planning of facilities, equipment and tooling to support large volume production.

- Lack of parts and vendor controls programs.

- Capitulation to schedule demands by accepting waivers and deviations not in the interest of the Army.

- Reliability requirements not being placed in contracts.

NEW POLICY

Since the credibility of DARCOM depends on the quality and reliability, availability and maintainability performance of material at Initial Operational Capability and since equipment must work reliably and be supportable at high readiness levels, the deputy commander for Research, Development and Acquisition has issued new policy on management of reliability and maintainability which recognizes the responsibility of all DARCOM scientists and engineers in this critical area.

This policy emphasizes basic engineering design, growth management and testing through system development and production, and directs that command principles and project managers be rated on their attainment of RAM requirements.

Starting with the development of requirements, the policy

insists that quantitative reliability requirements be established for all programs. These requirements must meet user needs and be consistent with the state-of-the-art of technology. These requirements should consider both hardware and operations and recognize that RAM will grow as design changes are implemented and troops gain training and experience. These requirements then become the operational reliability and maintainability requirements expressed in the Required Operational Capability. Before proceeding to the next milestone, these requirements must be met.

The policy next insists on basic reliability, availability, and maintainability engineering and design practices to meet these operational requirements. The reliability and maintainability designed into the hardware must exceed the minimum acceptable value expressed in the ROC. There must be a safety margin in the design to compensate for the degradation commonly experienced during systems integration. As such, development programs must impement parts control and include reliability parts in accordance with MIL-M-38510, MIL-STD-883 and other established military specifications.

The policy demands that reliability requirements be established during advanced development and implemented at the start of full-scale engineering development.

In addition, reliability and maintainability apportionment, tolerance analyses, failure modes and criticality analyses, and development of manufacturing process controls and inspection equipment are to be accomplished during the engineering design process. The policy also calls for the use of environmental stress screening, which employs thermally cycling and random vibration at all levels of assembly. This screening is accomplished to precipitate failures resulting from poor workmanship and defective parts so that these failures occur during manufacturing rather than in the field.

The applications of environmental stress screening to date have resulted in significant improvements in reliability and reduction of manufacturing costs by reducing rework.

In calling for the development of manufacturing and process controls and acceptance during design, the policy on management of reliability and maintainability reiterates the DARCOM policy on producibility.

Recognizing the strong and potent role that technology affords in controlling costs and quality and improving reliability and producibility, the director of Technology Planning and Management issued instructions that each laboratory identify quality and producibility opportunities and develop programs to address these opportunities.

The policy also requires a planned reliability growth program for use during development, production and initial deployment to achieve operational reliability, availability and maintainability requirements. The program is to be conducted in accordance with MIL-STD-189 and must address the entire system and critical components and subsystems.

An essential part of the growth program is the test-analyze-and fix (TAAF) concept. A test period must be scheduled in conjuncture with each major milestone to identify design, software and manufacturing defects. Test time and resources must also be scheduled to correct deficiencies found during testing. There must be sufficient dedicated people, facilities and test units to identify the "root cause" and eliminate design and manufacturing defects.

In addition to TAAF, achievement of reliability growth results from other processes which identify defects such as environmental stress screening, reliability predictions, failure modes and effect analyses, and component testing.

As a control and check on the acquisition process, the policy insists that approved reliability, availability and maintainability and supportability requirements for each major milestone be met before proceeding to the next phase or Initial Operational Capability.

Recent policies reemphasize that the responsibility for quality and reliability of Army weapon systems include all DARCOM engineers and scientists. It is only through the teamwork of all involved in the weapons acquisition process that we can fulfill the DARCOM goal of "Quality Equipment and Support for an Excellent Army."

TOWARDS A COMPUTERIZED MATERIALS CHARACTERIZATION EXPERT SYSTEM

Volker Weiss

Syracuse University
Syracuse, NY   13210

ABSTRACT

The feasibility of using logic programming for materials characterization in connection with an information base on materials is demonstrated. The previously described LOGLISP knowledge base for materials selection, which contains properties and rules commonly used in materials science, served as the basis for the present study. It is shown that the laboratory notebook format presented in the earlier study can readily be adapted for use in materials characterization. Characterization requirements are obtained as a response to simple LOGLISP inquiries for the conditions to meet a certain set of (unique) properties.

INTRODUCTION

LOGLISP is an implementation of logic programming in LISP, developed by J. A. Robinson and E. E. Sibert[1]. Weiss and Aha[2] have shown that this new programming language offers a powerful tool for software development in the materials science field. In a feasibility study they demonstrated the applicability of LOGLISP to materials selection, alloy substitution, fatigue data evaluation, and a research data notebook.

The language LOGLISP is a combination of logic, i.e. lamda calculus, unification and resolution, and LISP, a well established computer language, i.e.

LOGLISP = LOGIC + LISP

A LOGLISP knowledge base consists of facts and rules, i.e.

KNOWLEDGE BASE = FACTS + RULES

Facts may have the format:

(Is ST4340 400 TS 272)

which is LOGLISP shorthand for: "the tensile strength of 4340 steel, tempered at 400F is 272 ksi." The above entry style requires one line per datapoint. To save computer storage space it is possible and desirable to group related data. In the present knowledge base this has been done through the predicate Iss, i.e.

(Iss material condition ((property1 val1)
                         (property2 val2)
                              ...
                         (propertyn valn)))

For 4340 tempered at 400 the data in the ISS format are shown below.

(¦- (Iss ST4340 400. ((RC 55.0) (TS 272.0)

(YS 243.0) (KIC 40.0))))

A typical example of a rule is the formula for determining the critical crack length c from the plane strain fracture toughness KIC and the yield strength Y, namely

$$c = \frac{1}{\Pi}\left(\frac{K_{IC}}{Y}\right)^2$$

which, in LOGLISP has the form

(PROCEDURE CL)

(¦- (CL matl cond (% (SQR (% kic ys)) 3.1415925) kic ys)
 <- (Is matl cond KIC kic)
 & (Is matl cond YS ys))

The " % " symbol represents division. SQR is a function defined in LISP and represents the squaring operation. Inserting the name and the condition of the material into this rule yields the critical crack length for the material in question, provided that the fracture toughness and the yield strength values are contained in the knowledge base. For example:
(THE x (CL ST 4340 400 x)) yields 0.0086 inches, or

```
(THE x (CL ST4340 400 x))
0.86249698E-2
```

The knowledge base in queried, as hinted above, through the deduction commands THE, ALL, ANY or SETOF. The entire knowledge base can be queried in an ad-hoc fashion, without the need for prepared search routines and menus. The resulting response can be readily formatted for each purpose to include only the information sought, uncluttered by extraneous facts. A typical example is the request for a list of materials having yield strengths above 20 ksi and below 30 ksi. Note that only the name of the material, its condition, its tensile strength and its yield strength were sought, as indicated in the answer template which follows the deduction command ALL, i.e.

Table 1.

```
(DISPLAY (QUICKSORT (ALL (mat cond ts ys)
                         (Is mat cond YS ys)
                         (> ys 20.)
                         (< ys 30.)
                         (Is mat cond TS ts))
                    (ALPHABETICALLY)
                    1.)
         (Material Condition Tens-Str Yield-Str))
```

Material	Condition	Tens-Str	Yield-Str
A2117	T4	43.	24.
A6061	T451	35.	21.
A6061	T4	35.	21.
A6101	T6	32.	28.
A6463	T5	27.	21.
A7001	AN	37.	22.
C44300	AS	55.	22.
C44400	AS	55.	22.
C44500	AS	55.	22.
FMFE	SIN	40.	26.
SS502	PL	65.	25.
SS502	BA	65.	25.

It is relatively simple to develop software programs that do not require the user to be knowledgeable in LOGLISP or LISP programming. Such a program is presented below. The task is to select materials for a tensile member which will support a given load, be designed to a certain safety factor, with a maximum diameter and weight per unit length limitation. The designer selects the constraints and obtains a listing of eligible materials by responding to the questions presented by the program which is written in a conversational mode.

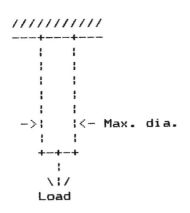

Load?*200

Safety factor?*3

Maximum diameter?*2.5

Max Wt/Length?*.75

Table 2.

Mat	Cond	Ys	Dia	Wt/L
CSIC-AL	VF50	262.	1.5957698	0.19000000
ST4340	400.	243.0	1.7730775	0.69135802
ST4340	QT400	243.	1.5957698	0.56000000
ST4340	600.	230.0	1.8224975	0.73043479
ST4340	QT600	230.	1.5957698	0.56000000
CSGLA-EPO	VF60	230.	1.5957698	0.14599999
ST4130	WQT400	212.	1.5957698	0.56000000
CBOR-EPO	VF60	180.	1.9544108	0.21900000
CKEV-EPO	VF60	180.	1.9544108	0.15000000
CEGLA-EPO	VF60	150.	2.2567593	0.30000000
CGRA-AL	VF40-1100	128.	2.2567593	0.36400000

The first demonstration of the use of LOGLISP in materials science and engineering[2] included examples for

1. Materials selection for critical applications, including the design of a tensile member under dimensional and weight constraints;

2. Alloy substitution, i.e. the listing of alternate alloys with equal or superior properties to the one currently used, but smaller contents of alloying elements that are in critical supply;

3. Fatigue data evaluation, including the construction of a Goodman diagram;

4. As a notebook, for storing and subsequent evaluation of data;

5. Determination of which properties a group of materials may have in common, given certain conditions, e.g. having yield strengths and fracture toughnesses in certain specified ranges.

The fourth example, a research notebook in LOGLISP, can serve to demonstrate the feasibility of using LOGLISP in connection with materials characterization.

APPLICATION TO MATERIALS CHARACTERIZATION

The flexibility of LOGLISP makes it very useful for storing experimental data. In this example, again for fictitious data, experimental results are entered under the procedure DPTS which has the format:

```
(DPTS  Test-material
       Condition
       (Descriptive Comments)
       Test-property
       (Test Results))
```

For example:

(DPTS AL7075 T651 (5 7 82) TS (75 80))

would indicate that two tensile tests were made on Al 7075 T651 on May 7, 1982 with the results of 75 and 80 ksi. If, at some future time a summary is desired, e.g. the individual results, the average and the standard deviation for all tensile strength results on Al 7075 T651, a LOGLISP function called STATS

```
(  ¦ -  (STATS mat cond com prop (vals avg stdd))
   <-  (= vals (FLATTEN (ALL x (DPTS mat cond com prop x))
   &  (= avg (AVERAGE vals))
   &  (= stdd (STANDARD-DEVIATION vals)))
```

will produce the desired printout. In this case

```
(SPRINT (THE x (STATS AL7075 T651 # TS x]
```

returns

```
((76.2 78.6 80.1 75.0 75.0 80.0 70.0 71.0 80.0 85.0)
 77.09
 4.33)
```

where the first group of numbers in parentheses represents all
data points in the "logbook" for the tensile strength of
AL 7075 T651. The second line is the average and the third the
standard deviation of these data points. Obviously the present
research logbook example is a very simple one. It is easy, however, to expand the format. Realizing that the results obtainable
from a tensile test include the tensile strength, yield strength,
elastic modulus, some characteristic relating to the strain
hardening behavior of the test material, uniform elongation, total
elongation and reduction in area, etc., one might choose a format
such as

```
(DPTS1 TMAT COND COMM ((TS (   ))
                       (YS (   ))
                       (RA (   ))
                        ...   ))
```

The desired printouts can then be obtained as the results of
readily programmed LOGLISP queries, similar to STATS discussed
above.

The fact that specification of several manufacturing and
processing parameters is required to achieve a unique set of
material properties has led to increased emphasis on characterization. An example of this lack of uniqueness is readily demonstrated
with the help of the data presented by McCauley[3]. His data for
Polyurethane samples have been added to the Materials knowledge
base in the above described laboratory notebook format. An
example of our entry is shown below.

(Iss PU S6 ((HS 66)   (HFP 9.4) (YS 30.0) (EL 400) (FS 33.0))))

When queried with

```
(DISPLAY (ALL (PU no hs ys fs el)
              (Is PU no HS hs)
              (Is? PU no YS ys)
              (Is? PU no FS fs)
              (Is? PU no EL el))
         '(MATL NO HARD-SEG YS FS EL))
```

we obtain the tabulated results

Table 3.

MATL	COND	HARD-SEG	INTL-MOD	ELONG	FR-STR	TOUGHNESS
PU	S1	10.	0.69999999	700.	1.7000000	83000000.0
PU	S2	21.	5.5	800.	7.8000000	3.3000000E+8
PU	S3	32.	3.4000000	1400.	22.	2.0999999E+9
PU	S4	43.	130.	1200.	33.	2.6000000E+9
PU	S5	55.	260.	700.	31.	1.6999998E+9
PU	S6	66.	520.	400.	33.	1.2000000E+9
PU	S7	77.	1400.	40.	38.	1.6000000E+8

The data, especially when plotted, clearly show that the elongation
is not a unique function of the hard segment content.

Table 4.

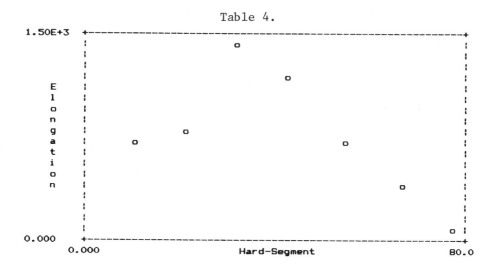

457

The "comment" part of the notebook entry format lends itself readily to indicate further how LOGLISP could be used for material characterization. To provide an example fictitious data for a ceramic, called CER1 in condition C1, were entered in the laboratory notebook. The data are the tensile strengths, the bend moduli and the fired densities. The tabulated data are shown below.

Table 5.

MATL	COND	COMMT	PROP	VALUE
CER1	C1	((Powder-den 60.) (SinT 2100.))	TS	(60.100.90. 120. 80. 70. 110.)
CER1	C1	((Powder-den 60.) (SinT 2100.))	IMOD	(20000000.0 22000000.0 24000000.0 29000000.0 15000000.0 22000000.0)
CER1	C1	((Powder-den 60.) (SinT 2100.))	DEN	(85. 86. 84. 83. 89. 83. 81.)
CER1	C1	((Powder-den 80.) (SinT 2100.))	TS	(120. 122. 123. 121. 118. 120. 124.)
CER1	C1	((Powder-den 80.) (SinT 2100.))	IMOD	(26000000.0 27000000.0 26500000.0 27000000.0 28000000.0 28500000.0 25800000.0 26500000.0)
CER1	C1	((Powder-den 80.) (SinT 2100.))	DEN	(97. 98. 96. 97. 98. 98.699995)

One might ask the question: What must be characterized to hold the scatter of tensile strength values below 3 ksi? This question has the form

```
(ALL (commt std)
     (DPTS CER1 # commt TS #)
     (STATS CER1 # commt TS (v a std))
     (< std 3.))
```

and results in

```
*(EVAL CHAR)
((((Powder-den 80.) (SinT 2100.00 1.8843840))
```

i.e. the green powder density as well as the sintering temperature.
The standard deviations of the tensile strength values as a function
of characterization parameters are obtained from

```
(SPRINT (ALL (CER1 c com stdd)
             (DPTS CER1 c com TS #)
             (STATS CER1 c com TS (v a stdd)))))
```

and result in

```
((CER1 C1 ((Powder-den 60.) (SinT 2100.)) 20.0)
 (CER1 C1 ((Powder-den 80.) (SinT 2100.)) 1.8843840))
NIL
```

Although the present example is rather trivial, it is clear
that the power and flexibility of LOGLISP will allow its application
to far more complicated practical applications in the field of
materials characterization. Especially the recently developed
methods in the QUALOG[4] system, for interfacing logic programming to
textual data, appear to have great potential. Specifications and
tables could be annotated. A logic query of these annotations
provides the text segments of concern. Text segments themselves
can be searched for words and phrases. The ease with which a
knowledge base can be modified or updated represents another
advantage for the development of "learning expert systems".

Among the potential applications of LOGLISP in connection with
materials characterization are:

* A directory of characterization techniques, their range of
applicability, accuracy, availability (location), cost etc.

* An expert system on characterization – properties relationships (reliability, cost/benefit, alternate methods..)

* An expert system for materials selection based on optimum
property – characterization requirements.

* A case histories knowledge base of characterization vs.
service experience information.

* General Expert systems e.g. for design or failure analysis,
possibly hybrid systems between LOGLISP and CAD/CAM systems.

CONCLUSIONS

The feasibility of using logic programming in the form of LOGLISP for materials characterization has been demonstrated. The previously developed research notebook format can readily be adapted to include characterization requirements. The search for unique sets of properties will yield the appropriate characterization requirements. Together with the text annotation capabiliites in QUALOG type systems, LOGLISP offers a powerful tool for software development in the field of materials characterization.

REFERENCES

1. J. A. Robinson and E. E. Sibert, "Logic Programming in LISP," CIS Report 8-80, Syracuse University, December (1980).
2. V. Weiss and D. Aha, "materials Selection with Logic Programming," Proc. 29th Annual SAMPE Symposium, Technology Vectors, Soc. for the Advancement of Material and Process Engineering, Vol. 29, p. 506, (1984).
3. J. McCauley, "Materials Characterization: Definitions, Philosopy and Overview of the Conference," 31st Sagamore Army Materials Research Conference, August (1984). (To be published by Plenum Press).
4. A. Shelly and E. Sibert, "QUALOG User Manual," CIS Technical Report 11-83, Syracuse University, November (1983).

ACKNOWLEDGEMENT

The author wishes to thank Professors E. Sibert and J. A. Robinson for their help and continued interest in the development of LOGLISP based expert system related to materials science. Mr. David Aha's contributions to the past and present efforts are also gratefully acknowledged.

A TOTAL QUALITY PROGRAM

3M CASE STUDIES

Robert E. Richards

New Products Department

3M Company

Historically, 3M has worked to maintain a quality image in its world-wide markets. Changing competition, raw materials evolutions, and the economic pressures of the late 1970s brought new recognition to the significance of quality as a key competitive factor. The new quality emphasis required a reworking of not only the techniques and priorities of quality management, but also a revitalizing of the quality commitment of senior management, and the development of a process for building organizational quality continuously.

COPY PAPER CASE HISTORY

This story starts in 1977 when 3M's Copying Products Division was facing growing Japanese competition of excellent quality and rapid inflation was creating financial pressures in both sales and costs. One of the prospects for quick relief was an improved copy paper for the line of 3M's VQC copy machines. In conjunction with the new paper, a new Quality System was desired to help reduce the increasing number of field recalls for paper that would not match up with the ever broadening base of customers' machines (see Figure 1).

Laboratory team members from Product Development, Quality Assurance, and Customer Service Engineering joined together to develop a proposal for a total quality program to be introduced with the new copy paper.

Customer comparisons with competitive product would be used to set final quality requirements (see Figure 2). Knowledge of the process variability would be used to work backwards through each step in the process to set statistically valid limits from the finished paper, to the in-process control standards, and then to the starting raw materials. Definitive material characterizations were to be developed, along with correlations to simulated use tests, to select and blend the raw materials. Target-centered controls, rather than two-sided limits, were planned to further reduce variability.

The Need:
  – Improved Performance
  – Reduced Cost

The Response:
  – A Minor Product Change
  – A Major Quality Program

Figure 1.

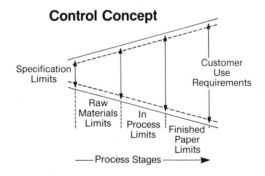

Figure 2.

The proposed plan was accepted and the work was started with enthusiasm. Rapid progress was made. The customer comparisons worked, the product changes worked, and in roughly one year the project moved into the factory. The project team's projections for the new product showed both a lower cost and a higher copy quality, with a better match to the aging field machines.

CONCEPT PROBLEMS

That ideal plan did not work out as we naively believed it would. As a result, valuable lessons in organization and management commitment were learned. The problems we encountered were in selling the new quality concepts (see Figure 3). The materials characterizations, even though they correlated well with final product properties, were resisted as a means of screening and selecting the starting materials. The new approaches to process control, especially correcting to a center target, were not accepted. Zero defects was considered an impossible and unaffordable goal; and the scrapping of out-of-standard materials, whether in-process or finished goods, was "unthinkable". We struggled for over eighteen months trying to implement the new product and its new Quality System before pulling back and rethinking the entire project. Today we understand that the attitudes toward compromise between cost and quality could not change without training and support throughout the organization. It took both good fortune and good timing to find a way to change those attitudes and to turn the project from an impending failure to a much needed success. Since those difficulties in 1978 and 1979, a major shift has evolved in 3M's approach to quality.

> Today – Quality at 3M is becoming A WAY OF LIFE
> and – WORLD CLASS QUALITY is a 3M COMMITMENT

BACKGROUND

3M is a multi-national enterprise. It is an organization which in 1983 had sales of over $7 billion, 87,000 employees, and operated forty-five product line businesses world-wide. Fourteen basic technologies form the basis for the present commercial products, and 3M research and development is busy on at least a half-dozen new ones. 3M markets over 40,000 products and serves ten major markets. In a word, 3M is diversity! The point is, that renewing and strengthening a quality emphasis, while facing stiff pressures for improved costs and ambitious goals for new products, is itself a challenge; but doing it inside a busy, successful, diverse organization that already believed in its existing quality program was even more difficult. The fact is, it is not done. It is still happening; there still is much to do!

Concept Problems:

Raw Materials Characterization
Process Control to Targets
Zero Defects
Scrapping Rejects

Figure 3.

Quality Emphasis Milestones

1979 – Crosby Programs – F. S. Webster
     – Quality College
1980 – Staff Quality – L. W. Lehr
     – All 3M Management Training
1981 – Supplier Quality Program – R. V. Peterson
     – Right the First Time
     – Zero Defects Day

Figure 4.

QUALITY EMPHASIS MILESTONES

By 1979 the copy paper project team's push for quality improvements had stalled; but fortune smiled as Division Vice-President, Francis S. Webster, discovered Philip Crosby's book "Quality is Free"[1] while on vacation. He was intrigued and carried his copy back to 3M. Soon the management team of the division was offered training at Quality College and slowly, but surely, changes in attitudes began to occur. Small groups, and later whole teams, began to implement the Crosby 14-step process.

In late 1979, more good fortune occurred. A request came from 3M's Chairman, Lewis W. Lehr, to 3M's Staff Quality organization to revitalize 3M's emphasis on quality. In early 1980, the Staff Quality organization began to organize to handle increased training demand, and cooperation with the Copying Products' effort grew and strengthened both.

Many, many 3Mers have become involved as leaders, trainers, and developers; their performance and contributions made possible the many steps and programs. Figure 4 lists the first major milestones in the development of the quality process.

By 1981 most Copying Products personnel had completed an internally developed training program based on and inspired by the Philip Crosby philosophies[2]. Quality Improvement teams, Zero Defects Day, a fledgling Cost of Quality measurement, and a major effort in expanding awareness to supplier and customers had been developed. R. V. Peterson, Copying Products Quality Assurance Manager, spearheaded the supplier program. Coordination with purchasing, top support from management, and a positive attitude toward assisting our suppliers "meet requirements the first time", like partners, was achieving remarkable results.

Ideas and elements for improving our approach to Total Quality were coming at us from many sources. We began to see them blended together to make up a comprehensive system. Quality training was moving to 3M Center. Hundreds from manufacturing, sales, laboratory, administration, and executive groups were involved.

concepts of a 3M strategy began to emerge[3], and emphasis was directed toward each (see Figure 5).

ESSENTIAL CONCEPTS OF 3M'S PROCESS

From these early efforts at systematic training, and the experiences of the first groups using the training, five essential

5 Essential Concepts

1. Definition of Quality:
   Conformance to Requirements
2. Prevention of Errors
3. Measurement of the Cost of
   Non-Conformance
4. Ultimate Goal: Zero Defects
5. Commitment: A Planned, Active Role
   for All Personnel

Figure 5.

Quality Emphasis Milestones

1982 – World Class Quality Theme
       – Stockholders Letter
       – Managing Change
1983 – Strategic Planning
       – Cost of Quality Reporting
1984 – Developing the Tools of Change
       – Do It Over Again

Figure 6.

1. Quality was defined as CONFORMANCE TO REQUIREMENTS. Crosby[1] uses the comparison between a Cadillac and a Pinto to illustrate the difference between a notion of goodness or luxury versus a difference in requirements. Organizational quality also receives emphasis at 3M. This stresses the relevance of setting and understanding the proper role by and for each individual and then performing the work so as to meet the requirements of that role.

2. PREVENTION pays dividends. Setting goals, making certain, and managing to prevent the errors is stressed. Crosby[1] demonstrates the costs of non-conformance are many times greater than the cost of prevention.

3. MEASUREMENT OF THE COSTS OF NON-CONFORMANCE. 3M's Cost of Quality is the sum of all the costs of failure, both internal (scrap, rework, etc.) and external (returns, warranty, customer service) as well as the traditional costs of testing, inspection, and training. Experience shows that for most companies the cost of non-conformance exceeds over 20% of sales, and inside 3M many believe that 20% number is a good estimate for many of the 3M product lines.

4. THE ULTIMATE GOAL - ZERO DEFECTS. No errors, doing the job right the first time. Much of our emphasis is that zero defects is a personal standard. It is an attitude that leads to a behavior that wins in business, in service work, or in life.

5. The last of the five essentials - COMMITMENT. Must start at the top of the organization and must recognize that quality improvement does not just happen; it has to be planned and actively managed if it is to become a way of life.

Returning to the milestones in the overall program development (see Figure 6), 1982 was the year the WORLD CLASS QUALITY theme was adopted. Training of 3M subsidiary leaders from around the world was started in St. Paul to enable each outside U.S. business unit to participate. A letter to 3M stockholders by L. W. Lehr, 3M's Chief Executive Officer, explained the new emphasis and

commited 3M to reduce the cost of quality by 50% by 1987. Based
on a general assumption of an original cost of quality level of
20% of sales, the savings could exceed $700 million.

In 1983, Quality and Cost of Quality became part of 3M's
business planning and strategic planning process. Real dollar
reductions were being documented, and positive effects on the business were being seen in over 20% of our ˈusiness units.

Quality Emphasis in
Copying Products Division

1979 – Start of Management Training
1980 – Manufacturing Staff Training (1st qtr)
     – Copy Paper Standards Republished (2nd qtr)
     – Process Controls Established (3rd qtr)
     – Full Production New Paper (September)

Figure 7.

Copy Products Functional Defects
Defective Lots of
In-Process Materials

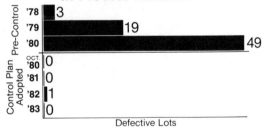

Defective Lots

Figure 8.

In 1984 the emphasis is on developing the tools of change.
For 3M, change is also a way of life. One of 3M's stated goals
is that new products sales, from products less than five years old,
should exceed 25% of our total sales. The need to Do It Right The
First Time is ever present, and the training and the updating of
the quality emphasis to "do it over again" for another pass at
improved performance and reduced errors was also underway.

COPY PAPER CASE RESULTS

These milestones and the five concepts give one an overview of
the philosophy that has developed since 1979. It is an ongoing
Total Quality Process; and into this changing, developing, corporate
quality emphasis, the Copying Products Division team became joined
as the pathfinder for implementing the 14-step program outlined
by Philip Crosby in "Quality is Free".

As the commitment to change increased, resistance to the earlier
quality concepts for the VQC Paper project began to decline.
Quarter-by-quarter during 1980, progress on the paper project
accelerated (see Figure 7). First the training of the manufacturing
staff, then the new Quality Standards, then the new process controls - all were accepted. By September 1980 the paper was in full
production and was operating with the new quality system.

The results were dramatic and unexpected (see Figure 8).
Functional defects stopped, field complaints stopped! From 1978
to 1980 the increased emphasis on customer comparisons and a
change in inspection diligence had resulted in a significant increase in the rate and number of product lots that were rejected.
Once the new product and the new quality system was installed,
rejects stopped. From 1980 until 1982 hundreds of millions of
yards of product were produced without a rejected lot. Zero defects
had been attained; or more specifically, the product was conforming
to requirements!

Dollar savings were documented (see Figure 9). From an early
estimate (before cost of quality procedures were available) of 25%
of sales for 1979-1980, the early 1984 cost of quality had declined
to less than 5% of sales. The benefits in improving the external
costs of failure were most appreciated by our customers.

Cost of Quality
Copy Products – VQC Paper Project

Internal Failure Costs Actual		External Failure Costs Estimated
Before Control (pre '80)	15%	+10%
After Control ('80-'82)	14%	–0–
Latest – 1st qtr '84	5%	–0–

Savings exceed 20% of selling prices

Figure 9.

Post-It Demo Program

1980 – Management Training
1981 – Program Team Training
    – Right the First Time Teams Formed
    – Program Goal Setting
    – International Management Training
1982 – Error Cause Removal
    – Prevention Emphasis
    – Zero Defects Day

Figure 10.

Cost of Quality
Type 665 – Post-It® – 3×5 Pads

1982	–	14.6%
1983	–	11.6%
1984 (1st qtr)	–	8.0%

Figure 11.

POST-IT® CASE HISTORY

The second case history starts in 1980 with the new product introduction of 3M's POST-IT® note pads (see Figure 10). In 1980, Francis Webster, who had been instrumental in starting the Copying Products Quality Emphasis process, had moved to and accepted the leadership of 3M's Commercial Tape Division. He began another Quality Emphasis program with that division's management. Following training, the management team was challenged to pick a program as a quality emphasis demo project. POST-IT® was chosen. As a result of their training and commitment, a promising new product opportunity rapidly became a major 3M success story. By every measure, the POST-IT® team moved ahead to meet and exceed their goals.

The process involved all elements of the division. Program introductions around the world were planned in detail; zero defects was installed as the universal goal. Targets were set for the cost of quality and quality improvement at each phase and level of the operation as a part of the new product introduction and scale-up. Organizational quality and individual performance became as much a part of the program as the product itself. Requirements were carefully set for each step of the products introduction. Announcement of the new product was actually delayed in one country's market when preparations for marketing did not meet the agreed requirements. In another instance, also international business, in-process improvements received crash emphasis in time to prevent the scale-up of a less than 100% quality process for a key raw material. The planning and performance show in the cost of quality results (see Figure 11). 20% to 30% cost of quality levels, typical of new product start-ups, were never observed for this program. Group by group throughout the Commercial Tape Division's organization - order entry, order filling, product quality, sales, marketing, advertising, dealer relations - better results were being obtained; each had a part in the new emphasis and made measurable reductions in their own cost of quality.

SUMMARY

By comparison, the POST-IT® example was a far more ideal case. The impact of training and the prior preparation of management enabled the introduction of a total quality program with a minimum of difficulty. The first case history had a far more developed technical effort for the characterization and control of each step in the materials processing; but without the organizational support and philosophy to "Make It Right The First Time", the benefits could not be achieved. This changed once the combination of both materials characterization and total management support was achieved, and the result was nothing short of spectacular. Zero defects, greatly reduced cost of quality, and outstanding customer acceptance were achieved immediately.

Summary

- Total Quality must be managed
  - Set specific goals
  - Compare to competition
  - Allocate for prevention
  - Prepare for change
  - Measure cost of quality
  - Use quality as a strategic tool
  - Nurture the commitment at
    all levels of the organization

Figure 12.

Corporate Quality Statement

Goal: To maintain or improve 3M quality by product and service performance that equals or exceeds that of any competitor in the same market, and meets all expectations of the user worldwide.

Policy: 3M will develop, produce and deliver, on time, products and services that conform to customers' requirements. These products and services must be useful, safe, reliable, environmentally acceptable and represented truthfully in advertising, packaging and sales promotions.

Implementation: Each business unit or function will be managed to provide personal, product and service performance exactly as required. All requirements must be clearly and completely stated and describe exactly what is needed. Any changes in requirements must be appropriately approved.

Figure 13.

The development of a Total Quality Emphasis is a process that needs to become continuing. One needs to involve not only the technical dimension, from raw materials characterization to finished product specifications, but also the management dimension, from senior management to all levels of the organization. These two cases illustrate the difference in one's ability to implement a superior product or quality system concept based on the degree of support existing within the organization.

In summary, Total Quality Emphasis involves not only the technical elements from raw materials characterization to finished product specifications, but also management commitment and all aspects of organizational performance.

So much attention is given to the technical tools and the mechanics of setting up the actual Quality System, but the management concepts in these cases dominated the technical aspects. Certainly one must also appreciate that in large organizations there are subtle factors to consider when attempting to combine new technology, human relations, and effective management. One must manage change within the culture of the existing organization. From these examples we conclude that the concepts Management needs to emphasize are shown in Figure 12. The commitments to improve and manage the key essentials need to combine to create an ongoing process. The nurture of the process cannot be a one-time activity but must be continuous as well. Strategists know the best products are those that truly meet the customers' expectations, and the best producers are the ones that supply those products time after time at lowest costs. We feel fortunate to have had these experiences to share with this conference. It was shown that given the proper supportive environment, a complex product can be produced without defects by using a process control system that started with material characterization, and that the benefits far outweighed the costs of the prevention system and preparation of the testing methods. For some time now, 3M has published the Corporate Quality Statement shown on Figure 13. It represents our commitment to a continuing Quality Emphasis.

REFERENCES

1. P. B. Crosby, Quality is Free, McGraw-Hill Book Company, ISBN 0-07-014512-1.
2. Quality Training Program - QTP-1, 3M Company Education and Training, (1980).
3. Douglas N. Anderson, The Quality Evolution, 3M Company Publication.

MATERIALS CHARACTERIZATION AT A UNIVERSITY

John B. Wachtman, Jr.

Director, Center for Ceramics Research
Rutgers, the State University of New Jersey

Materials characterization is a vitally important component of any university program dealing with materials. It is important both as an integral part of the materials research program of a university and as a part of the education of students. In the research area, characterization is essential as a support for research and is important as a research topic itself. In the education of university students characterization science is growing into a discipline rather than a fragmented collection of individual subjects. At Rutgers we have made materials characterization a central theme of our Ceramics Research Center. My comments draw upon our experiences with regard to ceramics but the conclusions are appropriate to any class of materials, in my opinion.

Characterization as a research area in itself is an intellectually challenging subject which can be rewarding both in a professional and a financial sense. However, some of the requirements pose special difficulties for university research. Advances in modern characterization techniques usually require sophisticated instrumentation which in turn requires many skills. Typically special sources, detectors and electronics are required as well as special environmental conditions such as high vacuum, electromagnetic fields, handling of reactive materials, etc. Computerized control and analysis of data is generally required to get the full benefits and special software is generally needed. Few university departments can put together the kind of team having all the required expertise although there are notable examples of some who do so very successfully. Despite these examples of entrepreneural success in building complete, state-of-the-art instrumentation in universities, the more common contribution to state of the art instrumentation by universities is likely to be through two other types of contribution: developing the principles of an improved characterization technique or adapting commercial instrumentation to special conditions. For example, adaptations of STEM seem a much more feasible line of work than building a STEM from the ground up for most of us. The apparent financial attractiveness of building a complete, complex instrument is likely to disappear if salaries and debugging costs are figured into the inhouse building program. Working with an instrument-building firm to extend the capabilities of some type of instrumental characterization through a university team contribution to selected features of its operation or analysis seems much more likely to be rewarding.

An area of this type which seems to me both promising and neglected is that of automated specimen preparation. For many measurements the specimen preparation phase is likely to be the weakest link. The same type of automation and feedback control which is today so successfully built into characterization instruments could be built into preparation equipment also. An individual STEM may cost over $500,000. Perhaps we need to accustom ourselves to the idea that the equipment to prepare specimens could reasonably cost over $100,000 and might be a good investment in terms of productivity and quality of measurements.

An area where universities should be especially well suited to contributing to the cutting edge of instrumentation is in improving our knowledge of what features of character are important to what properties. Somewhat related to this is the question of interpreting the results of instrumental measurements. In both cases the intellectual input (presumably a strength of universities) is likely to be the most important factor rather than the engineering of improved instruments.

An aspect of interpretation in some cases is·the need for a data base. A classic example of success is the X-ray diffraction data base maintained by the Joint Committee for X-ray diffraction. Similar data bases are needed to support computerized interpretation of Raman spectra, for example. There seems to be room for an entrepreneur here and for university participation or leadership.

Looking at trends toward future important uses of instrumental characterization of materials seems important for universities both in order to plan their research and to plan their teaching. One of the important trends is that toward use of in-process instrumentation with the associated special requirements for adaptability to process conditions and to realistic cost levels. This is especially significant because of its role in automated manufacturing. The leverage for good in process characterization is high because of the need for feed-back and feed-forward techniques to achieve zero rejects in manufacturing and to achieve the ability to hold tighter specifications at the same time. Other special trends include the need for micro-scale characterization (e.g. microelectronic devices), for special forms (e.g. surfaces of fibers), and for very short time-domain phenomena (e.g. details of reactive intermediates).

In preparing students for future materials characterization it seems important to recognize that many will be involved in processing and quality control as well as research. They must expect to use a variety of characterization techniques and to change the techniques which they use as the problems to which they are assigned change. Finally, they must expect extensive and continuing change in instrumental characterization techniques. Learning the principles underlying characterization techniques is more important than learning the hands-on practice of those other than the ones used in their thesis research.

Most universities with a serious interest in materials have some courses in materials characterization. A typical list would include a course in X-ray diffraction plus some instruction in electron microscopy. In addition materials students are some-times, but by no means always, encouraged to pick up some know-ledge of analytical chemistry. It appears that few have a systematic survey course on the principles of instrumental characterization of materials.

Our Ceramics Research Center and our Ceramics Department are working together to develop a better overall program of character-ization. We would welcome comments on what is needed and how to do it effectively and practically. Meanwhile a few remarks on our efforts may be of interest. We are putting a major effort into strengthening our in-house instrumental capability. We have set a goal of $5,000,000 for capital investment over a five-year period to do the job right and we have prioritized our needs and are filling them as we can. We also recognize that some instru-ments are beyond our reach and are trying to develop cooperative programs at national facilities in these areas such as synchro-ton radiation and small angle neutron scattering. I say coopera-tive programs rather than speaking of obtaining time on instru-ments because of the special skills required. We also use com-mercial service labs on contract and in one case we are working in a cooperative mode with a commercial service lab rather than paying them. Hopefully, our contributions to their instrumental development and adaptation to ceramics (and their subsequent sale of instruments and services) will repay them for the service they are providing to us.

Perhaps most important of all, we have budgeted for a supporting staff (including a very senior engineer to head this staff) and for maintenance and upgrading in our five-year plan.

Our students are expected to learn and use appropriate equipment in the performance of their research. Technicians are not supposed to do the work for them, but to maintain the equipment, give instruction, and monitor use.

We are instituting an overall materials characterization course to augment the special courses on selected aspects of characterization which are already given. The new course will concentrate on principles of characterization rather than details of practice.

To conclude these remarks some speculation is in order. The future of instrumental characterization of materials seems very bright to me. Advances in the components of characterization such as sources and detectors will surely continue. Perhaps completely new physical or chemical effects will become the basis for completely new methods of characterization. What does molecular biology have to offer? The extremely site-specific, chemical-species-specific nature of biological reactions may be put to work in materials characterization. Can we learn to characterize the surface chemistry of grain boundaries or the concentration differences at grain boundaries with something like the degree of specificity as that exhibited by the body's rejection mechanism for material foreign to itself? The fifteen years that have passed since the landmark NMAB Characterization of Materials report have been full of progress and excitement. The next fifteen years are likely to be even more so.

NITROGEN-15 SOLID-STATE NMR STUDY OF RDX AND HMX-POLYMORPHS

S. Bulusu

Energetic Materials Division, LCWSL
U.S. Army Armament Research and Development Center
Dover, NJ

ABSTRACT

The study of the molecular structure and conformation of explosive materials in the solid state is more relevant to the
explosive behavior than the structure in solution. The recently
developed high-resolution, solid-state NMR techniques, namely cross
polarization, magic angle rotation and high power proton decoupling,
have been applied to obtain N-15 NMR spectra of the important
nitramine explosives, RDX and HMX, enriched in N-15 content. The
spectra of RDX and the four polymorphs of HMX are clearly distinguishable and show promise of interesting correlations with the
charge densities on the nitrogens of each conformation in the
crystal structure derived from molecular orbital calculations. These
spectra represent some of the first few solid-state N-15 NMR spectra
available in the literature.

INTRODUCTION

The principal objective of this investigation was to explore
the potential of the most recent advances in nuclear magnetic
resonance (NMR) of solids for unravelling the molecular structure
and conformation, unique to the solid state, with special focus on

energetic materials such as explosives and propellants. For a vast majority of these materials, which are used as solids in their applications, it is the structural characteristics of the solid state that determines their properties such as thermal stability, initiability, sensitivity, and burning behavior. Experimental probes for the molecular structure in the solid state have been limited, however, and generally restrictive in handling solid materials like polymers. While high resolution nuclear magnetic resonance spectroscopy of liquids and solutions, especially of spin 1/2 nuclei (e.g. ^{1}H, ^{13}C, ^{15}N, ^{19}F, and ^{31}P), has been probably the most powerful technique during the last 25 years for molecular structure determinations, it has been quite ineffective in obtaining useful spectra of solids due to extensive line broadening. However, as a result of developments[1] in the last 7 or 8 years, it is now possible to obtain a resolution in NMR spectra of organic solids approaching that routinely achieved in solutions. Therefore, it was considered worthwhile to investigate the application of this technique to energetic materials by attempting a study of the nitramines, RDX (1,3,5-trinitro-1,3,5-triazacyclohexane) and HMX (1,3,5,7 tetranitro-1,3,5,7-tetra-azacyclooctane), which exhibit interesting polymorphism[2] and crystallographic properties[3-6]. In the initial studies, the results of which are presented here, the emphasis was of necessity on the NMR techniques and the quality of the spectra which were obtainable.

SOLID-STATE NMR PROBLEMS AND SOLUTIONS

The NMR spectra of solids obtained by conventional methods are characterized[7] by broad, featureless envelopes devoid of chemical shift and spin coupling information readily obtainable in liquids. There are three contributing reasons[1] for this: (1) Strong static magnetic (dipole-dipole) interactions among protons or between other nuclei under study (e.g. C-13 or N-15) and the surrounding protons having large magnetic moments. The result of this is a severe line broadening. (2) Chemical shift anisotropy, that is, chemical shift dependence on the orientation of chemical bonds to the applied field in amorphous samples contributing even more to line broadening and finally, (3) Long spin-lattice relaxation times of nuclei, such as C-13, causing poor signal intensities. These limitations have been largely overcome in the recently developed techniques[1,5] summarized here briefly. They consist of (1) high-power proton decoupling to eliminate the strong proton dipolar interactions, (2) spinning of the sample at the "magic angle" (54.7°) to the applied field to eliminate chemical shift anisotropy, and (3) a method of cross polarization whereby the strong spin polarization of protons is used to generate spin polarization in the less abundant nuclei, such as C-13 and N-15, thus alleviating the effects of their long spinlattice (T_1) relaxation times on signal intensities.

EXPERIMENTAL

The NMR spectra reported in this work were obtained using a Varian Associates, Model XL-200 NMR Spectrometer system, which uses

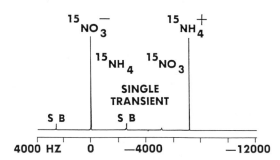

Figure 1.   Solid-State N-15 CP/MAS/HPD Spectrum of $^{15}NH_4$ $^{15}NO_3$.

a 47.5 kGauss superconducting magnet. Besides the conventional 5 mm, $^1H/^{19}F$ probe and a 10 mm 145-81 MHz broad-band probe for solutions, two fixed frequency C-13 and N-15 solid sample probes with magic angle spinning are provided. Samples are held in 8 mm i.d. x 9 mm high cylindrical Kel-F rotors with 2500 Hz routinely achievable spinning speeds. A Varian-supplied XPOL-pulse sequence for solids uses ^1H-cross polarization, magic angle spinning and ^1H-high power decoupling (CP/MAS/HPD) giving the high resolution spectra included in this paper.

The single pulse, N-15 CP/MAS/HPD spectrum of $^{15}NH_4$ $^{15}NO_3$ (Figure 1) illustrates the effectiveness of this technique and represents a particularly favorable case. The N-15 chemical shifts are reported with reference to $^{15}NO_3^- = 0$ ppm and spinning side bands are labeled SB. The linewidths are of the order of 8 Hz.

Unlabeled RDX and HMX were military grade samples recrystallized from acetone. The synthesis of N-15 labeled RDX and HMX was previously published[8]. Using the N-15 HMX thus prepared, the various polymorphs were made by the method reported by Cady[2]. The $^{15}NH_4$ $^{15}NO_3$ sample was obtained from Stohler Isotopic Chemicals, Waltham, MA.

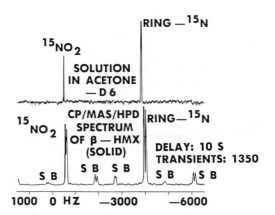

Figure 2.   Comparison of HMX N-15 Spectra in Solution and in the Solid-State.

NITROGEN-15 NMR SPECTRA OF RDX AND HMX-POLYMORPHS

The N-15 labeled RDX and HMX in acetone-d_6 solution give two singlets from NO_2 and ring-N, respectively, in the N-15 NMR, and the two spectra are indistinguishable from each other as is expected.

This spectrum is compared in Figure 2 with the CP/MAS/HPD N-15
spectrum of solid β-HMX. The latter shows two clear doublets indicating two inequivalent lattice sites for each nitrogen, shifted
by 50 Hz and 85 Hz respectively, for the NO_2 and ring-N.

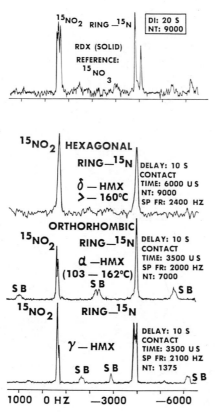

Figure 3.   N-15 NMR, CP/MAS/HPD Spectra of RDX and Polymorphs of HMX.

Figure 3 gives the solid state N-15 NMR spectra (CP/MAS/HPD) of RDX and three remaining polymorphs (γ, α, and δ) of HMX. It is readily seen that the spectra are strikingly different and distinguishable from each other as well as from the spectrum of β-HMX (Figure 2). These spectra represent accumulation times ranging from 4 to 50 hours indicating the limitations on signal intensities achievable in spite of the CP/MAS/HPD technique. It is believed that this is due to residual dipolar couplings among N-15 nuclei, which are, in fact, equal to protons in number. It is noteworthy that the highest temperature phase of HMX, δ-HMX gives two broad singlets indicating high motional freedom of the molecule nearly as in solution.

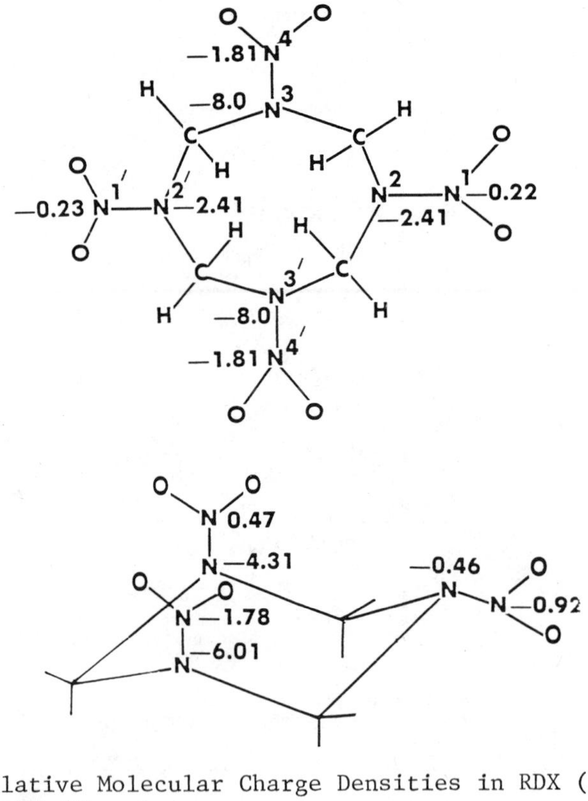

Figure 4. Relative Molecular Charge Densities in RDX (Below) and β-HMX (Above).

One can attempt to rationalize the observed multiplets in the above spectra by correlations with the crystallographic parameters[2]. However, this correlation will be discussed in a future paper pending determination of a complete crystallographic structure of γ-HMX missing in the current literature. The crystallographic structures, in turn, will enable molecular orbital calculations of the electronic charge densities on various nuclei in the structure. Figure 4 gives such calculations for RDX and β-HMX from a recent paper by Delpuech and Cherville[9]. An inspection of these numbers in Figure 4 suggests multiplet absorptions for N-15's as indeed, were observed in the CP/MAS/HPD spectra of RDX and β-HMX. Work is currently in progress to refine these calculations and extend them to the other polymorphs of HMX.

CONCLUSIONS

In summary, some of the first few solid-state N-15 NMR spectra of organic solids available in the literature are being reported. These spectra clearly reveal striking differences related to crystallographic and conformational forms of HMX-polymorphs. One can, therefore, easily envision the importance of solid-state NMR as a valuable new tool in the elucidation of the structure in crystalline organic solids with excellent applications to the generally nitrogen-rich energetic materials.

REFERENCES

1. R. G. Griffen, Anal. Chem., 49, p. 11, 951A (1977).
    R. E. Wasylishen and C. A. Fyfe, "High-Resolution NMR of Solids," in Annual Reports on NMR Spectroscopy, Vol., 12, Ed. G. A. Webb, Academic Press London (1982).
2. H. H. Cady and L. C. Smith, "Studies on the Polymorphs of HMX," Technical Report No. LAMS-2652, Los Alamos Scientific Laboratory, Los Alamos, NM (1962).
3. C. S. Choi and E. Prince, Acta Cryst., B28, p. 2857 (1972).
4. C. S. Choi and H. P. Boutin, Acta Cryst., b26, p. 1235 (1970).
5. H. H. Cady, A. C. Larson, and D. T. Cromer, Acta Cryst., 16, p. 617 (1963).
6. R. E. Cobbledick and R. W. H. Small, Acta Cryst., B&D, p. 1918 (1974).
7. V. J. Bartuska, G. E. Maciel, J. Schaefer, and E. O. Stejskal, Fuel, 56, p. 354 (1977).
    G. E. Maciel et al, Fuel, 58, p. 391 (1979).
8. S. Bulusu, J. R. Autera, and T. Axenrod, J. Labelled Compounds and Radio-Pharmaceuticals, XII(5), p. 707 (1980).
9. A. Delpuech and J. Cherville, Propellants and Explosives, 4, p. 61 (1979).

CARBON-13 AND NITROGEN-15 SOLID STATE NMR INVESTIGATION OF THE STRUCTURE AND CHEMICAL INTERACTIONS AMONG NITROCELLULOSE-NITRAMINE PROPELLANT FORMULATIONS

S. Bulusu

Energetic Materials Division, LCWSL
U.S. Army Armament Research and Development Center
Dover, NJ

ABSTRACT

The recently developed high resolution, solid-state NMR techniques, namely cross-polarization, magic angle rotation and high power proton decoupling have been applied to obtain C-13 and N-15 NMR spectra of nitrocellulose, RDX and HMX, both separately and in admixture as in typical propellant formulations. While the C-13 spectra were obtained at natural abundance, the N-15 spectra were obtained from N-15 enriched materials. The potential of this technique for detecting chemical interactions is explored by the spectra presented. C-13 spectra of cellulose with increasing degree of nitration (from 0 to 13% nitrogen content) illustrate the potential for analytical applications.

INTRODUCTION

The microstructure and morphology of nitrocellulose exhibit differences[1-3] based on methods of preparation and degree of nitration. Whereas ESCA has been a powerful tool[4] in recent years to probe the surface structure of nitrocellulose, it would be desirable to develop a rapid spectroscopic method to elucidate the bulk structural characteristics of nitrocellulose and related polymers. The purpose of this investigation is to explore the application of the recent advances in solid-state NMR firstly, to study the structure and morphology of nitrocellulose and secondly, to detect possible chemical interactions[5,6] within the propellant formulations, which use both nitrocellulose and nitramines in combination. The previous paper[7] in this Proceedings discusses the application of Nitrogen-15 solid-state NMR to the structural studies of RDX (1,3, 5-trinitro-1,3,5-triazacyclohexane) and HMX (1,3,5,7-tetranitro-1,3,5,7-tetrazacyclooctane).

Until 7 or 8 years ago, Carbon-13 and Nitrogen-15 NMR Spectra of solids have been characterized by very broad absorptions due to strong dipolar couplings to the protons, chemical shift anisotropy and long spin-lattice relaxation times. These problems have been remedied in the last few years by a combination of three techniques: [1]H-cross-polarization, magic-angle spinning and high-power proton decoupling (CP/MAS/HPD). These techniques are briefly reviewed in the preceding paper[7]. This paper describes the preliminary results of the application of these techniques to cellulose, nitrocellulose and mixtures of nitrocellulose and the nitramines, RDX and HMX.

EXPERIMENTAL

Details of the instrumentation used for this study are de-scribed in the preceding paper[7]. The C-13 spectra reported in this paper are referenced to hexamethylbenzene aromatic peak as 132.1 ppm from TMS.

Materials

RDX and HMX used were military grade samples recrystallized from acetone. Nitrocellulose samples were made and analysed by well-known methods[3]. N-15 labelled NC was made by using $H^{15}NO_3$ and H_2SO_4 to nitrate cellulose and the N-15 labeled HMX was prepared by a method already published[8].

C-13 NMR SPECTRA OF RDX, HMX, CELLULOSE AND NITROCELLULOSE

The natural abundance of C-13 being 1.1%, its only significant dipolar interactions in many organic materials are with protons (^{1}H). These interactions are effectively eliminated by the high-power proton decoupling used for obtaining solid-state spectra. However, in the highly nitrogenous materials studied in this work, a large number of N-14's and N-15's still contribute significantly to the dipolar broadening of C-13 resonances. As a result, the C-13 spectra seen in this paper are relatively broad. The C-13 CP/MAS/HPD spectra of RDX and HMX (Figure 1) show singlets are expected, but the line widths are of the order of 300 Hz indicating a spread in the chemical shifts of the carbons.

As a preparation for the study of C-13 NMR of nitrocellulose, a CP/MAS/HPD spectrum of pure cellulose (Whatman #2) was obtained. This spectrum (Figure 2), which agrees well with one recently ob-tained by Atalla[9] and Earl and VanderHart[10], shows a remarkable degree of fine structure and the peaks that are readily assignable. The broad absorptions on the high field side of C-4 and C-6 in the cellulose spectrum are attributed to the minor amorphous regions of otherwise crystalline cellulose polymer. It is interesting that the

multi-line spectrum of cellulose collapses to three broad absorption
bands in nitrocellulose . In addition to the dipolar broadening
mentioned below, it is believed that loss of crystallinity on
nitration contributes to the broadening.

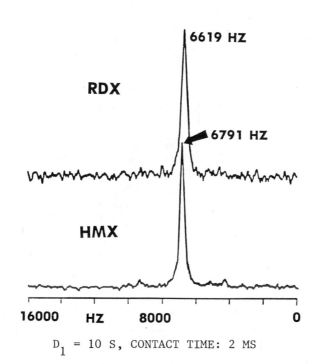

D_1 = 10 S, CONTACT TIME: 2 MS

Figure 1.   C-13 Solid-State, CP/MAS/HPD Spectra of RDX and HMX

C-13 SPECTRA OF NITROCELLULOSE AS A FUNCTION OF DEGREE OF NITRATION

Figure 3 shows the C-13 spectra of cellulose with progressive
nitration (0 to 13.5% N, fully nitrated % N = 14 · 14). These
spectra are helpful to make peak assignments starting with cellulose
peaks, to determine the order of preferential nitration, and to
analyze samples of unknown N-content. A detailed discussion of the
first two has to consider all the available solution spectra[1,2,11]
and they will be the subject of a separate paper. To make some
preliminary comments, it is readily seen that C-1 absorption at 105
ppm in the cellulose spectrum moves upfield to about 100 ppm in a
fully nitrated sample.

The C-4 absorption likewise moves upfield from 90 ppm to 80 ppm, while the C-2,3,5 absorptions undergo only minor shifts. It is also apparent that C-6 nitration precedes that of C-2 and C-3, implying

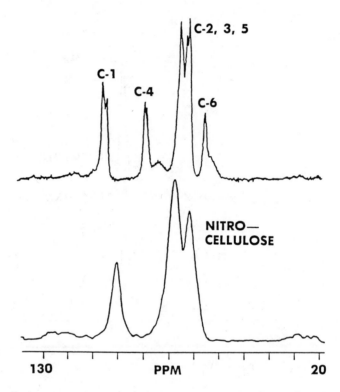

Figure 2. C-13 Solid-State NMR Spectra of Pure Cellulose and Nitrocellulose

a preference, thus enabling one to determine the nitration sequence of the hydroxyl groups.

The third application, namely analysis of samples, offers an elegant and rapid method of determining the degree of nitration of cellulose (%N) by quick spectral matching. Efforts are being made to put this technique on a quantitative basis.

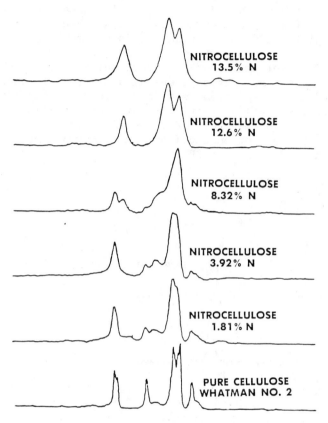

NITROCELLULOSE
13.5% N

NITROCELLULOSE
12.6% N

NITROCELLULOSE
8.32% N

NITROCELLULOSE
3.92% N

NITROCELLULOSE
1.81% N

PURE CELLULOSE
WHATMAN NO. 2

Figure 3. C-13 Spectra of Nitrocellulose with Increasing Degree of Nitration

PRELIMINARY DATA ON CHEMICAL INTERACTIONS BETWEEN NITROCELLULOSE AND NITRAMINES

Several double-base propellants use NC in combination with the nitramines, RDX or HMX. Knowledge about possible interactions among these and other additives, such as plasticizers, either initially or on storage, is important in predicting changes in the combustion characteristics. Nitramines are, in fact, well-known[12] for their propensity to complex with a variety of substrate molecules, and a recent IR study[5,6] strongly indicates hydrogen bonding between them and nitrocellulose chains.

Figure 4 shows the C-13 NMR spectra of NC and RDX, and those of two mixtures of the same. In one of these spectra, the two mixed and blended in a mortar and packed into the Kel-F sample holder. The other mixture was made by dissolution of the two in ethyl acetate, followed by removal of solvent under vacuum evaporation. A comparison of the NMR spectra indicates a clear reversal of intensities of the two upfield peaks of NC when it was mixed with RDX via precipitation from solution. This reversal was found to be reproducible in different mixtures including those which used N-15 labeled NC. The explanation of this reversal is most likely to be found in the changed relaxation properties of the C-13 nuclei responsible for those peaks because of close interactions with the additive molecules (RDX). Experiments are currently in progress to make a detailed measurement of the relaxation times of the C-13 nuclei from which further conclusions can be drawn about the nature of interaction between NC and RDX.

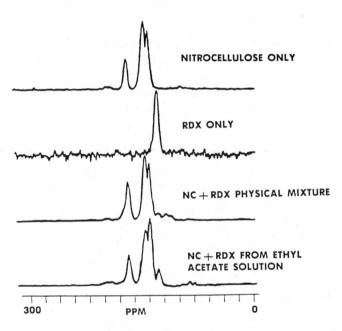

NITROCELLULOSE ONLY

RDX ONLY

NC + RDX PHYSICAL MIXTURE

NC + RDX FROM ETHYL ACETATE SOLUTION

300   PPM   0

Figure 4. C-13 CP/MAS/HPD NMR of NC-RDX Mixtures

Figure 5 shows the N-15 spectrum of a mixture of NC-N15 and HMX-N15 that was obtained by precipitation from solution. The two HMX nitrogen peaks (labeled HMX:NO$_2$ and HMX:Ring-N) are clearly singlets indicating that the HMX is probably in its δ-polymorphic form (see Figure 3). It is difficult to obtain this form by evaporation from any solvent. Therefore, it is concluded that the appearance of δ-HMX is the result of orientational preference of HMX alongside NC chains, which could involve interactions similar to those between NC and RDX discussed above. Here, too, relaxation time measurements of N-15 nuclei, currently in progress, are expected to shed light on the nature of the interactions.

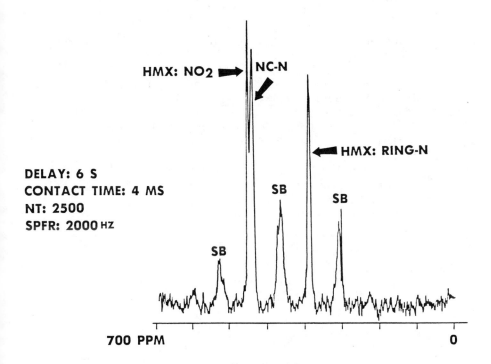

Figure 5. N-15 CP/MAS/HPD NMR Spectrum of NC-HMX Mixture

CONCLUSIONS

In summary, it may be stated that application of solid-state NMR to energetic materials gives valuable information about processes which may affect their performance. C-13 NMR spectra of partially-nitrated cellulose are helpful to develop a rapid analytical method for the estimation of the degree of nitration and to determine the sequence of nitration of the hydroxyl groups in cellulose. Both C-13 and N-15 spectra of mixtures give strong indications of interactions which are under further study. Thus, solid-state NMR spectroscopy promises to be a valuable new tool in the elucidation of the structure of organic solids, in general, with useful applications to the energetic materials studies.

REFERENCES

1. D. T. Clark and P. J. Stephenson, Polymer, 23, p. 1295 (1982).
2. D. T. Clark, P. J. Stephenson, and F. Heatley, Polymer, 22, p. 1112 (1981).
3. F. D. Miles, "Cellulose Nitrate," Interscience, Ch. 2, NY (1955).
4. D. T. Clark, "Photon, Electron, and Ion Probes of Polymer Structure and Properties," Ch. 17, Ed. D. W. Dwight, T. J. Fabish, and H. Ronald Thomas, ACS Symposium Series, #162, Washington, DC (1981).
5. B. W. Broadman, M. P. Devine, and S. Schwartz, J. Appl. Polymer Sci., 20, p. 2607 (1976).
6. B. W. Broadman, M. P. Devine, and M. T. Gurbarg, J. Macromol. Sci. - Chem., A8(4), p. 837 (1974).
7. S. Bulusu, Preceding Paper, This proceedings.
8. S. Bulusu, J. R. Autera, and T. Axenrod, J. Labelled Compounds and Radio Pharmaceuticals, XVII (5), p. 707 (1980).
9. R. H. Atalla, J. C. Gast, D. W. Sindorf, V. J. Bartuska, and G. E. Maciel, J. Am. Chem. Soc., 102:9, p. 3249 (1980).
10. W. L. Earl and D. L. Vander Hart, J. Am. Chem. Soc., 102:9, p. 3252 (1980).
11. T. K. Wu, Macromolecules, 13, p. 74 (1980).
12. W. Selig, Explosivstoffe, 14 (8), 174 (1966); 15 (4), 76 (1967); 17 (4), 73 (1969).

A STUDY OF POLYMERS USING FLUORESCENT DYES

Catherine A. Byrne
Polymer Research Division
Army Materials and Mechanics Research Center
Watertown, MA 02172

Edward J. Poziomek
Research Division
Chemical Research and Development Center
Aberdeen Proving Ground, MD 21010

Orna I. Kutai*, Steven L. Suib and Samuel J. Huang
Department of Chemistry and Institute of Materials
Science
University of Connecticut
Storrs, CT 06268

INTRODUCTION

Fluorescence spectroscopy is often used to study polymer properties which are difficult to learn about or incompletely understood by the use of more traditional techniques. [1]Fluorescent dyes have been used in the study of solid homopolymers[1], polymer blends[2], polymer colloids[3] and polymers in solution[4]. Block copolymers are a specific group of polymers which have not been studied using fluorescence techniques. In this work polyurethane elastomer block copolymers have been chosen and two dyes, 1, a fluorescent probe which can be dissolved or blended in a polymer film, and 2, a fluorescent label, which can react through the hydroxy functional groups to become part of the polymer chain.

EXPERIMENTAL

Fluorescent compounds 1 and 2 were prepared by published methods[5]. The polyurethane elastomers were prepared by the

---

* Present Address: Union Carbide Corporation, Old Saw Mill River Road, Tarrytown, NY 10591

2-DIPHENYLACETYL-1,3-INDANEDIONE-1-
(P-DIMETHYLAMINOBENZALDAZINE)

1   compounds   2

prepolymer method using hydroxy terminated poly(tetramethylene
oxide)(PTMO), with molecular weights 600-3000. The diisocyanate
was methylene bis(4-cyclohexylisocyanate)(H_{12}MDI, Desmodur W,
Mobay) and the chain extender was 1,4-butanediol. Except where
mentioned in the text, the polymer composition was 2.625 moles
H_{12}MDI, 1 mole PTMO and 1.5 moles 1,4-BD, giving a 29 weight
percent hard segment.

    A Perkin-Elmer LS-5 fluorescence spectrophotometer was used.
Polymer films (0.05 inches) were placed in the front surface
accessory or where constant or elevated temperatures were required, triangular glass cuvettes were used in the water jacketed
sample holder.

RESULTS AND DISCUSSION

    Concentrations in the range of 2 to 4×10^{-6} moles 1 or 2 per
100g polymer were used in the experiments. The excitation and
emission spectra of 1 in a cured polyurethane exhibited maxima of
472 and 549 nm respectively. For 2 in the same polyurethane, the
maxima were 480 and 545 nm. The concentrations of the fluorescence compounds, $\approx 10^{-5}$ M were used because they could be conveniently detected in the fluorescence spectrometer and were in a
range where spectral effects due to dye aggregation (excimer
formation) were minimized. The dyes were added during polymerization in tetrahydrofuran solutions at the chain extension step,
with 1,4-butanediol. In a study of polymerization at 80°C, the
fluorescence intensity increased and then reached a plateau value
after 75 minutes. This value did not change during the 16 hour
typical cure time. The increases in intensity were different for
1 and 2. The latter dye is analogous to 1,4-butanediol in that it
can react with diisocyanate and become polymerized in the polymer.
The intensity increased during cure, 100 percent for the sample
with 2 and 40 percent for the sample with 1. A covalently bonded
dye could be used in lower concentration since a greater increase
in fluorescence is observed. To show that 2 is covalently bonded,
a swelling experiment was done on polymer samples in methyl ethyl

496

ketone. The dyes are soluble and the polymer swells. After immersion in solvent for 15 days, the fluorescence of the sample containing 1 decreased 98 percent, whereas that containing 2 exhibited only a 4 percent decrease in fluorescence intensity.

To compare with another technique by which extent of reaction is judged, the infrared spectrum of the sample was followed. At 75 minutes, when the fluorescence levels off, the isocyanate-N=C=O stretch at $2270 cm^{-1}$ had decreased to 15 percent of its original size.

Thermal Analysis

Thermal analysis using differential scanning calorimetry is a useful method of characterizing phase segregated block copolymers like polyurethane elastomers.

These polymers exhibit a soft segment glass transition temperature at $-76°C$ and another transition at $62.5°C$. This transition, shown in Figure 1, does not look like a T_g or a crystalline melting. It has been attributed[6] to a partial ordering of hard segment regions in the polymer[6]. A similar transition in rubber modified epoxies has been attributed to enthalpy relaxation[7]. The nature of this transition will be discussed in detail elsewhere[8]. In the present work a sample was held at ten temperatures from 20°-90°C for ten minutes and the fluorescence intensity was measured at each temperature. The results are shown in Figure 2. The fluorescence decreases linearly with an increase in temperature, exhibits a sharp drop when the ordering transition is reached and then de-

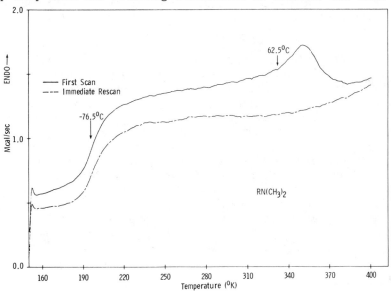

Figure 1. Dsc scan polyurethane. Scan rate was 20°C per minute. Sample size was 14.24 mg. Concentration of 1 was 3.7×10^{-6} moles per 100g polymer.

497

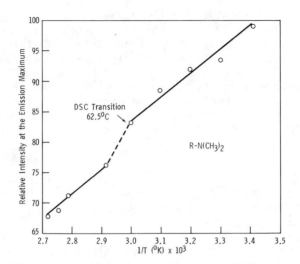

Figure 2. The change in fluorescence intensity in the polyurethane as the temperature is increased in the region of the DSC ordering transition.

creases linearly again above 70°C. The slopes of the lines are approximately parallel on either side of the transition. Loutfy and Teegarden[1] have observed a similar sharp drop at the T_g of the homopolymer isotactic poly (methyl methacylate), but not for atactic or syndiotactic polymer. For the isotactic sample, the slopes of the plot of I_F vs. T were different before and after the T_g. Activation energies for polymer motions above and below the T_g were different, meaning that different types of motion were occuring above and below the T_g.

The polymer motion indicated by the fluorescence change on either side of the ordering transition in the polyurethanes must be similar. The transition is one which involves the hard segment, which comprises 29 weight percent of the polymer. The fluorescent species in this case is the dimethyl dye, which is not covalently bonded in the hard segment. Dye 1 is believed to be uniformily distributed throughout the polymer sheet. It is probably not dissolved, because it slowly blooms to the polymer surface on standing. Optical microscopy on samples containing 1 using a magnification of 600x showed no crystals of 1 in the transparent polymer sheet. The effect measured on heating the fluorescent sample must be attributed to similar motion above and below the transition at about 60°, that is, motion in the soft segment.

Physical Aging

Physical aging in amorphous polymers is a phenomenon which goes on throughout the life of the polymer, but is most rapid during the weeks immediately following preparation, molding or

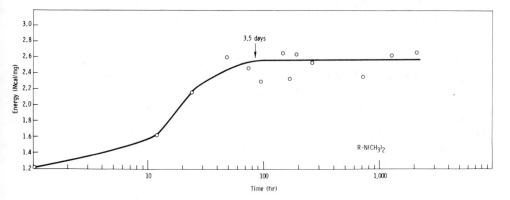

Figure 3. The effect of aging of the polyurethane on the size of the DSC transition with onset of 62.5 °C.

other thermal treatment. In phase segregated polyurethane elastomers, this process can be viewed as an improvement of the phase separation and also an ordering of the hard segment region[6]. The thermal transition associated with ordering has been discussed above and is shown in Figure 1.

The energy absorbed during the transition increases with time after polymerization and is shown in Figure 3. An apparent plateau occurs after three and one-half days. Polymer samples were cured for 16 hours, removed from the 100°C oven, cooled for one hour and placed in the fluorescence spectrometer at 30°C. The fluorescence intensity remains constant initially and then begins to increase, after ten hours for 1 in polymer and after three hours for 2 cova-

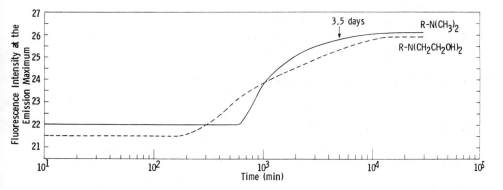

Figure 4. The change in fluoresence intensities of the polyurethanes as aging occurs.

lently bonded in the hard segment. At three and one-half days, the plateau region of the curve, shown in Figure 4 has not quite been reached. The increase in flourescence is small in both cases, about 18 percent. Note also that the effect is delayed slightly for 2 in the polymer. This suggests that the prescence of a fluorescent molecule like 2 in a polymer chain can be a sensitive probe of polymer chain motion during aging. Only a small amount of research has been done using a fluorescence technique to study aging. It is one area of polymer research using fluorescence which would be very fruitful, because experiments can be done at low temperatures and difficulties potentially associated with thermal decomposition of the dye for example, can be avoided.

Polyurethane Composition Variation

In another series of experiments, the weight percent hard segment (calculated as weight of diisocyanate plus weight of diol divided by total weight of sample) was varied. This was done in two ways, by changing the molecular weight of the PTMO and by changing the amounts of diisocyanate and diol added, as shown in Table I.

The samples all contained the same amount of 2 covalently bonded in the polymer hard segment, 2.2×10^{-6} g per 100g polymer. At ambient temperature as the weight percent hard segment increases, the fluorescence increases and then decreases for both series. At 80°C, above the ordering transition, the intensity increases monotonically in both cases. After two weeks aging again, the relative fluorescence intensities of the samples in both series are the same as they were before heating. The reasons for the behavior before heating are poorly understood.

Fluorescence intensity is expected to increase with a decrease in free volume and also to increase with an increase in rigidity of the polymers. This is due to a decrease in the amount of radiationless decay of the fluorescent dye. On a marcoscopic scale, both series of polymers are becoming more rigid as the weight percent hard segment is increased.

The free volume of a polyurethane varies directly with the distance between the temperature of the experiment and the soft segment T_g. Since the hard segment T_g is above the experimental temperature, the free volume in the hard segment is assumed to be very small. For the first series in Table I, the fluorescence increases with weight percent hard segment at room temperature, but is lower for the sample made with PTMO 650. This sample has the highest soft segment T_g and so should have the lowest free volume and the greatest fluorescence intensity at room tempera-

TABLE I. Effect of Amount of Hard Segment on Fluorescence

Length of Soft Segment Varied

Weight Percent Hard Segment	MW PTMO	Relative Fluorescence Intensity		$T_g(°C)$
		Ambient	80°	
22.1	2900	0.78	0.92	−76
29.6	2000	1.0*	1.0*	−73
45.7	1000	1.2	1.6	−60
56.1	650	0.60	2.0	+15

Length of Hard Segment Varied

Weight Percent Hard Segment	MW PTMO	Relative Fluorescence Intensity		$T_g(°C)$
		Ambient	80°	
29.6	2000	1.0*	1.0*	−73
34.2	2000	1.3	2.0	−74
38.0	2000	4.3	2.4	
41.5	2000	1.7	2.7	−75

*Taken as 1.0 for comparison, not the same absolute intensities

ture. Either the sample does not have the lowest free volume or another factor is more important. The elevation of the soft segment T_g above that of the pure soft segment (−85°C) indicates that the sample is highly phase mixed. The dye, covalently bonded in the hard segment, must experience a reduced local rigidity due to phase mixing and the fluorescence intensity is lower.

In the second series in Table I, the soft segment T_g's are constant and only slightly elevated, indicating that the poly-mers are phase separated. The free volumes at room temperature should be the same, on the basis of T_g elevation. The higher the weight percent hard segment, the more rigid the polymer should be and the greater the fluorescence intensity. This holds except for the last sample in the series. The hard segment of that polymer could be less rigid, due to mixing of hard and soft segment phase boundaries. The length of this hard segment is large relative to the soft segment length (compared to the others in the series), so perhaps there is mixing at the boundary. If the microscopic environment that the dye experiences is less rigid, then the fluorescence intensity would be lower. This explanation is not very satisfactory and so some commments about the location of dye 2 in the polymers will be given below.

First consider the case when the two series of samples are heated to 80°C. The fluorescence increases monotonically in both cases. This temperature is above the temperature of the DSC ordering transition shown in Fig. 2 which occurs at the same temperature for all the samples. The fluorescence increase is probably related simply to the increase in rigidity of the polymers when the weight percent hard segment is increased.

CONCLUSIONS

Fluorescence spectroscopy should be viewed as a simple, inexpensive characterization method for polymers. One of the difficulties encountered in the use of a fluorescent compound is an inexact knowledge of the location of the dye in a solid polymer. The location of the fluorescent probe 1 has been discussed earlier, in the section on thermal analysis. Label 2 is added at the chain extension step and reacts with the isocyanate endcap of the prepolymer or with free isocyanate (at least 0.625 moles excess) in the prepolymer mixture. It could be very close to the border of the hard and soft segments or it could be within the hard segment. This becomes more likely as the amount of diisocyanate used to make the polymer increases. The observed fluorescence results from the sum of the intensities at the various positions. Step-by-step syntheses of hard segments, with the dye located in several positions which are exactly known, would be of benefit. The fluorescence effects observed when the amount of hard segment is varied would then be interpretable in a more useful way.

ACKNOWLEDGMENT

One of us (OIK) would like to thank the Army Materials Research Center and the Chemical Research and Development Center for the opportunity to undertake this work as part of her doctoral research.

REFERENCES

1. R. O. Loutfy and D. M. Teegarden, Macromolecules 16, 452 (1983).
2. H. Morawetz, Polymer Eng. Sci. 23, 689 (1983).
3. O. Pekcan, M. A. Winnik and M. D. Croucher, J. Colloid and Interface Sci. 95, 420 (1983).
4. A. E. C. Redpath and M. A. Winnik, J. Amer. Chem. Soc. 104, 5604 (1982).
5. C. A. Byrne, E. J. Poziomek, O. I. Kutai, S. L. Suib and S. J. Huang, "Polymers as Biomaterials", S. W. Shalaby, Ed., Plenum Press, New York, 1984, in press.

6. J. W. C. Van Bogart, D. A. Bluemke and S. L. Cooper, Polymer, $\underline{22}$, 1428, (1981).
7. Z. H. Ophir, J. A. Emerson, G. L. Wilkes, J. Appl. Phys. $\underline{49}$, 5032, (1978).
8. A paper by C. A. Byrne, D. P. Mack and J. M. Sloan, discussing polyurethane elastomers prepared from $H_{12}MDI$, in preparation.
9. R. O. Loutfy, J. Polym. Sci., Polym. Phys. Ed. $\underline{20}$, 825 (1982).

MICROCOMPUTER BASED ULTRASONIC IMAGING

Jeffrey J. Gruber and James M. Smith

Army Materials and Mechanics Research Center

Watertown, Massachusetts

ABSTRACT

This report discusses an informational poster paper on acoustic imaging presented at the 1984 Sagamore Conference. Ultrasonic velocity, attenuation, defect, and contact C-scan maps are illustrated and discussed. This report also shows images of ultrasonic beam radiation patterns. All of the acoustic images presented in this paper are part of an on-going effort at AMMRC in the area of microcomputer based ultrasonic imaging.

INTRODUCTION

Ultrasonic inspection is a nondestructive method used to evaluate materials. Through the years, the vast majority of ultrasonic applications have been directed toward the detection of defects inside materials. Voids, inclusions, and delaminations are three examples of such defects. Maps of these internal defects (C-scans) are a commonly used inspection tool.

There are many material characteristics and properties other than flaws that can be detected using ultrasonic techniques. However, maps or images of such ultrasonic measurements have not been as common in the past as defect C-scans. Some ultrasonic information that was available from a test could not be extracted and processed for mapping without the aid of expensive, specialized, and dedicated instrumentation.

Most of the currently available instrumentation is microprocessor based, and is easily interfaced to computers or other instrumentation. This versatility has increased the accessibility of ultrasonic information. Various ultrasonic signal parameters that would have been difficult to measure in the past can now be mapped or imaged.

All of the ultrasonic images presented in this paper are part of an on-going effort at AMMRC in the area of microcomputer based ultrasonic imaging. These results were produced using systems that were developed in-house, utilizing off-the-shelf components. It is possible for these systems to use and share a small number of components as building blocks due to the common availability of intelligent electronic devices, coupled with a standardized instrument communication interface (IEEE-488). All of the systems share an inexpensive microcomputer to aquire, process, display, and store the ultrasonic data.

This report shows examples of ultrasonic attenuation, velocity, defect, and contact testing maps. This report also shows images of ultrasonic beam radiation patterns. All of the original images were displayed on a color RGB monitor or printed with a color printer. The photographs in this report are monochrome reproductions of the original color images.

CONTACT INSPECTION

Most ultrasonic mapping procedures require the utilization of a large electro-mechanical scanner, and require that the specimen which is being in inspected be immersed in water. A different technique is needed if the material to be tested is too large to fit in the water tank, can not be brought in from the field, or should not be immersed in water. To fill this need, a laboratory model of a field portable contact inspection system has been developed at AMMRC. Figure 1 is a block diagram of this system, and figure 2 is a photograph of the system's major components.

To test a material using this system, an operator "scrubs" a contact transducer across the material surface, while viewing the resulting C-scan on a color RGB monitor. The key element of this system is the L shaped microphone, which is used to determine the physical location of the transducer for mapping purposes. Figure 3a is a contact C-scan of a seven-ply graphite epoxy plate, and figure 3b is an immersion C-scan of the same material. The strip through the center of these images shows a region of delamination.

The contact testing method does not have the resolution of the immersion method. However, it is intended for use on large items where the critical defects are large. The contact method does have greater ultrasonic penetration capabiltity than the immersion technique. In one instance, the contact technique was capable of

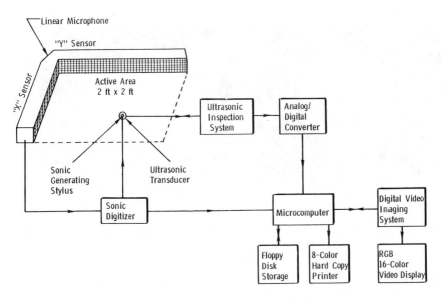

Figure 1. Block diagram of contact C-scan system.

Figure 2. Photograph of contact C-scan system components.

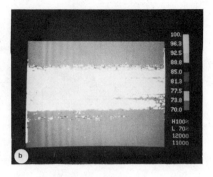

(a) Contact C-scan   (b) Immersion C-scan

Figure 3.   C-scans showing delamination in 7-ply graphite epoxy.

imaging a 1/4 inch diameter defect within a 64-ply glass epoxy material, whereas a 1 inch diameter defect within the same material could not be found using the immersion technique with its lesser penetration capability.

DEFECT MAGNIFICATION IMAGING

A defect scan, called a C-scan, is a map of significantly large ultrasonic reflectors inside the inspected material. Figure 4a is a defect scan of a 6 x 6 x 1/8 inch fiberglass epoxy specimen. Figure 4b is a defect scan of a 7 x 3 x 1/2 inch plastic test block containing various imbedded plastic shapes. Figures 5a - b are examples of magnified C-scans. Figure 5a is an 8x magnification of

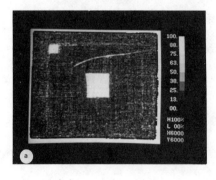

(a) Glass epoxy   (b) Plastic test block

Figure 4.   Examples of defect C-scans.

 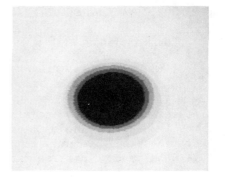

(a) 8x Magnification   (b) 40x Magnification

Figure 5.   Magnified C-scans of plastic block in figure 4b.

part of the test block in figure 4b, and figure 5b is a 40x magnification of the hole in figure 5a.   In figure 5b, the hole diameter is approximately equal to 1/16 inch.   Each pixel (picture element or dot) in figure 5b corresponds to 1/1000 inch length.

ULTRASONIC ATTENUATION IMAGING

Ultrasonic attenuation results from the scatter, reflection and absorption of ultrasonic energy.   Porosity within a material is one such scatterer of ultrasonic energy.   Figure 6 is an ultrasonic attenuation map of a 6 x 6 x 1/2 inch aluminum oxide material.   The black or dark grey areas of the map show regions of large ultrasonic attenuation.

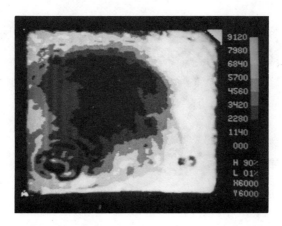

Figure 6.   Ultrasonic attenuation map.

ULTRASONIC WAVE VELOCITY IMAGING

For materials with a low Poisson's ratio,

$$VL \approx \sqrt{Y/p}$$

where VL equals longitudinal wave velocity, Y equals Young's modulus of elasticity, and p equals material density [1]. Generally, for a given type of material, the modulus of elasticity is usually dependent on the bulk density [2]. Therefore, an increase in material bulk density can increase rather than decrease the longitudinal wave velocity.

Figures 7 and 8 are ultrasonic time-of-flight maps for three types of ceramics. These maps can be interpreted as longitudinal wave velocity maps, since the inspected materials were of constant thickness and because of the inverse relationship between time and velocity.

Figure 7a is a velocity map of a silicon carbide material which cracked while under no external tension. Figure 7b is a velocity map of another silicon carbide sample. Both of these specimens have approximately a 4% change in longitudinal velocity, which is common for sintered silicon carbide. It is also common to have the smaller velocity (longer time-of-flight) near the center of the specimen.

Figure 8a is a velocity map of a titanium carbide material. This is the only type of material tested thus far at AMMRC that has this kind of velocity pattern. Figure 8b is a map of a 4 inch diameter, 1 inch thick zirconia based material. The roughly 10% change in wave velocity, as shown in this map, is untypically large if compared to the change that occurs for most ceramics.

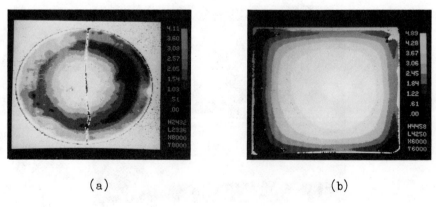

(a)   (b)

Figure 7.   Longitudinal wave velocity maps of silicon carbide.

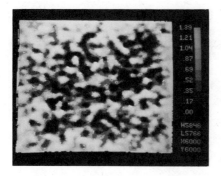 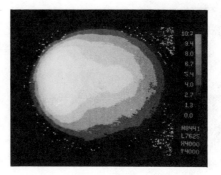

(a) Titanium Carbide   (b) Zirconia

Figure 8.   Longitudinal wave velocity maps.

ULTRASONIC BEAM IMAGING

The performance parameters of the ultrasonic transducer must be known and measured in order to assure the reliability and repeatability of test results.   One transducer characterization technique consits of the measurement and imaging of its spatial acoustic energy.   These images are known as beam profiles.

Figure 9a is an image of a 5 Mhz unfocused ultrasonic beam taken from a center plane parallel to the beam propagation direction.   In this image the beam length is geometrically compressed: The vertical scan length is 8 inches and the horizontal scan length is 0.6 inch. The first region of high acoustic intensity when approaching from

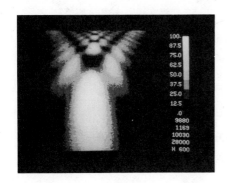

(a) unfocused beam image   (b) focused beam image

Figure 9.   Ultrasonic beam images.

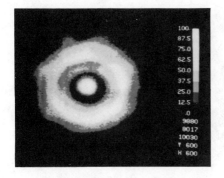 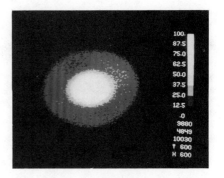

(a) Image at YO point   (b) Image at Y1 point

Figure 10. Unfocused beam cross-sectional images.

the far field is known as the YO point. The area between the YO
point and the transducer face is known as the near field. The second
point of high intensity is known as the Y1 point. Figure 9b is an
image of a 5 Mhz focused transducer. Note that this beam is
approximately one-third as wide as the unfocused beam in figure 9a.

Figure 10a is a cross-sectional image from the unfocused
transducer, taken at the YO point. Figure 10b is a cross-sectional
image from the beam Y1 point.

SUMMARY

This report has shown and discussed a variety of acoustic images
resulting from work underway at AMMRC. These maps or images have
proven to be useful in the nondestructive evaluation of materials.
The systems and techniques that are used to produce the images shown
in this report are made possible by easily interfaced microcomputers
and microprocessor based instrumentation.

REFERENCES

1. R. C. McMaster, Nondestructive Testing Handbook, Vol. II
   Section 43, p. 10, The Roland Press Co., NY, (1963).
2. Roy W. Rice, "Microstructure Dependence of Mechanical Behavior
   of Ceramics," Treatise on Materials Science and Technology,
   Vol. II, Academic Press, Inc. (1977).

MATERIAL DEGRADATION UNDER PULSED HIGH TEMPERATURE AND

HIGH PRESSURE

L. D. Jennings, S. Lin, and A. S. Marotta

Army Materials and Mechanics Research Center

Watertown, MA 02172

ABSTRACT

The AMMRC ballistic compressor is described as an instrument capable of adiabatically compressing gases to temperatures of several thousand Kelvins and pressures of several thousand atmospheres for a time of a fraction of a millisecond. Problems of measuring the gas conditions and its equations of state are discussed. Application to gun barrel erosion/corrosion is discussed through the example of various nitrogen/argon test gas experiments on 4340 steel. An interesting feature of an apparent melt and redeposit process in the formation of carbon containing nodules on the surface.

INTRODUCTION

Gases at temperatures of a few thousand Kelvins and pressures of a few thousand atmospheres are not readily produced and contained for long periods of time. Such conditions are obtained for short periods of time in the burning of a propellant, in a gun or a combustor for example. More controlled gas compositions, and also higher temperatures and pressures, may be obtained in a "ballistic compressor". This instrument uses a compressed (say 25 atmospheres) gas to drive a free ("ballistic") piston down a long cylinder. This action compresses a test gas approximately adiabatically to the desired high temperature and pressure. The AMMRC instrument utilizes a 4 kilogram piston which typically achieves 90% of maximum pressure for a time of about 0.5 milliseconds.

We consider three important classes of experiments which may be performed with a ballistic compressor:

513

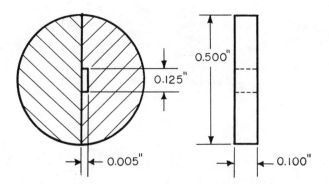

Figure 1.   Sample Dimesnsions, SHowing the Narrow Erosion/Corrosion
Channesl Milled in one of the Toe Pieces Which Make Up
The Disk

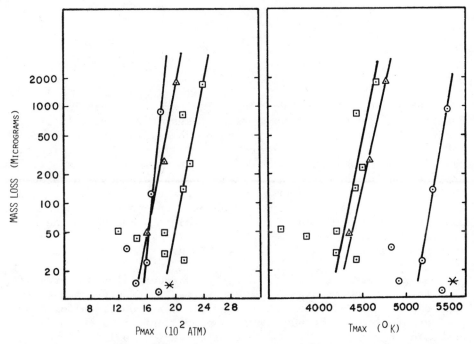

Figure 2.   Mass Loss (on a Logarithmic scale) vs. Maximum Pressure
(a) or Maximum Temperature (b) Note that Even on a
Logarithmic Scale the Onset of Appreciable Loss is
Sudden and Depends on a Combination of Temperature and
Pressure.   The Point Marked with an * Refers to a sample
that was Protected by a "Corrosion" Layer from a Previous
Shot. P_0 is the Initial Pressure of the 10% N in Ar Test
Gas.

1. The hot, high pressure gases can interact chemically (corrosion) and/or physically (erosion) with a sample. In general, some mechanism for increasing the heat transfer, such as flow through an orifice, is required. The mechanism of the "degradation" of the sample may then be studied.

2. The properties of the compressed gas, its equation of state for example, may not be well known at these high pressures and temperatures. If adequate instrumentation is available, these properties may be determined.

3. The operation of the instrument itself may not be well documented. Thus studies may be required to determine the conditions which obtain, during corrosion/erosion studies for example.

Clearly these three classes of experiments are interdependent and we indicate here progress in classes (2) and (3) and report some results on erosion/corrosion, the work with the most immediate Army relevance.

COMPRESSOR OPERATION

A fast-acting piezoelectric transducer is able to follow the pressure variation in the test section, so this parameter is easily measured. A magnetic pickup unit monitors the position of the piston via alternate magnetic and non-magnetic rings. By counting the number of pulses detected and by examination of the shape of the induced voltage near turn-around, the position of the piston may be determined within about 0.2 mm. The volume at maximum pressure or temperature may then be calculated. Knowing the initial mass of test gas, an estimate of the density might then be made. Unfortunately this is only an estimate; during the initial portion of the stroke, driver gas seeps by the piston; during the high pressure portion of the stroke, the blow-by has the opposite sign.

The details of the pressure and position vs. time curves may be used to estimate deviations from the ideal behavior of an isentropic, constant mass, compression. In particular, the reduction of the observed maximum pressure from the ideal and the time difference between the point of minimum volume and of maximum pressure give two parameters which may be used to model the operation of the compressor. We have done such modelling based on the descriptions of Lalos and Hammond[1], Takeo, Holmes, and Ch'en[2] and Alkidas, Plett, and Summerfield[3]. Unfortunately, all these approximations are of uncertain validity and further experiments are desirable to define the conditions obtained in the ballistic compressor.

Figure 4. SEM Photograph of the Nodules Which Appear Most Prominently in the Region Between Melted Regions (left of Figure 3) and Relatively Undisturbed Regions (right of Figure 3). The Width of the Imaged Region is 750 Microns.

Figure 3. A Typical Sample Exposed to "Intermediate" Temperature/Pressure Conditions. The Width of the sample is 2.5 mm.

TEMPERATURE DIAGNOSTICS

In addition to the pressure, the temperature would define the state of the test gas. If the equation of state and the density were known, the temperature would also be known. As stated above, these things are only approximately known, so the temperature is also only approximately known. Spectroscopic methods offer a promising method of determining the temperature in the required millisecond time frame. We have started development of a high repetition rate Coherent Anti-Stokes Raman Spectroscopy (CARS) system through which the temperature may be determined by measuring the population of vibrational levels of appropriate molecules in the test gas. Results are not yet available.

EROSION/CORROSION STUDIES

Semi-circular specimens of 4340 steel were prepared as shown in Figure 1. One of each pair has a notched surface so that an escape channel about 0.2 mm thick by 3 mm wide by 2.5 mm long is presented to the hot, high pressure gases. Test gases consisted of various mixtures of nitrogen and argon, both of which are comparatively inert. Because of the different heat capacities of nitrogen and argon, and by varying the initial pressure of the test gas and of the driver gas, a wide variety of final conditions were obtained.

Using the simplest of observations, visual appearance and weight loss, we may distinguish three regions of erosion/corrosion. In the lowest temperature region, we find surface chemical reaction, resulting in a tarnished appearance, with no appreciable mass change. Analysis by Auger spectroscopy shows that the reaction appears to be primarily with residual oxygen.

In the intermediate region there is evidence of melting at the leading edge of the orifice with redeposition downstream. There is then little net mass loss.

In the highest temperature/pressure region there is significant mass removal as shown in Figure 2. Note that there is a quite sharp onset of this effect. This onset is not determined by temperature or pressure alone, but by a combination.

The point labeled with an * in Figure 2 illustrates that some protection is available from a pre-existing, thin corroded surface. This sample was previously exposed to the same test gas, but under "corrosion" rather than "erosion" conditions.

An optical photograph of a typical specimen from the "intermediate" region is shown in Figure 3. On the left is the region of appreciable melting and on the right can be seen the essentially undisturbed machining marks on the original specimen. In between these two regions small nodules project from the surface as shown in the scanning electron micrograph (SEM) of Figure 4. In spite of all this activity on the surface of the specimen, the net mass

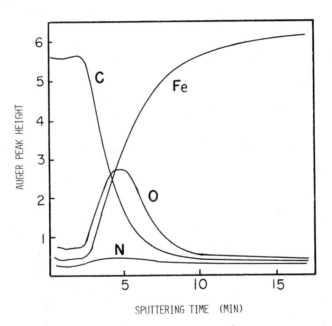

Figure 5.   The Depth Profile of the Major Constituents by Auger Analysis.

loss is only about 40 micrograms, barely more than our limit of detection and inappreciable by the standards of Figure 2.

The surface of the sample was observed with Auger spectroscopy, which showed strong oxygen and carbon lines. Presumably these arose from impurity gases in the compressor. Scanning Auger microscopy (SAM) showed that the nodules contained carbon and, in some but not all cases, oxygen. If the scanning beam was applied to the nodule for some time, the oxygen could be burned off.

The depth profile of the overall surface (as opposed to a nodule) was revealed by ion sputtering. The details of this profile depended on firing conditions, but a typical profile is shown in Figure 5.

After the sputtering off about 2 micrometers, the nodules were not visible in an SEM but there was still evidence of the high carbon concentration.

CONCLUSIONS

The ballistic compressor may be used to yield a wide variety of surface degradation mechanisms under controlled conditions of pressure and temperature. Much study remains to be done both with regard to such degradation and also to the definition of the "controlled" conditions; although they are well controlled, complex diagnostics are required to measure them.

REFERENCES

1. G. T. Lalos and G. L. Hammond, "The Ballistic Compressor and High Temperature Properties of Dense Gases," Chapter 25 pp. 1193-1218 in "Experimental Thermodynamics, Volume II, Experimental Thermodynamics of Non-reacting Fluids" Edited by B. Le Neindre and B. Vodar: Butterworths, London, (1975).
2. M. Takeo, Q. A. Holmes and S. Y. Ch'en, "Thermodynamic Conditions of the Test Gas in a Ballistic Compressor," J. Appl. Phys 38, 3544-3550, (1967).
3. A. C. Alkidas, E. G. Plett and M. Summerfield, "Performance Study of a Ballistic Compressor," AIAA Journal 14, 1752-1758, (1976).

MATERIALS AND PROCESS CHARACTERIZATION TO DETERMINE THE

EFFECT OF HIGH INTENSITY INFRARED LASER RADIATION ON A

COMPOSITE MATERIAL

J. S. Perkins, P. W. Wong, D. J. Jaklitsch, L. Elandjian
A. F. Connolly, S-S Lin, T. P. Sheridan, R. M. Middleton
A. J. Zani, C. E. Dady, W. K. Chin, and A. J. Coates

Army Materials and Mechanics Research Center
Watertown, MA 02172

INTRODUCTION

A wide variety of characterization methods have been used to
determine the effect of irradiating a specific composite material
with high intensity infrared laser radiation. The material, being a
metal-doped carbon/carbon composite, was intractable to the usual
methods of analysis; the behavior under intense radiation was
unexpected and, at the time, unexplained; and the desirability of
computer modeling required knowledge of the principal mechanisms of
radiant energy absorption and redistribution.

This study of the effect of an intense infrared laser beam on a
high-temperature, metalated ceramic system became feasible by use of
relatively new and powerful analytical techniques.

The examination we performed can be divided into three areas of
concentration, these being Materials Characterization, Laser
Testing, and Post-burn Analyses. Study of these was correlated with
data obtained from compilations, reports of experiments in the
recent literature, and information privately communicated.

An outline of the principal stages in the laser beam/material
interaction disclosed by this study is included.

MATERIALS CHARACTERIZATION

Initial studies on the effect of laser beam interaction with a metal-doped carbon/carbon composite showed that laser-produced front and back face temperature measurements during irradiation were reproducible providing material from a single sample source was used but that substantial variation in the resistance of the samples to damage occurred as different lots of material were tested. It thus became obvious that only by characterization of the different test samples would it be possible to unravel the effect of an intense laser beam on this type of material.

The material was a laminate of carbon cloth layers preimpregnated with a metal-containing resin. Different samples contained three to as many as a hundred and forty layers. These were successively subjected to curing, carbonizing, and graphitizing cycles under pressure, a process which allowed for substantial variation in the final products. These composite samples were black, opaque, and insoluble in most solvents, and analyses of the constituents required the development or refinement of special analytical methods. Especially useful techniques proved to be:

X-Ray diffraction measurements that showed carbon cloth/resin interaction during fabrication to produce varying amounts of the tungsten carbide products that consolidate the samples, i.e., WC, W_2C, and $\beta\text{-}WC_{1-x}$. These were produced at high temperatures by removal of the hydrogen, oxygen, nitrogen, and some of the carbon from the initial carbon, hydrogen, nitrogen, oxygen, and tungsten containing resin (Figure 1).

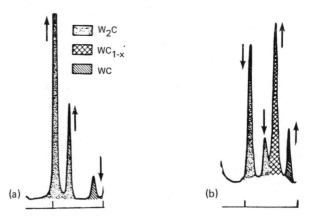

Figure 1. X-Ray diffraction patterns. (a) Resin fired to 2800°C showing initial WC converted to W_2C; (b) resin on carbon cloth similarly treated showing interaction of W_2C with cloth to produce $\beta\text{-}WC_{1-x}$ and WC in final composite, WC/C.

Elemental analyses that showed substantial deviations from the nominal W-content in a series of samples (Table 1) as well as uneven distribution of the metallic component throughout an individual sample.

Table 1. Variations in composition of eight nominally identical samples of 3-ply WC/C containing an added ingredient.

	C	W	X
11-1	86.1	8.7	4.4
11-2	83.6	9.1	4.9
11-3	85.1	9.3	5.0
11-4	83.4	10.1	4.8
11-5	86.4	8.7	4.5
11-6	76.7	12.7	6.0
11-7	76.5	12.5	5.0
11-8	80.2	6.8	5.5

X-Ray radiographs that also showed variations in the distribution of W in successive layers, the metal being concentrated near the top and bottom surfaces in some cases (Figure 2).

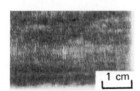

1 cm

Figure 2. X-Ray radiograph of 140-layer sample of WC/C showing concentration of WC's near top and bottom surfaces.

X-Ray fluorescence measurements that verified an uneven distribution of tungsten on the two surfaces, top and bottom, of a multilayered sample.

Apparent density measurements comparing samples cut from the surface and from the core.

Metallograhic mounting of a sample, which otherwise appeared nearly homogeneous, that revealed the known layered structure. It also revealed another phase that either interpenetrated the layers or was spread across several adjacent layers during extended polish of the surface (Figure 3).

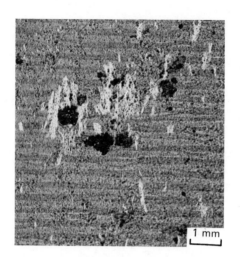

Figure 3.   Metallograph of a mounted sample showing 0°/90° cloth
            lay-up and a second phase (white).

LASER TEST DATA

During radiation with an intense continuous-wave (cw) CO_2-laser
beam of 10.6-µm wave length (a welding type laser), changes in the
sample were monitored with:

Thermocouples which were buried at different depths under the
area to be irradiated;

Optical pyrometers, Si (0.9 µm), Ge (1.45 µm), and/or Barnes
(4.6 to 7.4 µm) focused on the front and/or back face of the sample
in the burn area;

A circularly variable filter for the infrared region monitoring
a small area of the front face from which true front face
temperatures could be derived by subsequent computer-matching with
temperature-derived black body curves (Figures 4 and 5);

Video scans as well as high-speed cinematography, the former in
real time, the latter at 100, 200, or 400 frames per second, to
provide simultaneous records of each burn as viewed from the front,
side, and, in some cases, the back;

Spectroscopy over the near UV to near IR region (200 to 1200 nm)
either using an integrating monochrometer, a drum camera, or an
optical multichannel analyzer (OMA), the last triggered to give
eight instantaneous scans separated by selected equal time intervals
during the course of the laser runs which varied from three to ten
seconds in duration. These made possible the identification of
molecular or atomic species in the plume, both adjacent and at a

nominal distance from the irradiated surface, and, with the last two instruments, over a temporal period after initiation of laser radiation. A typical scan (Figure 6) shows the emission spectra of C_2 and C_3 molecules directly in front of the burn surface as viewed from the side.

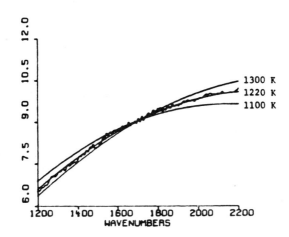

Figure 4.   Plots of corrected experimental test-case data compared with "best-fit" Planck function and ±100K Planck functions. (OptiMetrics, Inc.)

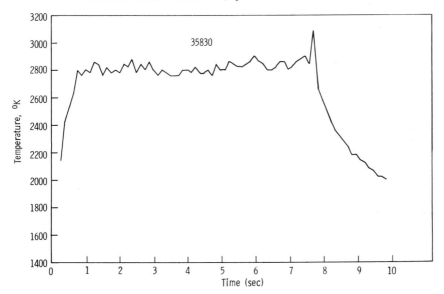

Figure 5.   True front-face temperatures of WC/C composite during irradiation with a 10.6-µm CO_2-laser in a 77-µm vacuum chamber. These were derived by Planck function fitting of data measured using a circularily variable filter.

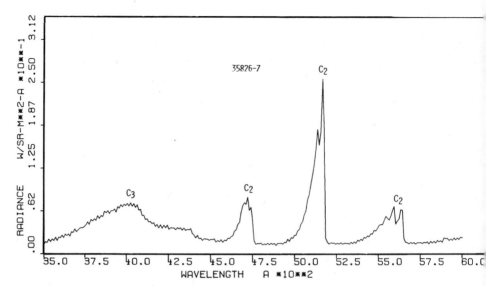

Figure 6. OMA scan of the C_2 Swann bands and the nearly
continuous C_3 emission bands seen from the side just
in front of the WC/C surface during laser irradiation.

A visicorder trace of concurrent events tied all data together,
i.e., the front, intermediate, and back face temperatures monitored
by pyrometers and thermocouples, the triggering of OMA signals, the
"on and off" timing of the laser, and any other measureable signal
such as the rise and fall of pressure during laser runs in vacuum.

These records permitted review of the "in-situ" laser beam-
sample interaction during post-radiation analysis of each laser run.
They provided revealing temporal evidence of physical and chemical
changes produced by the onset and course of steady laser irradiation
of the sample. In addition, they allowed a study of continued
changes during the immediate post-radiation period involving
relaxation and redistribution of the absorbed electromagnetic
energy.

We conclude that during irradiation the following stages are the
most significant:

Selective photo excitation, predominantly of the carbon
fabric by that portion of the beam absorbed and not
reflected;

Multiphoton absorption of the low-energy 10.6-μm photon
(3 kcal/mole) to activate rotational, vibrational, and
electronic transitions in carbon which produce
incandescence and, when the radiation is intense or
sufficiently prolonged, sublimation of the carbon;

Melting of the tungsten carbides in the matrix by the incandescent carbon cloth with release of postulated carbon atoms (C_1) from WC, forming tungsten-rich W_2C which then reacts with carbon in the matrix and cloth portions of the composite producing more molten WC that continues to release carbon.

This is a highly endothermic, cyclic process that causes the cloth to disappear (vaporize) if laser energy is sufficiently intense.

As the layer of liquid tungsten carbides becomes thicker, due to the disappearance of the carbon substrate, the endothermic reaction proceeds more rapidly. Simultaneously, the molten carbide surface layer generally becomes smoother and reflects more of the incident laser beam. These can counteract the accelerating effect and the reaction slows down to a steady state.

The most significant aspect of this mechanism, we believe, is the highly endothermic removal of monoatomic carbon (C_1) by the following mechanism:

At the liquid/vapor surface:

$$2\ WC(l) \longrightarrow W_2C(l) + C(g) \qquad \Delta H = -5\ \text{kcal} \qquad (1)$$

At the liquid/cloth surface:

$$W_2C(l) + C(cloth) \longrightarrow 2\ WC(l) \qquad \Delta H = +175\ \text{kcal} \qquad (2)$$

$$C(cloth) \longrightarrow C(g) \qquad \Delta H = +170\ \text{kcal} \qquad (3)$$

which is equivalent to an energy absorption of 60 kJ/g of carbon released.

POST-BURN ANALYSES

Post-burn analyses strengthened our understanding of the mechanisms and were accomplished by use of:

Scanning electron micrographs (SEM) and associated tungsten-mapping by energy dispersive analysis of X-rays (EDAX) (Figure 7);

Scanning electron detector (SED) micrographs and scanning Auger micrographs (SAM) which additionally map carbon, oxygen, and other light elements as well as tungsten (Figure 8);

X-Ray diffraction that shows the varying content of W_2C, β-WC_{1-x} and WC on quenched burns resulting from different amounts of energy deposition (Figure 9);

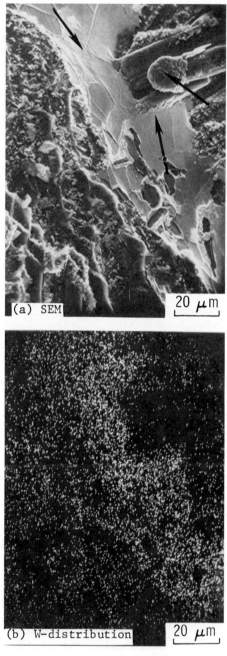

(a) SEM

20 μm

(b) W-distribution

20 μm

Figure 7. (a) Scanning electron micrograph and (b) an EDAX W-map of the area near a laser burn on WC/C. Note the liquid matrix phase, the attack on the side and end of one carbon fiber, and the growth of carbon on the end of a cooler nearby fiber.

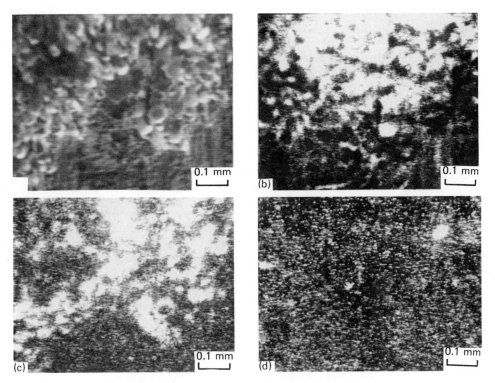

Figure 8.   (a) SED-produced micrograph of the edge of a laser burn on WC/C and SAM-produced (b) C-map, (c) O-map, and (d) W-map of the same area.

LASER-HEATED MATERIAL

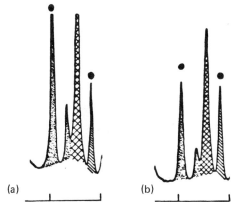

(a)   (b)

Figure 9. X-Ray diffraction patterns from burn surfaces produced by different intensities of a CO_2-laser beam interacting with WC/C. (a) 15 kW/cm^2, (b) 30 kW/cm^2. The higher energy flux intensifies the reaction of W_2C with carbon cloth to produce WC.

Optical photomicroscopy from metallographically prepared samples (Figure 10);

X-Ray radiography that reveals the movement of molten metal carbides within the burn craters (Figure 11) and can be used to compare the relative amounts and distributions of W in the various samples.

These tell a considerable amount about the course of mass movement as a result of laser-beam action. Most importantly the information gained verifies the liquifaction of the tungsten carbides, the changing composition of this liquid with increased radiant energy input, the segregation of the tungsten alloy at the edges of the burns, and the removal of carbon in the vapor state.

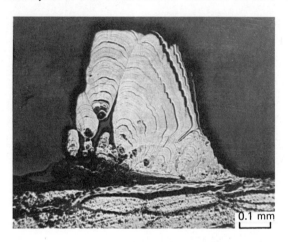

Figure 10. Optical photomicrograph of the cross-section of a carbon wall formed downwind (to left) of a deeply penetrating laser-burn on WC/C in a 0.1 Mach wind tunnel.

Figure 11. X-Ray radiograph of a laser-burned sample of WC/C showing ring of WC's pushed to the edge during radiation and droplets separating on the surface of the burn during cooling.

That the last named occurs is indicated by the spectra of C_2 and C_3 found immediately in front of the burn surface during laser irradiation, the large amount of carbon trapped in the burn surface when no burn-through occurs, and the redistribution of carbon downwind or in cooler parts of the carbon fiber composite. The wall downwind from the laser-produced cavity in Figure 10, for example, is pyrolytic in character (i.e., layered) which occurs when carbon is deposited from the vapor state.

The heavy ring of tungsten carbides about the burn in Figure 10 is indicative of mass flow during irradiation. The pressure exerted on the molten tungsten-carbon "alloy" due to gas evolution is partly the cause of the ring appearing. The cohesive character of this high melting fluid causes the thin residual film over the burn surface to break into droplets and also to add material to the ring.

The ring and the thin film remaining in the burn area frequently crack on cooling to relieve stresses but it is believed this arrests further shattering of the material that survives. It is also believed that thermal conduction through the cloth layers and through some modified matrix compositions further limits structural damage.

CORRELATION WITH OTHER EXPERIMENTS AND DATA

The final phase of this work involved the correlation of our work with other experiments and data. This entailed examination of the published literature, private communication with other scientists, and personal observations during experiments in other programs. This phase led to the use of the following:

Thermodynamic data[1] leads to equations (1), (2), and (3) which explain the high-energy absorption or "sink" that causes this material to survive a heavy energy flux.

The phase diagram of the tungsten-carbon alloy system[2] (Figure 12) shows the peritectic character of WC. On melting, the tungsten-carbon alloy becomes enriched in tungsten as it releases carbon, probably atom by atom.

The concept of jet cooling applied to some plasmas explains the fleeting disappearance of radiation due to dropping of radiation producing electrons to the ground level as vapor particles approach supersonic speeds in high vacuum[3]. Reradiation becomes possible as this supersonic kinetic energy is converted into internal electronic excitation with sudden deceleration to the velocity of one Mach. Figure 13 illustrates this effect.

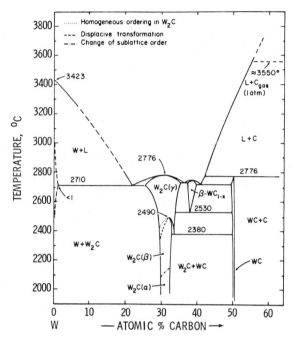

Figure 12.   Phase diagram of the tungsten-carbon system[2].

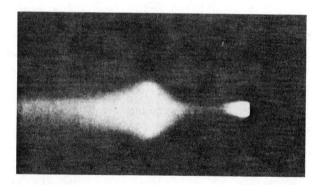

Figure 13.   Free-jet expansion from vaporizing WC/C surface.
Radiating C_2 and C_3 species close to the surface cool
as supersonic speeds are reached, become invisible,
and then reradiate strongly on deceleration to 1 Mach
producing a "Mach-disk".

This final concept in the literature has let us infer the locale and significance of C_1 formation at high energy inputs. We propose that release of C_1 followed by formation of C_2 and C_3 is the dominant mechanism absorbing energy under such conditions.

Let us examine the structure of the radiant plume during CO_2-laser radiation of our composite in vacuum. Directly in front of the irradiated area is a small bright plume that disappears farther from the surface and then blossoms into a large radiant "Mach disk" trailing into a diminishing tail "normal" to the irradiated surface, even though the angle of the impinging beam may fall at any angle up to 45° to the normal and produce the same effect. There is a thin barrel-shaped shock-wave connecting the perimeter of the burn area to the Mach disk not visible in this photograph. The disappearance of the original radiating plume is attributed to jet cooling.

We propose that the occurrence of the small radiating plume attached to the burn surface is related to release of C_1 from molten WC in this surface. This mode of vaporizing carbon absorbs the maximum amount of energy since six bonds connecting the graphitic carbon atoms must be severed instead of three or two of the C-C bonds to release C_2 or C_3 moities. If two C_1's were to combine without escaping from the surface, the released energy of combination could produce one photon of UV-energy detectable only in the vacuum UV-region. However, if the combination energy is distributed between both electron activation, which can result in radiation, and kinetic energy as the new particles escape from the surface and approach supersonic speeds, the formation of C_2 and C_3 molecules can be occurring and produce this first bright plume. The new molecules could then produce the spectra observed in Figure 6.

Consequently, an examination of these plumes as near to the surface as possible in order to detect and measure any C_1 as well as the C_2 and C_3 already found there, would contribute to understanding the ablation mechanism and determine the relative importance of reflection and ablation for blocking the damage very high intensity lasers are capable of producing.

SUMMARY

In this paper we have attempted to show the means we have employed to reveal the nature of the interaction of a high-intensity infrared laser beam with a specific material system.

We have divided this study into four phases:

(i)   Initial material characterization
(ii) "In situ" mass and energy changes during laser irradiation of the sample
(iii) Post-radiation analyses of burns and surrounding areas
(iv) Correlation with other experiments and data recorded in the literature, directly observed, or reported in private communications

which are illustrated by descriptions of the analytical techniques used and brief indications of the interpretations made of the acquired data.

We have presented an outline of the principal stages in the interaction of a particular laser (a cw 10.6-μm CO_2-laser, welding type) with a particular composite material (a carbon cloth laminate with a pyrolysis-produced carbon and sintered mixed tungsten carbide matrix), and described a proposal for understanding the ablation mechanism involved.

REFERENCES

1. Thermodynamic data taken from "JANAF Thermochemical Tables," 2nd Ed., D. R. Stull and H. Prophet (Project Directors), NSRDS-NBS 37, 1971; and H. L. Schick, "Thermodynamics of Certain Refractory Compounds," Vol.2, Academic Press, New York, 1966, pp. 83, 85.
2. E. Rudy, "Ternary Phase Equilibria in Transition Metal-Boron-Carbon-Silicon Systems," Part V, AFML-TR-65-2, Air Force Materials Laboratory, 1969, p. 192.
3. M. A. Covington, G. N. Liu, and K. A. Lincoln, Free-Jet Expansions from Laser-Vaporized Planar Surfaces, AIAA J., 15:1174 (1977).

SUB-SURFACE RESIDUAL STRESS MEASUREMENTS BY MEANS OF NEUTRON

DIFFRACTION: ALUMINUM, STEEL, AND DEPLETED URANIUM

H. J. Prask and C. S. Choi
DRSMC-LCE-P(D)
U.S. Army Armament, Munitions, and Chemical Command
Dover, NJ   07801
and
Reactor Radiation Division
National Bureau of Standards
Gaithersburg, MD   20899 301/921-3634

ABSTRACT

Neutron diffraction closely parallels x-ray diffraction in methodology and analytical formalism. However, in the normal diffraction range ($\lambda \sim 1$ Å) neutrons are nondestructive and are generally about a thousand times more penetrating than x-rays. We have made use of these properties, and an energy-dispersive experimental mode, to measure sub-surface residual stress, nondestructively, in several samples: an aluminum shrink-fit ring-plug specimen, an elastically bent steel bar, and two depleted uranium cylindrical rods. The former two specimens were calibration samples for the technique and were successfully characterized. The depleted uranium rods were of unknown stress distributions but differed significantly in their thermomechanical histories: slow furnace cooled vs. quick-quenched and rotary straightened. With the energy-dispersive neutron diffraction technique we have determined radial, tangential, and axial residual stress components within depleted uranium specimens and find them to be consistent with the heat treatments.

INTRODUCTION

Although limited to a research reactor or an accelerator neutron source, neutron diffraction offers distinct advantages relative to x-ray diffraction in certain applications. For example, neutron penetration for most elements is typically on the order of centimeters, whereas x-rays (in the normal diffraction range: ~ 1.5Å) typically penetrate about $10^{-2} - 10^{-3}$ cm. Furthermore, since neutron

535

diffraction involves a neutron-nucleus interaction, scattering sensitivity varies irregularly from element to element (in fact, isotope to isotope). In x-ray diffraction, scattering sensitivity

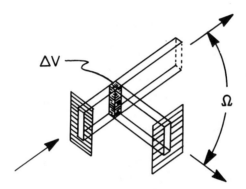

Figure 1a. Perspective Schematic of How Absorbing Masks (e.g. Cd) Can be Used with Bragg's Law to Define the Examined Volume.

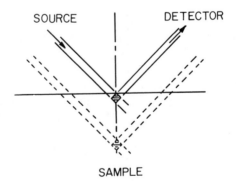

Figure 1b. Plan View of Movement of Examined Volume Through the (translated) Sample.

increases almost monotonically with atomic number. In addition, neutron diffraction is truly nondestructive.

These properties of neutron diffraction led us to use this technique for subsurface residual stress measurement employing a scattering geometry illustrated schematically in Figure 1. Some success was achieved for large "differential" volumes[1]. However, in many samples texture and path length effects caused anomalous shifts in peak positions when more realistic differential volumes were examined ($\sim$40 mm^3), and a conventional θ-2θ scan was employed.

In approximately the last three years, other research groups have reported the successful use of neutron diffraction for residual-stress gradient measurement[2-4]. In particular, Pintschovius et. al[2]. have utilized instrumentation which appeared promising for highly textured, highly attenuating samples of interest in many army applications. In the Pintschovius et. al. measurements, wavelength is scanned rather than Bragg (or scattering) angle from the sample. This is achieved by employing a "triple-axis" neutron spectrometer which utilizes both a monochromater crystal and analyzer crystal - preceding and following, respectively, the sample - rather than a "two-axis" diffractometer with no analyzer crystal.

EXPERIMENTAL

Instrumental

One great advantage of the wavelength scan possible with a three-axis spectrometer - although not emphasized by Pintschovius et. al. - is that movement need only occur in the monochromator and analyzer systems. With reference to Figure 1, the angle between incident (source) and scattered (detector) neutrons, and between neutron directions and sample remains fixed. In this configuration, path length in the sample does not change during a scan, so that beam attenuation is, essentially, constant. Equally important is the fact that because of the fixed geometry the examined "differential" volume does not change in content or orientation during a scan, minimizing the effects of texture.

Stress measurements reported in this work were performed on a triple-axis spectrometer installed at Beam Tube 6, at the NBS Research Reactor. In general, measurements are made at a scattering angle of 90^0 to minimize the differential volume for a given collimator (slit) geometry. Pyrolitic graphite is used in the mono-chromator (002) and analyzer (004) systems.

To test the effectiveness of the technique on highly-textured, highly-attenuating samples, two calibration samples were studied. The first was a highly textured 2024-T351 aluminum shrink-fit, ring/plug sample (3 cm high, 7.6 cm O.D., 2.54 cm diameter plug) fabricated to be an ultrasonics calibration sample for residual stress[5].

TABLE 1. U^{238}-0.75 wt% Ti Samples

Property/Treatment	DU1	DU2
Extrusion (10X R.A.)	X	X
Solutionizing (2 hrs., 850°C, vacuum)	X	X
Furnace Cooled	X	
Rapid Quench		X
Rotary Straightened		X
Aged		X
α-U + TiU$_2$ Structure	X	
α'-U Structure		X

The second was a warm-worked, heat-treated, mild steel bar (30.5 X 5.1 X 0.95 cm^3). The bar was clamped at one end in a fixture, and strained elastically by screws attached to the fixture at the other end[3]. Steel being highly attenuating, the measurements were made through the 0.95 cm dimension.

Two cylindrical specimens of depleted uranium (DU), each 10.6 cm long by 2.54 cm in diameter, with unknown stress distributions but quite different thermo-mechanical histories, were also studied. These samples differed as shown in Table 1. From its treatment, it was expected that DU1 should be very nearly stress free.

RESULTS

Calibration Samples

Measurements were made on the aluminum and steel calibration samples with examined differential volumes of 1.5 X 1.5 X 20 mm^3 and 1.0 X 1.0 X 30 mm^3, respectively. In the case of the aluminum sample, the measurements were in excellent quantitative agreement with the expected stress distribution, both for ($\sigma_r - \sigma_\theta$) which is measured directly and for ($_6\sigma_r$) and (σ_θ) which are obtained after application of equilibrium conditions.

In the case of the steel bar, some difficulty was encountered in measurement of the macroscopic strain applied to the bar. Nevertheless, for an applied (unknown) strain the diffraction-determined stress exhibits the expected dependence on position through the bar, changing linearly from compression to tension.

These results on samples of known stress distribution, along with those of Pintschovius et. al.[2] and others, gave us confidence that the neutron diffraction technique can indeed provide a quantitative measure of subsurface residual stress, even in textured, highly-attenuating materials such as uranium.

Depleted Uranium Samples

Shown in Figure 2 are the region of examination and differential-volume geometries for R-θ and axial-direction scans in the DU rods. The solid dots represent the center points of the differential volumes examined. The two orthogonal lines in the plane of examination were chosen arbitrarily and designated X and Y axes. Typically, measurements were made at each position (e.g. X = +4mm, Y=0) for two R-θ plane probe directions and one measurement with

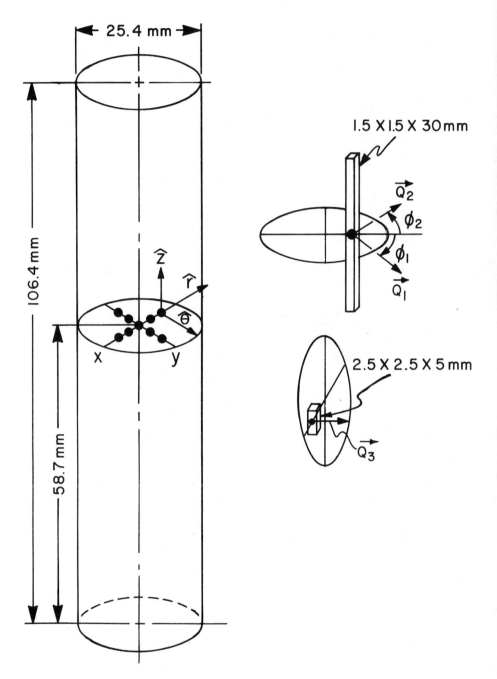

Figure 2. Overall Geometry of the DU Samples and Points at Which Measurements were Made. Also Shown are the Differential Volumes Examined for $\vec{Q}$ in the r-θ Plane (upper right) and Parallel to the Cylinder Axis ($\vec{Q}_3$).

the probe direction parallel to the cylinder axis, centered at the
same point. By "probe direction" is meant the wave-vector transfer
direction in the scattering process, generally denoted $\vec{s}$, $\vec{\kappa}$, or $\vec{Q}$
in diffraction measurements. With reference to Figure 1, it is
the bisector of the incoming-outgoing radiation angle. It should
be noted that it would have been preferable for the R-θ and Z-
direction measurements to include exactly identical volumes (e.g.,
spheres); however, since our measurements on DU were very exploratory
in nature, not all aspects were completely taken into account at
the beginning. In the results presented below, R-θ and axial
direction data are treated as if they include identical volumes.

The relation between stress and strain applicable to diffraction measurements is summarized in the following way[7]. With
reference to Figure 3, the $\vec{P}_i$ represent specimen axes, the $\vec{L}_i$
represent laboratory axes, and the strain $\varepsilon_{\phi\psi}$ is measured along
$\vec{L}_3'$

$$\varepsilon_{\phi\psi}' = (d_{\phi\psi} - d_o)/d_o \qquad (1)$$

where $d_{\phi\psi}$ is the lattice spacing along $\vec{L}_3'$ and d_o is the unstressed
lattice spacing. The stresses are related to the measured strains
through

$$\varepsilon_{\phi\psi}' = 1/2\ S_2(hkl)\,[\sigma_{11}\cos^2\phi\sin^2\psi + \sigma_{22}\sin\phi^2\sin^2\psi$$

$$+ \sigma_{33}\cos^2\psi + \sigma_{12}\sin2\phi\sin^2\psi + \sigma_{13}\cos\phi\sin2\psi \qquad (2)$$

$$+ \sigma_{23}\sin\phi\sin2\psi]\ +\ S_1(hkl)\,[\sigma_{11} + \sigma_{22} + \sigma_{33}].$$

For the cylindrical samples, we assume that the principal axes
of stress in the samples coincide with the cylindrical geometry so
that $\sigma_{12} = {_J\sigma}_{13} = \sigma_{23} = 0$. The $S_1(hkl)$ are diffraction elastic
constants which depend on the elastic coefficients, the coupling of
the crystals within the matrix, the (hkl)-plane, and on the
(ϕ,ψ) of the measurement. For elastically isotropic solids,

$$1/2\ S_2(hkl) = (1 + \mu)/E$$

and

$$S_1(hkl) = -\mu/E \qquad (3)$$

where μ, E are Poisson's ratio and Young's modulus, respectively.

As in x-ray residual stress measurements, in the neutron
diffraction technique what is actually measured is the strain
produced by residual stress, as manifested by the change in "d"
spacing of a Bragg reflection relative to the unstressed "d^o" value.
In Figure 4 the measured d-spacings for the (111) reflection are

shown for points along a single radial direction for both the furnace-cooled (DU1) and rotary straightened (DU2) samples. Because of mechanical limits in the spectrometer, in this case a scattering angle of 72.4° was used rather than the preferred 90°.

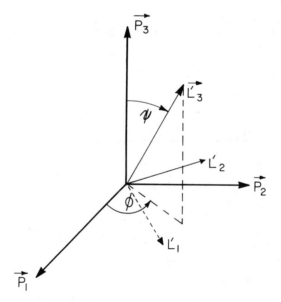

Figure 3. Specimen and Laboratory Coordinates at a Point in the Specimen. The $\vec{L_3'}$ (or probe) Direction Coincides with the Momentum Transfer, $\vec{Q}$, Direction.

In the DU1 sample, only seven measurements were made at a total of three positions. However, the uniformity of d-spacing obtained shows clearly that the sample is indeed essentially stress free. This, in turn, provides an excellent value for the unstressed lattice spacing, d^0, for α-uranium. In contrast, the d_{111} spacings measured for DU2 vary enormously compared to DU1. Nine of twenty-seven measurements are shown, but these are representative of the complete data set. It is clear from this that regardless of the approximate nature of our present analysis, the sensitivity of neutron diffraction to stress in DU is demonstrated.

As indicated in Table 1, the DU1 and DU2 specimens differ in that DU1 is a two-phase system, α-u and TiU_2, and DU2 is a one phase solid-solution, α'-U (.0075 Ti). Although the α- and α' -U structures are virtually identical, it is known that they differ sufficiently to be of significance in residual stress/strain measurements. We have measured the absolute values of the lattice parameters of α-U and α'-U by means of total-profile refinement of data from DU1 and DU2 samples taken with a high-resolution neutron powder diffractometer. The d^0_{111} of DU2 for the energy-dispersive residual stress measurements was obtained by appropriate normaliza-tion using d_{111} (α-U), d_{111} (α'-U) and d^0_{111} (DU1)--the last value taken from the data in Figure 3. Equations (1)-(3) were then used with the isotropic-elastic constants, E = 165480 MPa and μ = 0.21, to obtain the stress distribution of Figure 4 for quick-quenched, rotary-straightened DU2. Several aspects of the measured stress distribution of DU2 are of interest:

1. These results represent the first nondestructive determina-tion of residual stresses in uranium by any technique;

2. Qualitatively, the measured distribution agrees well with that expected for a quenched cylinder (see, for example, ref. 2);

3. Quantitatively, the measured stress distributions appear to have steeper downward gradients (toward compressive stress) than the stress profiles for a quenched cylinder;

4. The stress distribution for each principal axis shows some degree of deviation from cylindrical symmetry. This may arise from the rotary straightening or from other more fundamental anisotropic properties, such as texture.

SUMMARY

We have described the successful application of neutron diffraction to the nondestructive determination of sub-surface residual stress in metallurgical samples. In the case of depleted uranium, neutron diffraction may have considerable importance as

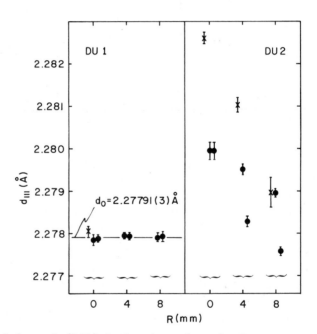

Figure 4. Selected (111) D-Spacings for the Furnace-Cooled (DU1) and Quenched (DU2) Specimens. Measurements with $\vec{Q} \parallel Z$ are Shown as Xs.

a means for testing various thermo-mechanical processes to improve production. In addition to being nondestructive, neutron diffraction contrasts with the Sachs boring out method by not requiring an assumption of cylindrical symmetry for analysis.

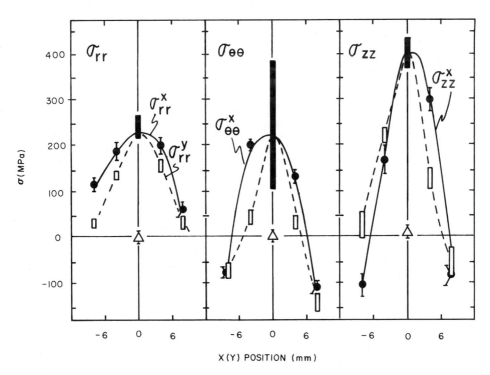

Figure 5. Residual Stresses in the Two Depleted Uranium Samples [DU1 Results are Shown as Triangles at x(y) = 0].

REFERENCES

1. H. J. Prask, C. S. Choi, S. F. Trevino and H. A. Alperin, "Advances in the Application of Neutron Diffraction to Non-Destructive Testing Problems," Proc. 26th Def. Conf. on NDT, p. 158 (1977).
2. L. Pintschovius, V. Jung, E. Macherauch and O. Vöhringer, "Residual Stress Measurements by Means of Neutron Diffraction," Matls. Sci. Eng. 61, 43 (1983).
3. A. Allen, C. Andreani, M. T. Hutchings and C. G. Windsor, "Measurement of Internal Stress Within Bulk Materials Using Neutron Diffraction, NDT Intnl., 251 October, (1981).
4. M. J. Schmank and A. D. Krawitz, "Measurement of a Stress Gradient through the Bulk of an Aluminum Alloy Using Neutrons," Met. Trans. 13A, 1069 (1982).
5. N. N. Hsu, T. M. Proctor, Jr., and G. V. Blessing, "An Analytical Approach to Reference Samples for Ultrasonic Residual Stress Measurement," J. Test. Eval. 10, 230 (1982).
6. H. J. Prask and C. S. Choi, "NDE of Residual Stress in Uranium by Means of Neutron Diffraction," J. Nucl. Matls., in press.
7. Reviewed by V. M. Hauk, in Residual Stress and Stress Relaxation (ed. E. Kula and V. Weiss), Plenum Press, New York and London, pp. 117-138, (1982).

THERMAL WAVE IMAGING OF DEFECTS IN OPAQUE SOLIDS

D. N. Rose, D. C. Bryk and D. J. Thomas
US Army TACOM, Warren, Michigan

R. L. Thomas, L. D. Favro, P. K. Kuo, L. J. Ingelhart
and M. J. Lin
Department of Physics
Wayne State University, Detroit, Michigan

K. O. Legg
School of Physics
Georgia Tech University, Atlanta, Georgia

ABSTRACT

      Thermal wave imaging of defects in opaque solids is carried
out by focusing the periodically modulated intensity of a heat
source (conventionally a laser, electron beam, or ion beam) at the
surface of the solid. Solution of the heat equation shows that the
ac temperature in the solid is wave-like (thermal waves), and
critically damped. These waves can be used to probe the thermal
properties of the subsurface of the solid, and their reflections
from discontinuities in thermal impedance can be used to image
subsurface defects such as inclusions, voids, cracks, and
delaminations. Various techniques have been developed to detect
the resulting ac temperature variations at the sample surface,
including photoacoustic (gas-cell) detection, thermoacoustic
(piezoelectric transducer) detection, optical probe beams, and ac
infrared detection. Thermal wave imaging is particularly useful to
probe subsurfaces from depths of about 1 micron to 300 microns,
with a maximum depth of about 2 mm. The effective depth is about
one thermal diffusion length, and can be varied systematically by
varying the heat source modulation frequency.

Examples will be given of thermal wave images of ceramics and metals, as well as an image of a turbine blade. These images were taken with laser beam excitation and gas cell or optical probe beam detection. An additional "ion-acoustic" image of implanted areas in a single crystal slab of zirconia will be presented as an example of the use of ion beam excitation of thermal waves.

INTRODUCTION

Thermal wave imaging, or photoacoustic microscopy as it is also called, offers a different view of materials through its dependence on thermal properties. Its history began with a report by Alexander Graham Bell in 1880. There was an initial flurry of work by scientists of the day including Roentgen and Lord Rayleigh, but after that, the effect lay dormant for 50 years until it was revived for work with gases by Viengerov in the Soviet Union. It was not reapplied to solids until 10 years ago[1] and the first photoacoustic image was produced only 5 years ago using a borrowed laser and an O-ring for a drive belt between a motor and a stage micrometer[2].

PHOTOACOUSTIC PRINCIPLE

Figure 1 schematically describes the effect as applied to solids. Periodic light, which was chopped sunshine in one of Bell's later experiments, shines on a material in a closed vessel. As the surface of the material heats up, it heats the air above it. As this air is heated, the pressure in the vessel rises. When the light is off, the material's surface cools, so the air above it cools, so the pressure drops. This periodic pressure variation or sound was detected using a listening tube in Bell's day. It was found that the sound was stronger when the sunshine was stronger and was also stronger when the material had a darker, more absorbent color[1].

The first and most widespread use of the effect has been in spectroscopy where the interest lies in the variation of the signal with excitation light frequency. This has permitted spectra to be taken of absorbing and highly scattering materials. While the interest here is microscopy, where point to point differences over a surface are imaged, there are good illustrations from spectroscopic applications. The photoacoustic signal has a phase with respect to the light modulation as well as a magnitude. The phase can be related to depth. For example, the red in an apple peel shows up in a photoacoustic spectrum of the peel taken with a phase delay because the natural wax coating on an apple delays the appearance of the thermal response at the surface[3]. Utilizing this, it was shown that a lobster shell which is black optically had different colored dyes at different depths[4]. The phase is also dependent to an extent on surface profile though it is relatively insensitive to surface reflectivity.

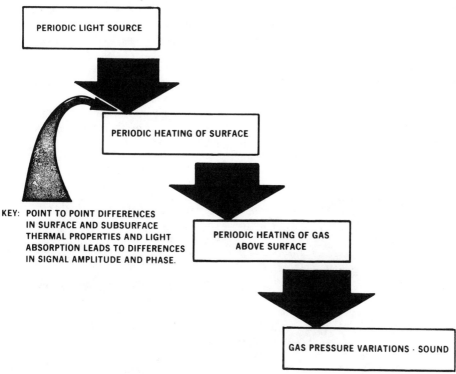

Figure 1.   Principle of the photoacoustic effect using gas cell
            detection.

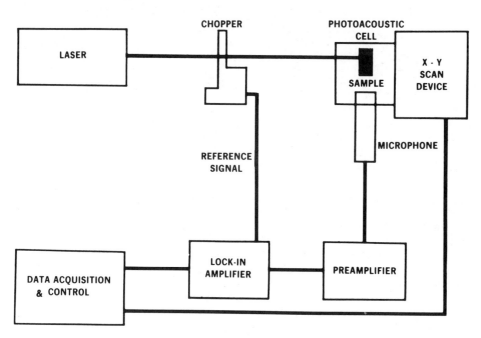

Figure 2.   Schematic of a photoacoustic microscope using gas cell
            detection.

Figure 2 shows a schematic layout of a modern imaging system utilizing gas cell detection. The sound is detected with a sensitive microphone and the electronics that are now available to extract small signals from considerable noise. Lack of these hindered nineteenth century researchers. In addition, the advent of the laser gave easily directed beams with which to interrogate different spots on a surface. Figure 3 shows top and bottom views of a modern gas cell. Note the window to allow the radiation to reach the surface and the shallow channel to the hearing aid microphone. In use, the cell is sealed to the surface to be examined with a thin bead of sticky wax.

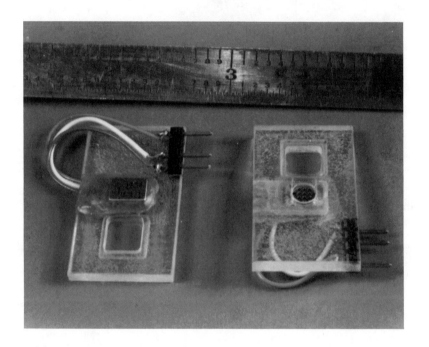

Figure 3.   Photoacoustic microscopy gas cell.

Figure 4 shows the photoacoustic effect in a more general form. The excited spot may be heated by energetic beams other than a laser[5], for example, by electrons[6] or by ions[7]. The periodic heating of the spot generates a thermal wave which is not depicted in Figure 4. This is where the other name for the field, thermal wave imaging arises. The interaction of this wave with the surface, subsurface and gas above the surface, if any, determines the gas pressure variations, gives rise to periodic IR radiation and, through thermal expansion and contraction of the area around the spot, launches an acoustic wave. If we model the basic situation in an opaque material as a one dimensional case of the heat equation, on a semi-infinite solid, a solution for the temperature T into the material can be written as

$$T = T_o \exp [i (qx - \omega t)]$$

where $q = (1 + i) (\omega \rho c/2K)^{\frac{1}{2}}$, $f = \omega/2\pi$ is the modulation frequency, ρ is the mass density, c is the specific heat capacity and K is the thermal conductivity. This solution is periodic in space and time but is highly damped, dropping by a factor of 535 in one thermal wavelength which, in turn, equals $2\pi(2K/\omega\rho c)^{\frac{1}{2}}$. It is customary to take a thermal diffusion length, defined as $(2K/\omega\rho c)^{\frac{1}{2}}$, as a measure of the depth for thermal wave imaging for a particular material at a particular modulation frequency f. At this point, the wave has dropped to 1/e of its initial amplitude. This corresponds to skin depth for electromagnetic waves. This illustrates that this technique is a near surface tool being useable in aluminum down to roughly 1 mm maximum using the magnitude of the signal and 2 mm using the phase. More details of the theory using gas filled cells are given elsewhere[8,9,10].

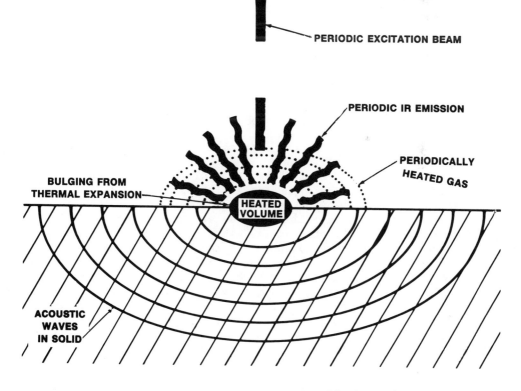

Figure 4.   Effects at a periodically heated spot.

DETECTION TECHNIQUES

Figure 5 shows 5 of the most common methods for detecting the thermal wave signal. The first one, gas cell detection, is what we now call the approach of Bell which has been described.

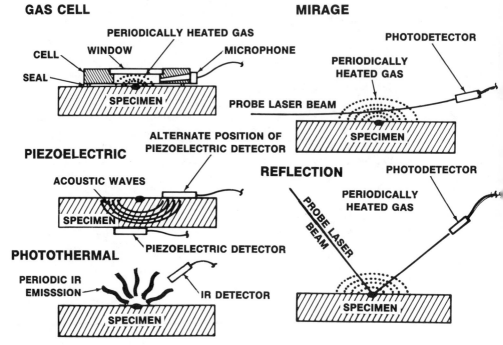

Figure 5. Common photoacoustic detection techniques.

The "mirage" technique, also called optical beam deflection[11], was pioneered by the French[12]. In this approach, a second laser beam called the probe beam skims within 50 microns of the surface. The periodically heated air above the excited spot deflects the probe beam. The principle is the same as that responsible for what appears to be water ahead of us on a hot highway. With this technique one can go to lower modulation frequencies and thus see deeper with longer thermal waves. Unfortunately, this makes image acquisition slower because of the longer diffusion time. If the probe beam is slightly displaced from the excitation beam, very characteristic signatures are seen of cracks oriented along the probe beam direction as will be shown. The mirage technique can obviously not be applied to concave surfaces. It also requires maintaining the relative position of the excitation beam focal spot, the probe beam and the sample surface during a scan. With this technique local variations in surface height can produce signal variations in the magnitude and phase that are not easily separated from subsurface thermal wave interactions[13]. It is, however, a non-contract method.

Piezoelectric detection relies on the acoustic waves that are generated as the excited spot on the sample that is effected by the thermal waves periodically expands and contracts. The conversion efficiency from thermal to acoustic energy is typically quite low but this technique can be faster since one does not have to wait for the thermal information to come back to the surface. Thermal wavelengths are typically 0.3 microns in insulators and 3 microns in metals at 1 MHz dropping to 0.03 microns and 0.3 microns respectively at 100 MHz. This approach can be used where the exciting beam is a chopped electron or ion beam in which case the sample must be in a vacuum. Using this technique and a modified electron microscope, images have been obtained in the same length of time that it takes to get an electron microscope picture. Subsurface flaws in integrated circuits, the doping pattern in semiconductors, the ion implantation pattern in GaAs and the crystalline structure in welds and also in wire bonds to integrated circuit chips have been imaged in this way[14]. This technique does require bonding a transducer to the sample. Theoretically, the interpretation is more complicated since the material's elastic properties, generally expressed as a tensor, become important and acoustic wave reflections must be considered. Theoretically, by themselves, thermal waves are nice because they die out so quickly that reflections, if they come into consideration are simpler and wave mechanics approximations give accurate descriptions[15].

In photothermal imaging, the periodic infrared emissions from the heated spot or the emissions from the back of thin specimens from thermal waves propagating through the specimen are detected. This is a point sampling approach but it relies on thermal diffusion so it is not as inherently fast as piezoelectric detection. It is however, a non-contact method and can also be used in a vacuum. Using this approach, 0.5 micron changes in a 50 micron layer of paint on a 0.5 mm thick sheet of metal have been detected[13,16]. Also the thickness of diffusion hardened surface layers in steel has been measured to an accuracy of 0.15mm[17]. It is adaptable to complex geometries and its resolution is at least 0.001 Celsius degrees.

In the reflection approach, a probe laser is reflected directly off the excited spot. It is felt now that changes down to at least 10 nm can be detected (for reference, green light has a wavelength of 550 nm). While this is very good, state of the art ultrasonic transducers can do better. Both thermal lensing, as in the mirage technique, and thermal expansion, as in the piezoelectric approach, affect the signal[18]. This technique does not require a particularly well polished surface. It does require maintaining the relative positions of the excitation beam focal spot and the spot at which the probe beam is reflected which is usually slightly displaced from the excitation beam's focal spot. It is a non-contact method and high modulation frequencies are possible giving fine resolution.

A sixth approach not shown in Figure 5 uses interferometry to measure the height variation of the heated spot[19]. If this technique, also known as photodisplacement, were used, it should be possible to simultaneously profile the surface.

Yet another approach utilizes an ultrasonic beam to interrogate the periodic gas pressure variations by detecting the periodic phase shift in the ultrasonic wave in the gas as it is beamed down to the heated spot and reflected to another transducer[20]. A variation of this is to monitor the phase of an ultrasonic surface wave in the solid as it traverses the heated area. Changes corresponding to 0.001 Celius degree have been detected. This approach was used to show poor bonding between a silicon wafer and an aluminum backing.

This variety in detection approaches has been given to emphasize the generality of the effect and to show the range of options for applying thermal wave imaging in particular situations. Examples of gas cell, mirage and piezoelectric detection are given in the following examples.

NEAR SURFACE RESOLUTION

Figure 6 shows how another sample was constructed which illustrates the photoacoustic principle further. Two slots, approximately 70 microns deep, were milled in aluminum and came together forming a sharp point. The slots were then filled with plastic and the surface was polished smooth. Next a coat of silver

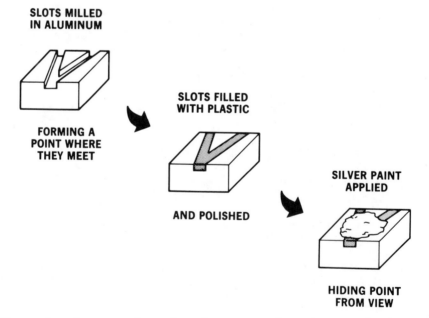

Figure 6. Construction of a photoacoustic point resolution test specimen.

paint was applied, hiding the structure from optical view. Figure 7 shows the photoacoustic picture at two modulation frequencies. This presentation is in inverse video as shown by the reference bar at the bottom so that higher thermal responses are darker. The thermal wavelength in the upper image is 2.2 mm, while the thermal wavelength of 1 cm at 11 Hz for the lower image is longer than the entire image and yet the point is sharp. This feature is in the near field for thermal waves so geometrical shadowing holds[21]. Under these conditions though, the phase shift is essentially zero so no depth informaton can be obtained from the phase.

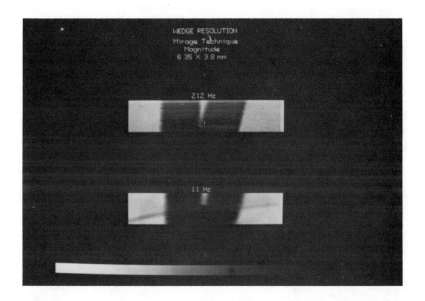

Figure 7. Mirage image of point resolution test specimen.

BURN-IN CARRIER

A burn-in carrier for integrated circuits made of glass bonded mica was imaged for Figure 8. The metal connecting leads are the broad vertical strips angling in at the top to where the integrated circuit would be. Optically, the matrix was white. Under a microscope, the matrix had small, irregular clear glassy areas in a white matrix. Included with the magnitude and phase images is a concurrently collected image formed by monitoring the light reflected from each point sampled. This was inverted to serve as a visual reference of the relative amount of light energy absorbed. As expected, more of the light was reflected from the metal leads and the thermal wave signal from the leads is

correspondingly lower. However, the matrix shows areas of strong
thermal response for which there is no corresponding feature in the
absorbed light. This illustrates that thermal wave imaging is a
different way of seeing. A crack can be seen slanting vertically
on the right side of the thermal wave images. This crack could be
seen optically under the right conditions. With gas cell
detection, perfectly vertical, closed cracks can not be seen[10];
however, "vertical" cracks in practice, such as this one, are
frequently neither perfectly vertical nor perfectly closed so they
can be detected.

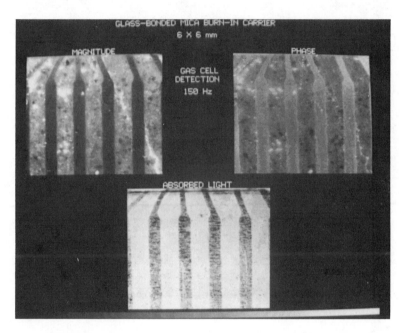

Figure 8. Gas cell and corresponding scanned optical absorption
images of a glass-bonded mica burn-in carrier. A crack
may be seen slanting vertically on the right side of the
thermal wave images.

OPEN FATIGUE CRACK IN ALUMINUM

Figure 9 demonstrates the distinctive signature that can be
obtained from a crack using mirage detection. The upper image
shows the crack using gas cell detection while the lower is the
same crack using mirage detection with the probe beam slightly
displaced from the heating beam and also parallel to the crack[13].
As a first approximation, one can think of the crack as

556

blocking the heat flow when the crack is between the heated spot and the probe beam so that the thermal response the probe beam sees is low in this case, but, when the probe beam is between the crack and the heated spot, the crack reflects the heat so that the probe beam is more strongly deflected.

When the probe beam is displaced from the excitation beam, there is a transverse deflection of the probe beam as well as a deflection normal to the surface. This transverse deflection can be used for extracting the thermal diffusivity of the spot under examination[22].

In the mirage image, there is a crown-like structure over the crack and a vertical line in the upper half of the image. We first noticed this in a pseudocolor image. We have found a pseudocolor capability useful for revealing subtle detail in the images which we subsequently verify in the gray scale images.

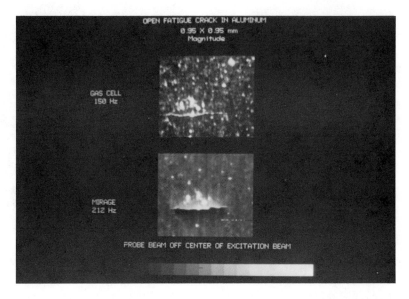

Figure 9. Fatigue crack in aluminum: Gas Cell vs mirage detection.

Figure 10, in inverse video, shows a comparison of thermal wave images of Knoop indentations on silicon carbide obtained using gas cell detection and mirage detection. Though the images are of different indentations, they illustrate the two approaches. The visible indentation, 140 x 20 microns, is somewhat less than half the height and 6 % of the width in the field of view in these images. There is a background signal, minimal for the 90 degree images and highest for the 0 degree images that has been subtracted from the data.

According to the theory, the effect of cracks that are shallow compared to the thermal wavelength should show up at − 45 degrees phase angle[9,23]. Accordingly, the smallest indications can be seen in the + 45 degree images and the biggest indications in the − 45 degree images. The residual indication in the + 45 degree gas cell image corresponds in size to the visible surface deformation of the indentation. In contrast, mirage detection responds to vertical cracks and a much larger response is seen for mirage detection at + 45 degrees. The + 45 degree mirage response however, is still less than the − 45 degree mirage response as can be seen most clearly in Figure 11 which is a

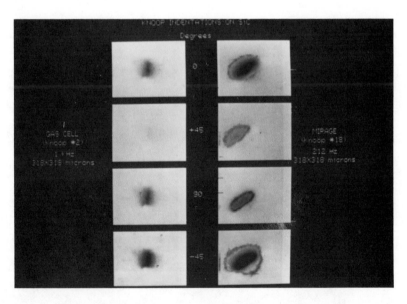

Figure 10. Knoop indentations in sintered silicon carbide: Gas Cell vs Mirage detection.

comparison of these two images. Figure 11 also illustrates that a combination of data presentations helps to bring out all of the features of the data.

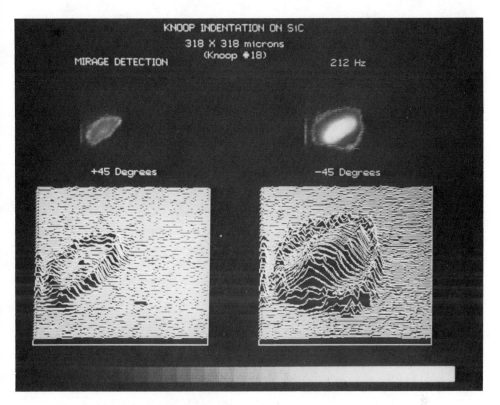

Figure 11. Knoop indentation in sintered silicon carbide: Mirage images at + 45 degrees and – 45 degrees

TIGHT VERTICAL CRACK IN SiC

For this demonstration, a Knoop indentation was again used. The indentation was made in hot pressed silicon carbide with a 2.7 Kg load but then, the visible indentation was machined off, leaving only the vertical crack formed during indentation, estimated as 140 microns long and 70 microns deep. Figure 12 shows a comparison of a reflected light image obtained as in Figure 8 and a mirage image. The crack was not detectable by optical microscopy, by scanning laser acoustic microscopy (SLAM) or by an advanced eddy current probe. In this case, weaker bands can be seen on either side of the strong pair delineating the crack.

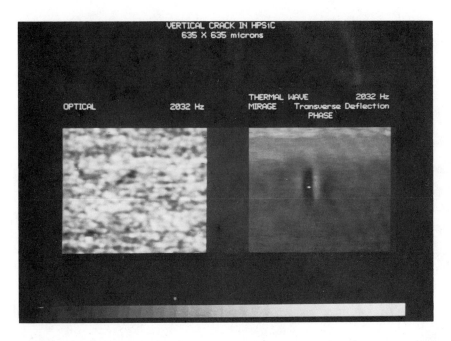

Figure 12. Tight vertical crack in hot pressed silicon carbide:
Scanned optical and mirage images.

ION ACOUSTIC IMAGE OF ZIRCONIA

For Figure 13, a modulated ion beam was used for excitation
with piezoelectric detection to image two nitrogen implanted
regions in a slab of single crystal zirconia. Only very faint
yellowish areas were visible to the eye at the implanted areas
using glancing light. We see the best potential for "ion
acoustics" in on-line control of ion implantation on complex or
inhomogeneous surfaces[7]. Similarly, the photoacoustic effect
can be useful for control of materials processing via lasers.

SUBSURFACE COOLING PASSAGES

Figure 14 is an image at the trailing edge of a TF30 first
stage turbine blade which has through cooling air holes of which
two are visible as broad horizontal bands in the image[24]. As
expected, the thermal signal is stronger where the heat flow is
blocked by the air holes as shown by the whiteness of the bands
compared to the reference bar at the bottom of the picture. This,
also illustrates that an absolutely flat surface is not required
for thermal wave imaging, even with mirage detection.

560

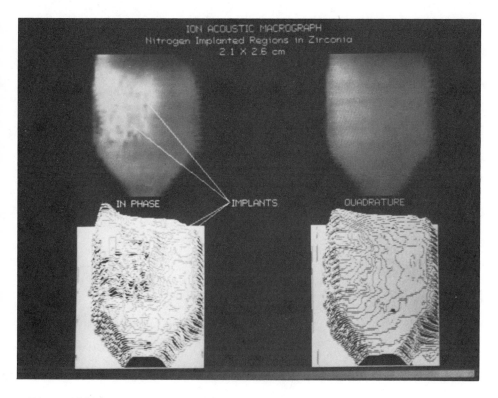

Figure 13. Ion acoustic image of nitrogen implanted areas in a
            zirconia crystal.

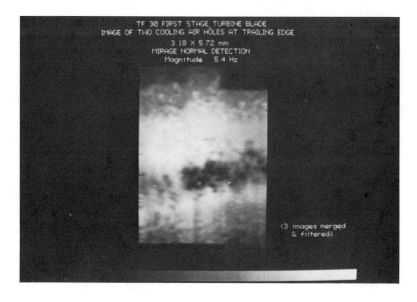

Figure 14. Image of subsurface turbine blade air cooling holes.

CONCLUSION

Thermal wave imaging offers a unique view of materials through its dependence on thermal properties. It is capable of probing opaque materials. It is better adapted to surfaces which can withstand more heat such as ceramics and metals but rubber has been successfully imaged[25].

ACKNOWLEDGEMENTS

This work was partially funded by Dr. Fred Schmideshof of the Army Research Office, the Track Elastomer program under Mr. Jacob Patt at the US Army TACOM, the Reliability for Readiness Program under Mr. Paul Doyle and Mr. Jim Kidd at the US Army AMMRC, and an MTT program under Mr. Robert Brockelman of the US Army AMMRC.

The burn-in carrier sample was furnished by Mr. Jack Liker of MYKROY/MYCALEX Division of Monogram Industries. The fatigue crack sample was part of the DoE Ames Laboratory NDE development program. The hot pressed silicon carbide sample was prepared by Dr. George Quinn of the US Army AMMRC. Mr. Kenneth Fizer of the Naval Air Rework Facility, Norfolk supplied the turbine blade and Xavier Mann prepared the zirconia crystal sample.

REFERENCES

1. A. Rosencwaig, "History of Photoacoustics", Chapter 2 in Photoacoustics and Photoacoustic Spectroscopy, John Wiley and Sons, New York, NY (1980).
2. Y. H. Wong, R. L. Thomas and G. F. Hawkins, "Surface and Subsurface Structure of Solids by Laser Photoacoustic Spectroscopy", Appl. Phys. Lett. 32, 538 (1978).
3. J. A. Noonan and D. M. Munroe, "What is Photoacoustic Spectroscopy", Optical Spectra, February (1979).
4. T. A. Moore, R. Tom, E. P. O'Harra, D. M. Anjo, and D. Benin, "In Vivo and In Vitro Studies of Biological Materials by Photoacoustic Techniques", presented at AAAS meeting, 26-31 May, Detroit, Michigan (1983).
5. A. Rosencwaig, Photoacoustics and Photoacoustic Spectroscopy, op. cit, p. 297.
6. E. S. Cargill, III, "Electron-acoustic Microscopy", Phys. Today, 34, 27 (1981).
7. K. O. Legg and D. N. Rose, "Scanning Ion Acoustic Microscopy for Analyzing Ceramic Surfaces", presented at the 8th Annual Conference on Composites and Advanced Ceramic Materials, 15-18 January, Cocoa Beach, Florida (1984).

8. R. L. Thomas, J. J. Pouch, Y. H. Wong, L. D. Favro, P. K. Kuo and A. Rosencwaig, "Subsurface Flaw Detection in Metals by Photoacoustic Microscopy", J. Appl. Phys. 51, 1152 (1980).
9. P. K. Kuo, L. D. Favro, L. J. Inglehart, R. L. Thomas and M. Srinivasan, "Photoacoustic Phase Signatures of Closed Cracks", J. Appl. Phys. 53, 1258 (1982).
10. P. K. Kuo and L. D. Favro, "A Simplified Approach to Computations of Photoacoustic Signals in Gas Filled Cells", Appl. Phys. Lett. 40, 1012 (1982).
11. J. C. Murphy and L. C. Aamodt, "Signal Enhancement in Photothermal Imaging Produced by Three Dimensional Heat Flow", Appl. Phys. Lett. 39, 519 (1981); L. C. Aamodt and J. C. Murphy, "Photothermal Measurements Using a Localized Excitation Source:, J. Appl. Phys. 52, 4903 (1981).
12. A. C. Boccara, D. Fournier and J. Badoz, "Thermo-optical Spectroscopy: Detection by the Mirage Effect", Appl. Phys. Lett., 36, 130 (1980); A. C. Boccara, D. Fournier, N. Jackson and N. Amer, "Sensitive Photothermal Deflection Technique for Measuring Absorption in Optically Thin Media", Opt. Lett. 5, 377 (1980).
13. R. L. Thomas, L. D. Favro, K. R. Grice, L. J. Inglehart, P. K. Kuo, J. Lhota and G. Busse, "Thermal Wave Imaging for Nondestructive Evaluation", p. 586, Proceedings of the 1982 IEEE Ultrasonics Symposium, B. R. McAvoy, ed., IEEE Press, New York, NY (1982).
14. A. Rosencwaig, "High Resolution Thermal Wave Imaging" presented at AAAS meeting 26-31 May, Detroit, Michigan (1983).
15. L. D. Favro, K. R. Grice, L. J. Inglehart, P. K. Kuo and R. L. Thomas, "Gas-Cell Scanning Photoacoustic Microscopy (SPAM)", presented at AAAS meeting Detroit, Michigan, 26-31 May (1983).
16. G. Busse, "Photothermal Remote Nondestructive Material Inspection", presented at AAAS meeting Detroit, Michigan, 26-31 May (1983).
17. M. Luukkala, J. Jaarinen and A. Lehto, "Photothermal Measurement of the Thickness of Diffusion Hardened Surface Layers in Steel", Proceedings of the 1983 IEEE Ultrasonics Symposium, B. R. McAvoy, ed., IEEE Press, NY, NY (1983).
18. J. C. Murphy and L. C. Aamodt, "Photothermal Deflection Imaging and Microstructural Characterization of Solids", presented at AAAS meeting Detroit, Michigan, 26-31 May (1983).
19. Y. Martin, H. K. Wickramasinghe and E. A. Ash, "Thermo and Photo Displacement Microscopy", p. 563, Proceedings of the 1982 IEEE Ultrasonics Symposium, B. R. McAvoy, ed., IEEE Press, New York, NY; L. C. M. Miranda, "Photodisplacement Spectroscopy of Solids: Theory", Appl. Optics, 22, 2882 (1983).

20. R. G. Stearns, B. T. Khuri-Yakub and G. S. Kino, "Measurements of Thermal-Elastic Interactions with Acoustic Waves", p. 595, Proceedings of the 1982 IEEE Ultrasonics Symposium, B. R. McAvoy, ed., IEEE Press, New York, NY; R. Stearns, B. T. Khuri-Yakub and G. S. Kino, "Phase Modulated Photoacoustics", p. 649, Proceedings of the 1983 IEEE Ultrasonics Symposium, B. R. McAvoy, Editor, IEEE Press, New York, NY (1983).

21. L. J. Inglehart, M. J. Lin, L. D. Favro, P. K. Kuo and R. L. Thomas, "Spatial Resolution of Thermal Wave Microscopies", Proceedings of the 1983 Ultrasonics Symposium, B. R. McAvoy, ed., IEEE Press, New York, NY (1983) L. J. Inglehart, D. J. Thomas, M. J. Lin, L. D. Favro, P. K. Kuo and R. L. Thomas, "Resolution Studies for Thermal Wave Imaging", Review of Progress in Quantitative NDE, Vol. 4, D. O. Thompson and D. Chimenti, ed., Plenum Press, New York, NY to be published; L. J. Inglehart, K. R. Grice, L. D. Favro, P. K. Kuo and R. L. Thomas, "Spatial Resolution of Thermal Wave Microscopes", Appl. Phys. Lett. $\underline{43}$, 446 (1983).

22. R. L. Thomas, L. J. Inglehart, M. J. Lin, L. D. Favro and P. K. Kuo, "Thermal Diffusivity in Pure and Coated Materials", Review of Progress in Quantitative NDE, Vol. 4, D. O. Thompson and D. Chimenti, ed., Plenum Press, New York, NY to be published.

23. P. K. Kuo, L. J. Inglehart, L. D. Favro and R. L. Thomas, "Experimental and Theoretical Characterization of Near Surface Cracks in Solids by Photoacoustic Microscopy", P. 837, Proceedings of the 1981 IEEE Ultrasonics Symposium, B. R. McAvoy, ed., IEEE Press, New York, NY (1981).

24. D. N. Rose, "Photoacoustic Microscopy - An Emerging Tool", p. 201, Proceedings of the 32nd Defense Conference on Nondestructive Testing, Wright Patterson Air Force Base, Ohio 1-3 November (1983).

25. R. L. Thomas, L. D. Favro, P. K. Kuo and D. N. Rose, "Scanning Photoacoustic Microscopy of Aluminum with Aluminum Oxide, Roughness Standards and Rubber", US Army Tank-Automotive Command Research and Development Center, Warren, Michigan, Technical Report no. 12668 (1982); R. L. Thomas, L. D. Favro, P. K. Kuo, D. N. Rose, D. Bryk, M. Chaika and J. Patt, "Scanning Photoacoustic Microscopy of Aluminum with Aluminum Oxide, Roughness Standards and Rubber", US Army Tank-Automotive Command, Research and Development Center, Warren, Michigan, Technical Report No. 12957 (1984).

THE RELATION BETWEEN CRYSTAL STRUCTURE AND IONIC CONDUCTIVITY

IN THE NASICON SOLID SOLUTION SYSTEM, $Na_{1+x}Zr_2Si_xP_{3-x}O_{12}$ $(0 \leq x \leq 3)$

L. J. Schioler
Ceramics Research Division
Army Materials and Mechanics Research Center
Watertown, MA 02172

B. J. Wuensch
Department of Materials Science and Engineering
Massachusetts Institute of Technology
Cambridge, MA 02139

E. Prince
Reactor Radiation Division
National Bureau of Standards
Washington, DC 20234

The NASICON system is a solid solution system that can be represented by the formula $Na_{1+x}Zr_2Si_xP_{3-x}O_{12}$, with x between 0 and 3. The word NASICON is an acronym for Na super-ionic conductor and is the name given to a specific composition in the solid solution system, that with x = 2.0.

Prior to this work (Schioler, 1983), only the structures of the end members had been determined due to the lack of single crystals (Hagman & Kierkegaard, 1968; Tranquietal., 1981b). Because of this, the exact changes which occur in the structure as the compositions changes were unknown. In the NASICON system, the ionic conductivity shows a strong dependence on the composition, changing some three orders of magnitude between x = 0 and x = 2.0 at 300°C. At room temperature, the change is not so great, but still follows the same trend as seen at 300°C, Figure 1.

When the changes in the lattice parameters with compositions are examined, Figure 2, it is seen that the a parameter changes linearly with x, but the c parameter does not. This anomalous change in c causes the unit cell volume to pass through a maximum at a composition close to x = 2.0. That the ionic conductivity reaches a maximum at a composition close to that where the cell

Table I. Crystallographic Data for Compounds in the System $Na_{1+x} Zr_2 Si_x P_{3-x} O_{12}$

	X = 0*	X = 1.0	X = 1.6	X = 2.0	X = 2.5	X=3.0**
SPACE GROUP	R3̄c	R3̄c	C2/c	C2/c	R3̄c	R3̄c
RHOMBOHEDRAL AXES						
a (Å)	8.815(1)	8.9346(1)	8.9954	9.0351	9.1115(3)	9.198
c (Å)	22.746(7)	22.8363(7)	22.9716	22.9870	22.6573(16)	22.210
α	90.	90.	90.445	90.953	90.	90.
β	90.	90.	89.555	89.047	90.	90.
γ	120.	120.	119.999	119.892	120.	120.
VOLUME (Å³)	1530.7	1578.7	1609.8	1626.7	1628.6	1627.3
MONOCLINIC AXES						
a (Å)	15.268	15.4752	15.5805(9)	15.6407(6)	15.7816	15.931
b (Å)	8.815	8.9346	8.9956(5)	9.0498(3)	9.1115	9.198
c (Å)	9.1317	9.1953	9.2137(5)	9.2102(4)	9.2023	9.1110
β	123.871	124.124	123.795(4)	123.709(3)	124.866	125.652
VOLUME (Å³)	1020.4	1052.5	1073.2	1084.4	1085.7	1084.9

* From Hong 1976
** From TranQui Et Al. 1981b

Table II. Na Occupancies

	X=0*	X=1.0	X=1.6	X=2.0	X=2.5	X=3.0**
Na(1)	1	1.002(30)	1.00	1.00	.987(30)	1
Na(2)	0	.332(10)	.813(31) [.533]	.732(24) [.667]	.838(10)	1
Na(3)	–	–	.393(15) [.533]	.634(12) [.667]	–	–

* From Hong, 1976
** From TranQui Et Al. 1979

567

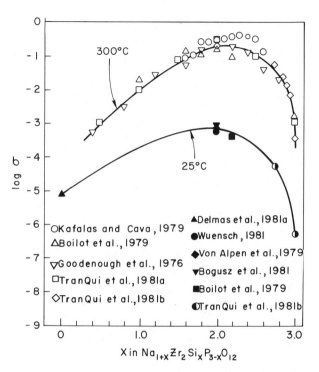

Figure 1.   Ionic Conductivity at Room Temperature and at 300°C of Compositions in the NASICON Solid Solution System.

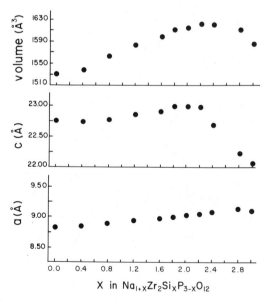

Figure 2.   Unit Cell Parameters of Compositions in the NASICON Solid Solution System (from Hong, 1976).

volume is maximized leads one to believe that there is a strong correlation between the changes in the structure and the changes in the conductivity.

Structure refinements were performed on four compositions in the solid solution range, those with x = 1.0, 1.6, 2.0 and 2.5, using neutron powder diffraction and the Rietveld technique for profile refinement (Schioler, 1983). It was found that the range of monoclinic compositions was larger than previously supposed, from x equal to 1.6 to 2.2 rather than from 1.8 and 2.2, and might be larger on the high x side (Table 1). The main change in the structure was found to be a distortion and rotation of the coordination polyhedra of the tetrahedral sites in the structure.

The structure is based upon a lantern unit. Each lantern is made up of two zirconia octahedra linked at the corners by three Si or P tetrahedra. In the rhombohedral compositions, all the tetrahedra are alike, but in the monoclinic compositions, there are two types of tetrahedral sites and thus two types of tetrahedra. The central tetrahedra in the lanterns are linked to two other lanterns: one corner is linked to the bottom of one lantern and the other is linked to the top of another lantern. The distortion and rotation of the tetrahedra is subtle and is most easily seen in the tetrahedra along the right side of the projection, Figure 3. A measure of the distortion is the angle the oxygen-oxygen separation in the central tetrahedra makes with the horizontal. This angle was found to increase to a maximum at x = 2.0 and then decrease. When this tilt angle and the c lattice parameter are plotted as a function of the composition, it is seen that their variation with x is almost identical, Figure 4.

Two factors affect the ionic conductivity of a material: the concentration of the conducting species and the mobility of the conducting species. In the NASICON system, the concentration is determined by the composition and the mobility by the structure, which determines the site energies and the window size of the pathways for transport.

The ionic conductivity of material is given by:

$$\sigma = \frac{zC(1-C) \, e^2 a^2 \nu}{kT} \, \exp(-\Delta G/kT)$$

where z is the concentration of the sodium sites, C is the concentration of the sodium, e is the charge of the sodium, a is the jump distance, ν is the jump frequency, and ΔG is the free energy for the jump. At any given temperature the conductivity is maximized where the concentration of the conducting species, C, is one half. In the NASICON system, if all the sodium sites are considered, this maximum occurs at a composition with x = 1.0. If only the Na(2) sites are considered, the maximum occurs for x = 1.5. Since

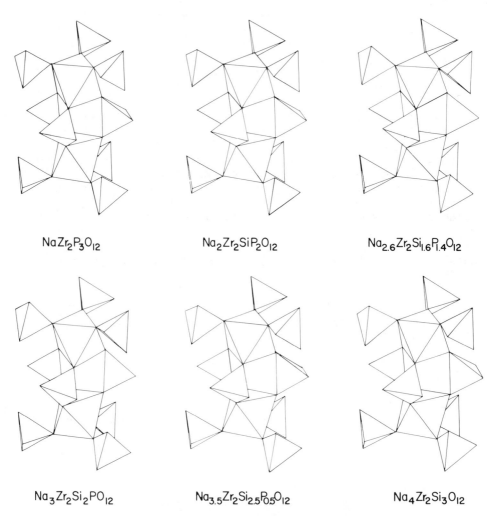

$NaZr_2P_3O_{12}$

$Na_2Zr_2SiP_2O_{12}$

$Na_{2.6}Zr_2Si_{1.6}P_{1.4}O_{12}$

$Na_3Zr_2Si_2PO_{12}$

$Na_{3.5}Zr_2Si_{2.5}P_{0.5}O_{12}$

$Na_4Zr_2Si_3O_{12}$

Figure 3. Orthographic Projections of the Lantern Units of Six Compositions in the NASICON Solid Solution System.

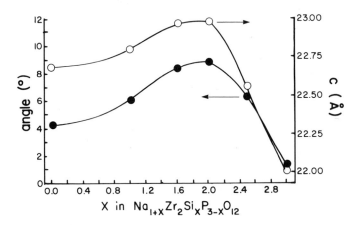

Figure 4.   The Tilt Angle of the Tetrahedra as a Function of
Composition.

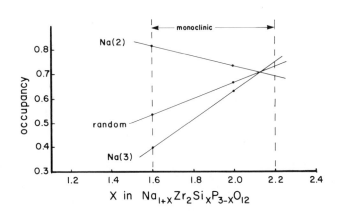

Figure 5.   Occupancy of the Na(2) and Na(3) Sites as a Function
of Composition.

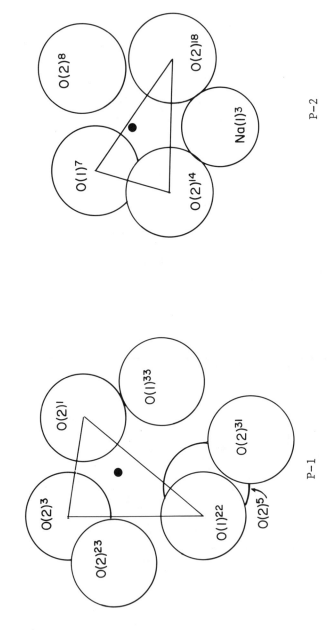

P-1

P-2

Figure 6. Projections of Two Pathways for Transport Between Sodium Sites.

572

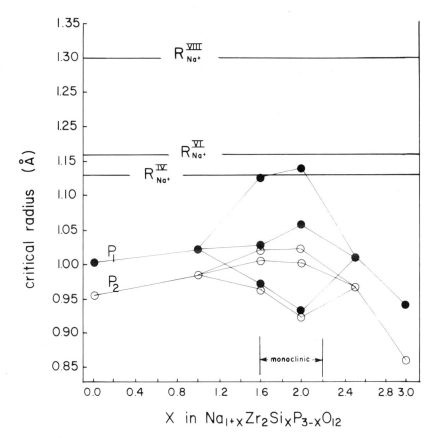

Figure 7.   The Critical Radius of the Pathways P1 and P2 as a
Function of Composition.

the observed maximum in the conductivity at room temperature and at $300^{0}C$ occurs at about x = 2.0, it would imply that the Na(2) concentration is the important factor and that the pathway for transport is between Na(2) sites.

The observed changes in the structure affect the mobility by affecting the site energies of the sodium sites. In the rhombohedral structure there are two types of sodium sites, Na(1) and Na(2), while in the monoclinic structure, the Na(2) site splits into two independent sites, Na(2) and Na(3), due to the decrease in the symmetry. In all compositions, the Na(1) site was found to be fully occupied, while the occupancy of the Na(2,3) sites increased as the concentration of the sodium in the composition increased, Table 2. (The values in the square brackets are those for random occupancy.) As can be seen, the sodium are ordered over the Na(2) and Na(3) sites in the monoclinic compositions. This ordering is less in the composition with x = 2.0 than in that with x = 1.6.

When the occupancy of the Na(2) and Na(3) sites are plotted as a function of x in the composition, Figure 5, it is seen that the occupancy of the two sites is equal at a composition with x about 2.1. This implies that the site energy difference at this composition is zero, while that between the Na(2,3) and the Na(1) sites is much larger, and that the pathway for transport is between Na(2,3) sites.

The final factor in the ionic conductivity is the size of the window the sodium must pass through between sites. Two pathways must be considered: one between a Na(1) and a Na(2,3) site, P1, and one between two Na(2,3) sites, P2, Figure 6. In the monoclinic compositions, these pathways split into three independent pathways. The window size can be calculated for each of the possible pathways. When this calculated size is plotted as a function of x in the composition, it is seen that they, on average, reach a maximum size at about x = 2.0, and that the window size of P1 is larger than that of P2, Figure 7. Note that the size of both is smaller than the radius of a sodium ion!

In summary, the concentration term is maximized at a composition with x = 1.5, the side energy difference between Na(2,3) sites is minimized at x = 2.1, and the size of the windows for transport is maximized at a composition with x = 2.0. From these three factors, it was concluded that transport at room temperature is via Na(2,3) sites and that in the NASICON solid solution system the ionic conductivity is at a maximum at a composition where the concentration and the mobility of the sodium are optimized.

574

REFERENCES

1. W. Bogusz, F. Krok and W. Jakubowski, "Bulk and Grain Boundary Electrical Conductivities of NASICON," Solid State Ionics, $\underline{2}$, 171-174, (1981).
2. J. P. Boilot, J. P. Salanie, G. Desplanches and D. LePotier, "Phase Transformation in $Na_{1+x}Si_xZr_2P_{3-x}O_{12}$ Compounds," Mat. Res. Bull., $\underline{14}$, 1469-1477, (1979).
3. C. Delmas, J. C. Viala, R. Olazcuaga, G. LeFlem, P. Hagenmuller, F. Cherkaoui and R. Brochu, "Ionic Conductivities in NASICON-Type Phases $Na_{1+x}Zr_{2-x}L_x(PO_4)_3$ (L = Cr, In, Yb)," Solid State Ionics, 3/4, 209-214, (1981).
4. J. B. Goodenough, H. Y.-P. Hong and J. A. Kafalas, "Fast Na^+-Ion Transport in Skeleton Structures," Mat. Res. Bull., $\underline{11}$, 203-220, (1976).
5. L.-O. Hagman and P. Kierkegaard, "The Crystal Structure of $NaMe_2^{IV}$ $(PO_4)_3$; Me^{IV} = Ge, Ti, Zr," Acta Chem. Scand., $\underline{22}$, 1822-1832, (1968).
6. H. Y.-P. Hong, "Crystal Structures and Crystal Chemistry in the System $Na_{1+x}Zr_2Si_xP_{3-x}O_{12}$," Mat. Res. Bull., $\underline{11}$, 173-182, (1976).
7. J. A. Kafalas and R. J. Cava, "Effect of Pressure and Composition on Fast Na^+-Ion Transport in the System $Na_{1+x}Zr_2Si_x$ $P_{3-x}O_{12}$," Fast Ion Transport in Solids, (Vashishta, Mundy Shenoy, Editors), Elsevier North Holland, Inc., NY, 419-422, (1979).
8. L. J. Schioler, "Relation Between Structural Change and Conductivity in the Fast-Ion Conducting NASICON Solid Solution System, $Na_{1+x}Zr_2Si_xP_{3-x}O_{12}$," ScD. Thesis, MIT, (1983).
9. D. TranQui, J. J. Capponi, M. Gondrand, M. Saib and J. C. Joubert, "Thermal Expansion of the Framework in NASICON-Type Structure and its Relation to Na^+ Mobility," Solid State Ionics, 3/4, 219-222, (1981a).
10. D. TranQui, J. J. Capponi, J. C. Joubert and R. D. Shannon, "Crystal Structure and Ionic Conductivity in $Na_4Zr_2Si_3O_{12}$," J. Sol. State Chem. $\underline{39}$, 219-229, (1981b).
11. U. Von Alpen, M. F. Bell and H. H. Hofer, "Compositional Dependence of the Electrochemical and Structural Parameters in the NASICON System $(Na_{1+x}Si_xZr_2P_{3-x}O_{12})$," Solid State Ionics, 3/4, 215-218, (1981).
12. B. J. Wuensch, "Principles of Superionic Conduction," Annual Report, Department of Energy contract #450-7910, July, (1981).

APPENDIX

On the final morning of the Conference a Workshop Panel Discussion was convened to discuss various experiences and problems associated with Materials Characterization being carried out in universities, industry, and government laboratories. A representative cross section of experienced practitioners were selected to present short summaries of their experiences and views and to field questions from the audience. The panel consisted of the following with Dr. James W. McCauley acting as Chairman:

Mr. Charles Craig, U. S. Department of Energy
Dr. Robert Green, The Johns Hopkins University
Dr. Rustum Roy, The Pennsylvania State University
Dr. John Wachtman, Jr., Rutgers University
Dr. Ram Kossowsky, Westinghouse R&D Center
Dr. Volker Weiss, Syracuse University
Mr. Harry Light, U. S. Army Materiel Development and
    Readiness Command

The general thrust of the workshop focused on "how to manage Materials Characterization in a Materials Science and Engineering context with particular focus on the manufacture of advanced materials in the 21st century." In order to stimulate the audience to think of issues in a constructive way prior to the workshop, a questionnaire was prepared by Drs. John Wachtman, Jr. and Donald G. Groves and distributed at the beginning of the Conference. The results of the questionnaire were analyzed and summarized at the workshop by Dr. John Wachtman, Jr. The questionnaire is reproduced in chapter 22.

The views presented in the questionnaire essentially confirmed that there is a fundamental lack of good characterization and a lack of use of existing equipment. Education of future materials scientists and engineers and current decision makers was another major issue. The consensus was that there is a fundamental lack of good materials characterization courses and textbooks for undergraduate and graduate curricula. Further, there did not seem to be a viable national means (society, committee, etc.) to help create a continuing dialogue in this area. The whole area of standardization of advanced techniques by various means, including well-planned round robins received considerable attention during the workshop. Finally, the management problems associated with having highly trained scientists carry out characterization support was discussed.

PARTICIPANTS AND STAFF

1. Fran Adar
    Instruments S. A.
    173 Essex Avenue
    Metuchen, NJ 08840

2. John J. Antal
    AMMRC
    DRXMR-OM
    Watertown, MA 02172

3. Ray Blajda
    PLASTEC
    Armament R&D Center
    Building 351N
    Dover, NJ 07804

4. Cyril M. Brown
    USA DARCOM
    Material Readiness Spt Avtivity
    DRXMD-MO, AOAP Branch
    Lexington, KY 40511-5101

5. Alfred L. Broz
    AMMRC
    DRXMR-STN
    Watertown, MA 02172

6. Gordon A. Bruggeman
    AMMRC
    DRXMR-MD
    Watertown, MA 02172

7. Darryl Bryk
    U.S. Army Tank-Automotive Command
    DRSTA-ZSA
    Warren, MI 48090

8. S. Bulusu
    Sarge Caliber Weapons Systems Lab
    Armament R&D Center
    Building 3028, EMD
    Dover, NJ 07801

9. Douglas P. Burum
    Bruker Instruments, Inc.
    Manning Park
    Billerica, MA 01821

10. Catherine A. Byrne
    AMMRC
    DRXMR-OP
    Watertown, MA 02172

11. C. S. Choi
    Large Caliber Weapons Systems Lab
    Armament R&D Center
    Reactor Rad Div, NBS
    Gaithersburg, MD 20899

12. Frederic Cohen-Tenoudji
    Rockwell International Science Center
    1049 Camino Dos Rios
    Thousand Oaks, CA 91360

13. James R. Cook
    ARMCO Research
    5788 Trenton-Franklin Road
    Middletown, OH 45042

14. Charles H. Craig
    U.S. Department of Energy
    Office of Veh & Eng R&D
    Code CE131
    Washington, DC 20585

15. Joseph R. Crisci
    David Taylor Naval Ship R&D Center
    Asst. Dept. Head for Research
    Code 2801
    Annapolis, MD 21402

16. Wenzel E. Davidsohn
    AMMRC
    DRXMR-OP, Polymer Research Div.
    Watertown, MA 02172

17. Robert T. DeHoff
    University of Florida
    Dept. of Mat Sci & Eng.
    Gainesville, FL 32611

18. C. Richard Desper
    AMMRC
    DRXMR-OM
    Watertown, MA 02172

19. R. Judd Diefendorf
    Rensselaer Polytechnic Institute
    Dept. of Mat Eng.
    Troy, NY 12181-3590

20. Bruno Fanconi
    National Bureau of Standards
    A303/224
    Gaithersburg, MD 20899

21. Renee Ford, Editor
    Hi-Tech Materials Alert
    P. O. Box 1304
    Fort Lee, NJ 07024

22. Charles P. Gazzara
    AMMRC
    DRXMR-OM
    Matls. Char. Div.
    Watertown, MA 02172

23. Robert E. Green, Jr.
    The Johns Hopkins University
    Materials Science & Engineering
    Baltimore, MD 21218

24. Donald G. Groves
    National Academy of Sciences, NMAB
    2101 Constitution Avenue
    Washington, DC 20418

25. Gary L. Hagnauer
    AMMRC
    DRXMR-OP
    Polymer Research Division
    Watertown, MA 02172

26. Kay Hardman-Rhyne
    National Bureau of Standards
    Building 223, Room A-331
    Gaithersburg, MD 20899

27. Ralph J. Harrison
    AMMRC
    DRXMR-SMM
    Watertown, MA 02172

28. Gerald J. Hoag
    AFPRO Hughes Aircraft Company/QA
    P. O. Box 92463
    Los Angeles, CA 90009

29. Albert S. Ingram
    U. S. Army Missile Command
    DRSMI-RLM
    Redstone Arsenal, AL 35898

30. Larry D. Jennings
    AMMRC
    DRXMR-OM
    Watertown, MA 02172

31. Rebecca Jurta
    AMMRC
    DRXMR-OM
    Watertown, MA 02172

32. George B. Kenney
    Massachusetts Institute of Technology
    Matls Proc. Center, Room 12-007
    77 Massachusetts Avenue
    Cambridge, MA 02139

33. Michael J. Koczak
    Drexel University
    Dept. of Materials Engineering
    32nd & Chestnut Streets
    Philadelphia, PA 19014

34. Ram Kossowsky
    Westinghouse R&D Center
    1310 Beulah Road
    Pittsburgh, PA 15235

35. Robert W. Lewis
    AMMRC
    DRXMR-DA
    Watertown, MA 02172

36. Harry L. Light
    Headquarter
    U. S. Army Materiel Command
    AMCQA-EQ
    5001 Eisenhower Avenue
    Alexandria, VA 22333

37. James W. McCauley
    AMMRC
    DRXMR-OM
    Watertown, MA 02172

38. William McClane
    U. S. Army Aviation Systems Command
    4300 Goodfellow Boulevard
    St. Louis, MO 63120-1798

39. Robert W. McClung
    Oak Ridge National Laboratory
    4500-S, D-61
    Box X
    Oak Ridge, TN 37831

40. William Meyer
    U. S. Army Armament R&D Center
    DRSMC-LCU-CT
    Dover, NJ 07801

41. Andrus Niiler
    Ballistic Research Laboratory
    DRDAR-BLF (A)
    Building 120
    Aberdeen Proving Ground, MD 21005

42. Gregory B. Olson
    Massachusetts Institute of Technology
    Room 13-5050
    77 Massachusetts Avenue
    Cambridge, MA 02139

43. Frank C. Palilla
    GTE Laboratories, Inc.
    40 Sylvan Road
    Waltham, MA 02254

44. Carlo G. Pantano
    The Pennsylvania State University
    Department of Materials Science & Eng.
    123 Steidle Building
    University Park, PA 16802

45. Phillip A. Parrish
    U. S. Army Research Office
    DRXRO-MS
    P. O. Box 12211
    Research Triangle Park, NC 27709

46. Janet Perkins
    AMMRC
    AMXMR-OC
    Watertown, MA 02172

47. Roger S. Porter
    University of Massachusetts
    Polymer Science & Engineering Dept.
    Amherst, MA 01002

48. Robert E. Richards
    3M Company
    New Products Dept.
    219-01-01
    3M Center
    St. Paul, MN 55144

49. James Michael Rigsbee
    University of Illinois
    Department of Metallurgy
    201C MMB
    1304 W. Green Street
    Urbana, IL 61801

50. Michael A. Riley
    Ballistic Research Lab
    Aberdeen Proving Ground, MD 21005

51. Douglas Rose
    U. S. Army Tank-Automotive Command
    DRSTA-ZSA
    Warren, MI 48090

52. Alan R. Rosenfield
    Battelle Columbus Laboratories
    Physical Metallurgy Section
    505 King Avenue
    Columbus, OH 43201

53. Rustum Roy
    The Pennsylvania State University
    Materials Research Laboratory
    University Park, PA 16802

54. Clay O. Ruud
    The Pennsylvania State University
    159 MRL
    University Park, PA 16802

55. L. J. Schioler
    AMMRC
    DRXMR-MCS
    Watertown, MA 02172

56. Robert B. Schulz
    U. S. Department of Energy
    Program Manager
    Veh Eng R&D
    CE-131, Forrestal Building
    Washington, DC 20585

57. Wendel Shuely
    Commander, Chemical R&D Center
    DRSMC-CLB-CP (A)
    Building E3300
    Aberdeen Proving Ground, MD 21010

58. Victor I. Siele
    Commander, Large Cal Weapons Sys.
    Lab ARDC
    DRSMC-LCP-SL
    Building 94
    Dover, NJ 07801

59. Donald M. Smyth
    Lehigh University
    Materials Research Center #32
    Bethlehem, PA 18015

60. W. Evan Strobelt
    Boeing Aerospace Company
    11029 Woutheast 291
    Auburn, WA 98002

61. Wayne K. Stuckey
    The Aerospace Corporation
    P. O. Box 92957
    MS M2/250
    Los Angeles, CA 90009

62. John D. Venables
    Martin-Marietta Labs
    Assoc. Dir
    1450 South Rolling Road
    Baltimore, MD 21227-3898

63. John B. Wachtman, Jr.
    Rutgers University
    Director, Center for Ceramics Research
    Room A274
    P. O. Box 909
    Piscataway, NJ 08854

64. Volker Weiss
    Syracuse University
    304 Administration Building
    Syracuse, NY 13210

65. James A. Whiteside
    AMMRC
    DRXMR-MPM
    Watertown, MA 02172

    STAFF

1. Mr. Larry Felton
   Syracuse University
   Chemical Eng/Mat Science Dept.
   Syracuse, NY 13210

2. Mrs. Mary Ann Holmquist
   Syracuse University
   Office of Sponsored Programs
   Skytop Office Building
   Syracuse, NY 13210

3. Ms. Karen Kaloostian
   AMMRC
   Watertown, MA 02172

4. Mr. Robert Sell
   Syracuse University
   203 Link Hall
   Syracuse, NY 13210

584

Bright-field transmission
electron micrograph,
of steel alloy, 110,
113
Brittle-fracture resistance of
steel, 413-422
Buckley, A., 80, 81
Built-in equipment for aircraft oil analysis,
406-407
Bulk density of yttria powders,
151, 161-162, 169, 170
Bunning, E. J., 374, 381
Burn-in carrier, thermal wave
imaging, 555-556

Californium-252, neutron radiography, 396
Capiati, N., 73
Carbide particles, and
toughness of steel, 415
Carbon, on glass surfaces,
132-134
Carbon content, and thoughness
of steel, 418-419, 422
Carbon double bond concentration, IR monitoring,
278-279
Carbon fibers, 71
Carbon-13 NMR spectra, 487,
488-491
CARS (Coherent Anti-Stokes
Raman Spectroscopy),
517
Cellulose
C-13 NMR spectra, 487,
488-494
silicon wafer polluted by,
347-351
Cementite, 114, 117
Ceramic powders
dynamic compaction of,
441-442
processing of, 443
Ceramics, advanced, 16-25, 28
and neutron radiography, 393
markets for, 15-18, 27-29
materials characterization,
354-355
small-angle neutron
scattering, 257-269

Ceramics, advanced (continued)
x-ray stress measurement,
361-383
wave velocity maps, 510-511
processing, 438-439
studies of, 427
Ceramics Processing Research
Laboratory, MIT, 25
Chain structure of polymer, 196
Characterization, *See* Materials
characterization
Characterization laboratories,
319-321
Charge, of glass surfaces, 136
Charpy ductile-brittle
transition curves, 118
Charpy Impact Test, 414
Chemical purity requirements
of yttria powders,
152, 154, 172-174
and sintering outcomes,
180-185
Cheverton, R. D., 415
Chromatographic analysis of
polymers, 220
Chuah, H. H., 96
Classification of polymers,
192-195
Cleavage behavior of iron,
122-124
Cleavage fracture resistance
of steel, 414-422
Cluster methods, appied to
grain boundaries, 122
Coextrusion, 75-76, 92
Cohen, Morris H., 95-96, 109
Coherent Anti-Stokes Raman
Spectroscopy, (CARS),
517
Coherent scattering length,
258
Cohesive properties of steels,
122
Cold rolled steel, detection
of surface defects,
308
Collimation systems for SANS,
258-259
Combat vehicles, oil analysis,
404, 407

Crystallinity (continued)
  relative, measured by NMR,
      313-316
Crystallites
  polyethylene resin branch
      points, 323, 324
  of sintered yttria, 157, 159,
      160
    and pore size, 167
  sizes, and extrusion draw
      ratio, 87-89
Crystal structure, and ionic con-
      ductivity, 565-575
C-scans, 505-509
C-shear process, 81
Cure chemistry, 277-280
Cured sealants, change in
      formulation, 244-245
Cure monitoring for polymer
      matrix composites,
      275-291
Cutting tool applications of ce-
      ramics, 18-20

Dangling bonds, 128
Dark field electron microscopy,
      86, 88-89
Data bases, 476
Debye-Scherrer x-ray diffraction
      patterns, 35-36, 38
Decane, 198
Defects, 505
  in advanced ceramics, 21, 23
    microstructural, 267-268
  identification of, in
      polymers, 357
  in metal structures, 427
  study recommendations, 436
  thermal wave imaging, 547-562
Defect scans, 508-509
Deformability of metals, 427
Deformation
  of high density polyethylenes,
      86-91
  of low density polyethylenes,
      80-85
Degradation of materials,
      513-519
  of polymers, 196, 210
Degree of polymerization, 190
Delhaye, M., 344

DeMicheli, R. E., 96
Deming, W. E., 1
Density
  of polyethylene, and draw ratio,
      87, 96
  of polymers, 217
  of polymer solvents, 199-200
  of sintered yttria-alumina,
      155-157, 161-166,
      168-172, 178
  of translucent body, 149
Department of Defense
  Army Oil Analysis Program,
      403-412
  testing requirements, 396
Depleted uranium, diffraction
      testing, 535, 539-546
Desorption, electron or photon
      stimulated, 132
Destructive property measurement,
      10
Determinate sampling errors, of
      polymers, 209
Dewhurst, R. J., 47
Dielectric constant
  of polymers, 208
  of solvents, 200
Dielectric oxides, 59-68
Dielectric spectroscopy,
      286-288
  to test epoxy resins, 230
Differential scanning calorimetry
      (DSC), 248, 249, 253
  of block copolymers, 497-498
  of epoxy resins, 203, 233
  of polymers, 214
Differential thermal analysis (DTA)
  of polymers, 214
Diffraction patterns, WAXS
      measurement of, 321-322
Diffuse transmittance of
      yttria, and density
      175-176, 179
Diffusion coefficient, 94-95
Dilute solution characterization
      of polymers, 217
Dimensional Resonance Profiling,
      302
Dislocation theory, 430
Donor impurities, 62

587

Nylon thin films, 76

Oil analysis program, 403-412
On-line process monitoring
  dielectric spectroscopy,
    286-288
  fluorescence spectroscopy,
    281-286
  ultrasonic techniques, 288-291
Opaque solids, thermal wave
    imaging of, 547-564
Optical beam deflection, 552
Optical birefringence of
    polymers, 358
Optical fibers
  characterization of, 352-354
  and fluorescence spectroscopy,
    282-286
Optical imaging techniques,
    47-51
Optical photomicrograph of
    laser irradiated
    composite, 530
Optical properties of sintered
    yttria, 178
Organic chemistry, 427
Organic contaminants on silicon
    wafers, 344
Organic polymer character-
    ization, NMAB study,
    439
Orientation function of poly-
    mers, 322, 332, 336
Oscillating sphere viscometer,
    405
Osmometry, 224
Oxidation-type semiconductors,
    61-62, 66-68
Oxygen steelmaking processes,
    304

Pacific Northwest Laboratory,
    294
Peel plies, and composite
    bonding, 272
Pennsylvania State University,
    Materials Research Lab-
    oratory, 363
Performance levels
  of advanced ceramics, 23
  and process control,
    27

Perkin-Elmer LS-5 fluorescence
    spectrophotometer, 496
Permeability of polymers,
    92-101
  and draw ratio, 96-97, 99
Permeation flux, in extrusion-
    drawn polystyrene,
    100, 101
PET (polyethylene terepthalate)
    chips, 347
Peterlin, A., 96
Peterlin's model, 87, 91
Peterson, R. V., 465
Phase diagram of tungsten-carbon
    alloy system, 531, 532
Photoacoustic microscopy,
    548-564
Photodisplacement, 554
Photon stimulated desorption,
    132
Photothermal imaging, 53, 54,
    552, 553
Photovoltaic detectors, 441
Physical aging, in amorphous
    polymers, 498-500
Physical change, in polymer
    matrix composites, 276
Physical property test
  of lubricants, 411-412
  procedures for oil analysis,
    404-405
Physical specifications for
    yttria powders, 154,
    172
Piezoelectric detection,
    552, 553
Pintschovius, L., 537
Plasmas, jet cooling of,
    531-533
Plastic deformation, absent in
    ceramics, 361
Plastics, 192, 194
Plastic shear instability, 114
Point sources of neutrons, 389
Polarity switching, 38, 39
Policies of quality control,
    U. S. Army, 447-449
Polished glass surfaces, 128
Polycrystalline metals
  texture of, 45
  steel, ultrasonic testing of,
    41, 43

Wide Angle X-ray Scattering
(WAXS), 319-337
Wilchinsky, Z. W., 335
Windshield glass, 145-146
Winter, J. M., 33
Wire contact, broken, micro-
analysis of, 351
World market, for ceramics,
15-18, 27-29
Wyball distribution of ad-
vanced ceramics, 23

X-ray diffraction (XRD),
32-39, 44, 321, 427
carbon cloth/resin inter-
actions, 522
data base, 476
of laser irradiated composite, 527, 529
and neutron diffraction,
535-537
of polymers, 358
of residual stresses,
363-382
X-ray fluorescence, 230
of metal-doped carbon
composite,
523

X-ray imaging techniques, 46
X-ray photoelectron spec-
troscopy (XPS), 130,
132, 133, 134, 136
in composite bonding, 272
of laminated glass, 145-146
X-ray radiography, 387
imaging failures, 393
of laser irradiated composite, 530
of metal-doped carbon composite, 523
X-ray stress measurement of
structural ceramics,
361-383

Yttria, translucent, 149-187

Zachariades, A. E., 96
Zero defects, 467
Zirconia, 355
ion acoustic image, 560, 561
velocity map, 510, 511